Series Editor: L. L. Langley
National Library of Medicine

Published Volumes and Other Volumes in Preparation

HOMEOSTASIS: Origins of the Concept
L. L. Langley
THE PHYSIOLOGY OF RESPIRATION
Julius H. Comroe, Jr.
ELECTROPHYSIOLOGY
Hebbel E. Hoff
CARDIOVASCULAR RESEARCH
James V. Warren
CONTRACEPTION
L. L. Langley
MICROCIRCULATION
Mary P. Wiedeman

Benchmark Papers in Human Physiology

HOMEOSTASIS
Origins of the Concept

Edited by
L. L. LANGLEY
National Library of Medicine

Stroudsburg, Pennsylvania

Library of Congress Catalog Card Number: 72–79805
ISBN: 0–87933–007–4

Manufactured in the United States of America.

Exclusive distributor outside the United States and Canada:
JOHN WILEY & SONS, INC.

Acknowledgments and Permissions

ACKNOWLEDGMENT

The Congress of American Physicians and Surgeons—*Transactions of the Congress of American Physicians & Surgeons*
"Some General Features of Endocrine Influence on Metabolism"

PERMISSIONS

The following papers have been reprinted with the permission of the authors and copyright owners.

American Association for the Advancement of Science—*Science*
"Claude Bernard's Conception of the Internal Environment"

The Royal Society of London—*Proceedings of the Royal Society, London*
"The Physiological Basis of Thirst"

American Physiological Society—*American Journal of Physiology*
"Studies on the Conditions of Activity in Endocrine Glands"

American Physiological Society—*Physiological Reviews*
"Organization for Physiological Homeostasis"

Charles C. Thomas, Publisher, Springfield, Illinois—*Selected Readings in the History of Physiology*
"Physiological Regulation of Normal States: Some Tentative Postulates Concerning Biological Homeostasis"

The Williams & Wilkins Co.—*Elements of Physical Biology*
"Displacement of Equilibrium"

Society for Experimental Biology—*Society for Experimental Biology Symposia*
"The Thermal Homeostasis of Man"
"Feedback Theory and Its Application to Biological Systems"

Preface

When the publishing firm of Dowden, Hutchinson and Ross announced the Benchmark Book Program, I responded as reflexively as a Dalmatian does to the clanging of fire engines. In 1965, I published a small volume entitled "Homeostasis." It was designed to offer the undergraduate student a means of easily obtaining fundamental knowledge of the very basic biological concept termed homeostasis. In that volume, the concept was explained and illustrated. The first chapter defined the term and discussed its origin. But there was space only for the briefest mention of those who were instrumental in the development of the concept, and certainly no space to publish more than a short quote or two. (That book was also one of a series and the ground rules called for a total of exactly 128 pages, including front material and index!)

To be sure, there are restrictive ground rules for the Benchmark Papers in Human Physiology (400–450) pages. The purpose of this series is to publish the original writings in a variety of fields which developed an important concept. So here is the opportunity to gather together in one place the publications that established homeostasis as such a solid biologic principle.

In this volume, the problem is not limitation of space; it is deciding where to begin, and, more troublesome, ascertaining where to stop. We now know that the living organism is the sum total of self-regulating mechanisms that maintain the constancy of the internal environment despite great variations in the external environment. And that is what homeostasis is all about. This means that a truly definitive volume should contain landmark papers which embrace the entire gamut of physiology. That is clearly not practical; therefore, just as the artist must at some stage step back from his painting, sign his name, and declare it finished, so have I had to carry this compilation to a certain point, and then cease.

Anyway, the real fun, I think, is to be found in the older literature. With libraries becoming inundated, with scientific journals becoming financially pinched, one can no longer publish long, rambling, but utterly delightful articles filled more with philosophy than with physiologic fact. And that, to me, is lamentable. Thus, this opportunity to savor those charming publications and to bring them together in a single volume could not be resisted.

Happily, I had all the resources and the knowledgeable and cooperative staff of the National Library of Medicine at my disposal. I was permitted to roam the stacks, a copying machine was made available, and, after selection had been made, microfilms were made so that the reader could have before him a true replication of the original publications, at least of those in English. Articles in other languages have been translated. In short, any errors of commission or omission are, all too obviously, mine alone.

L. L. Langley
Bethesda, Maryland
November, 1972

Contents

Contents by Author

Introduction

If the word homeostasis is analyzed it is found to consist of *homeo*, which means like or similar, and *stasis*, a standing still. A better idea of stasis can be obtained by considering how it is used in medicine. One speaks of stasis in reference to the circulation to indicate a stoppage of circulation. Likewise, it is used to connote retarded movement of the intestinal contents. Homeostasis, then, conveys the impression that something is stopped, unchanging. If we look in the dictionary, we find homeostasis defined as "the tendency of a system, especially the physiological system of higher animals, to maintain internal stability, owing to the coordinated response of its parts to any situation or stimulus tending to disturb its normal condition or function." There are thus two essential ideas therein: 1) internal stability, and 2) the coordinated response responsible for the maintenance of that internal stability.

One's body temperature, blood pressure, heart rate, respiratory rate, urine output, body weight, blood composition, total body composition, to name but a few, are static day after day, year after year. A whole battery of clinical tests have been developed which measure something. In each instance, the results are meaningful because we know that certain values are always obtained in a state of health and that those values may vary in sickness. We expect these values to be static. We have come to understand many of the physiological mechanisms that maintain this internal stability. When abnormal values are detected that fact alone tells us something about the responsible physiological mechanism. Thus, the importance of the concept of homeostasis to physiology and to medicine.

Without an appreciation of homeostasis and the comprehension of homeostatic mechanisms, medicine is empirical, not rational. Such was the state of the art prior to the "benchmark" papers that established this concept.

First, the means had to be developed for accurate measurement. The early physician depended upon his own senses. He could estimate body heat by virtue of his own temperature receptors, he could count the pulse, he could smell the breath, he could detect sugar in the urine by his sense of taste, he could gain much knowledge by using his eyes. But to measure specific body temperature, blood pressure, blood and urine sugar, the chemical makeup of the whole body and its components all awaited appropriate instrumentation. Just as our current knowledge of the biochemistry of inheritance could not have developed without radioisotope detectors and the amino acid analyzer, our understanding of body temperature required the thermometer.

It is with the thermometer and the measurement of body temperature that I have elected to begin this series of benchmark papers. Man, of course, simply by using his hands has known since antiquity that the body of certain species is warm. He also came to recognize that in death the body cools, and in sickness it becomes abnormally warm. With the thermometer, Charles Blagden and John Hunter in the eighteenth century in England not only measured body temperature, for that had been done by many others years before, but they established that body temperature remains amazingly constant over a remarkable range. Blagden raised the ambient temperature to 260°F! Such observations really began the development of the concept of internal stability.

In Part II, the idea of internal stability is seen to be firmly established in the nineteenth century primarily by the work and publications of Charles Robin, Claude Bernard, Leon Fredericq, and Charles Richet. And early in the twentieth century, there was an explosion of activity which resulted in an understanding of the coordinated responses responsible for the maintenance of that internal stability. As this knowledge grew and as Walter Cannon contributed to it he came to conceptualize what he termed homeostasis. Thus, Part III is devoted mostly to Cannon's landmark publications culminating with his 1929 article in *Physiological Reviews.*

Part IV should really contain articles which served to establish comprehension of the control mechanism of every physiological value. Obviously, that is impossible. Therefore, only two articles have been included. The first finishes where we began, namely with temperature control. The second embodies an extension of the concept of homeostasis by virtue of mathematical expressions. The span in time between the first and last paper is almost exactly 200 years.

Editor's Comments on Papers 1, 2, and 3

The Power of the Body to Destroy Heat

I

Early man, by means of his own senses, detected that certain living things are warm and others are not. He was also able to ascertain that the former maintained about the same warmness despite the ambient temperature whereas the latter varied with the ambient temperature. Obviously, however, progress depended upon the development of instrumentation; in this case, a means of actually measuring temperature.

The first thermometer is said to have been produced in 1592 by Galileo. Early instruments contained water, then wine, and finally, in 1670, mercury. But the inquiring mind could not wait for these refinements. An Italian physician with the marvelous name of Sanctorio Sanctorius, made his own clinical thermometer, determined body temperature with it, and concluded that man's temperature remains remarkably constant except during illness when it rises. He published his findings in 1614 in *De Statica Medicina.*

In 1714, Gabriel Fahrenheit, a German physicist, constructed a thermometer with mercury but took a rather arbitrary reference point for zero. That happened to be the lowest temperature observed in his hometown during a particular winter. And, it was not the air temperature but rather the temperature of a mixture of snow and sal ammoniac! The boiling point of water he set at 212°. Why, I do not know. At any rate, he did measure body temperature with this instrument and found it to be constant at 96°. About the same time, a Swedish astronomer, Anders Celsius, also constructed a thermometer. He decided to use the freezing point of water as 0° and the boiling point as 100° thereby producing a very logical instrument. Why the Fahrenheit thermometer became more popular than the Celsius one in some countries would be interesting to know. And why Antoine Reaumur in 1731 decided to make still another thermometer with only 80° between freezing and boiling could be made part of that investigation.

But whatever the scale, there was now a means of measuring temperature of the air as well as of the living body. Just where the instrument should be placed on, or in, the body was still to be resolved. At first, investigators pressed it against the skin, or in the armpit, or between the thighs. Dr. George Fordyce, about whom we shall read in the Blagden article, was the first to suggest that the bulb of the thermometer be placed under the tongue.

Charles Blagden (1748–1820) was born in Edinburgh. He obtained his M.D. in 1768 and, at age 24, became a Fellow of the Royal Society. His contributions were such that he later became Sir Charles Blagden.

His publications illustrate beautifully the depths of his inquiring mind. The origins of much of our knowledge of the physiology of temperature regulation are to be found therein. He states, for example, in an atmosphere of high temperature, that, "The external circulation was greatly increased; the veins had become very large, and a universal redness had diffused itself over the body."

In his second paper, he says that, ". . . it appears beyond all doubt, that the living powers were very much assisted by the perspiration, that cooling evaporation is a further provision of nature for enabling animals to support great heats." And, "Perhaps no experiments hitherto made furnish more remarkable instances of the cooling effect of evaporation than these last facts; a power which appears to be much greater than hath commonly been suspected."

John Hunter (1728–1793), another Englishman, in this case a surgeon, was born 20 years before Charles Blagden. His main work was done at St. George's Hospital in London, both as a surgeon and as an anatomist. But, as his article clearly shows, he was also a physiologist much concerned with body temperature.

Hunter appeared to be unaware of the work of Blagden; at least he never mentions his name. Still, this was not unusual in those days as the citing of previous work by others was not generally practiced. They both belonged to the Royal Society of London, read their papers at about the same time, and thus, they more than likely were acquainted.

John Hunter was not content with putting the thermometer on the skin, or in the armpit, or under the tongue. He wanted to know what the body temperature was in a variety of places. So he had relatively small thermometers made and inserted them everywhere; in the male urethra, the rectum, and in experimental animals in the body cavities and a variety of organs. He carried out hundreds of experiments, many of which are reported in this paper. His curiosity was obviously insatiable. Most importantly, he confirmed that body temperature remains constant despite broad variations of the external environmental temperature.

Blagden concentrated on high temperatures and was impressed with "the power of the body to destroy heat." Hunter found that the body could generate heat as well. The ability of the body to generate heat in a cold environment, and to dissipate heat when the ambient temperature rises are, of course, fundamental homeostatic mechanisms.

The reading of these papers is hindered somewhat by the representation of the letter "s" everywhere in a word except as the final consonant. They appear to us to be "f's." Yet, interestingly, after struggling through the first few pages, one makes the adjustment and is able to read on quite easily.

1

Experiments and Obſervations in an heated Room.
By Charles Blagden, *M. D. F. R. S.*

Redde, Feb. 16, 1774.

ABOUT the middle of January, ſeveral gentlemen and myſelf received an invitation from Dr. GEORGE FORDYCE, to obſerve the effects of air heated to a much higher degree than it was formerly thought any living creature could bear. We all rejoiced at the opportunity of being convinced, by our own experience, of the wonderful power with which the animal body is endued, of reſiſting an heat vaſtly greater than its own temperature; and our curioſity was not a little excited to obſerve the circumſtances attending this remarkable power. We knew, indeed, that of late ſeveral convincing arguments had been adduced, and obſervations made, to ſhew the error of the common opinions on this ſubeject; and that Dr. FORDYCE had himſelf proved the miſtake of Dr. BOERHAAVE [a] and moſt other authors, by ſupporting many times very high degrees of heat, in the courſe of a long train of important experiments; with which, and his moſt philoſophical concluſions from them, every lover of ſcience muſt earneſtly wiſh that he may ſoon favour the public. In the mean time time, I am happy in an opportunity of laying before this So-

(a) Elem. Chemiæ, tom. I. p. 277, 278.

ciety

ciety the following ſhort account of ſome of theſe experiments, and of the views with which they were undertaken; for the particulars of which I am obliged to Dr. FORDYCE himſelf.

DR. CULLEN long ago ſuggeſted many arguments to ſhew, that life itſelf had a power of generating heat, independent of any common chemical or mechanical means; for, before his time, the received opinions were, that the heat of animals aroſe either from friction or fermentation(b). Governor ELLIS in the year 1758 obſerved(c), that a man can live in air of a greater heat than that of his body; and that the body, in this ſituation, continues its own cold. The Abbé CHAPPE D'AUTEROCHE informs us, that the Ruſſians uſe their baths heated to 60°(d) of REAUMUR's thermometer, about 160 of FAHRENHEIT's, without taking notice, however, of the heat of their bodies when bathing. With a view to add further evidence to theſe extraordinary facts, and to aſcertain the real effects of ſuch great degrees of heat on

(b) To do further juſtice to the philoſophy of this moſt ingenious and reſpectable profeſſor, I muſt here declare, that during my ſtay in Edinburgh, from the year 1765 to 1769, the idea of a power in animals of *generating cold* (that was the expreſſion) when the heat of the atmoſphere exceeded the proper temperature of their bodies, was pretty generally received among the ſtudents of phyſic, from Dr. CULLEN's arguments; in conſequence of which I applied a thermometer, in a hot ſummer-day, to the belly of a frog, and found the quickſilver ſink ſeveral degrees: a rude experiment indeed, but ſerving to confirm the general fact, that the living body poſſeſſes a power of reſiſting the communication of heat.

(c) Philoſophical Tranſactions, vol. L. p. 755.

(d) Voy. en Siberie, tom. I. p. 51.

the

the human body, Dr. FORDYCE tried the following experiments.

He procured a *ſuite* of rooms, of which the hotteſt was heated by flues in the floor, and by pouring upon it boiling water; and the ſecond was heated by the ſame flues, which paſſed through its floor to the third. The firſt room was nearly circular, about ten or twelve feet in diameter and height, and covered with a dome, in the top of which was a ſmall window. The ſecond and third rooms were ſquare, and both furniſhed with a ſky-light. There was no chimney in theſe rooms, nor any vent for the air, excepting through crevices at the door. In the firſt room were placed three thermometers; one in the hotteſt part of it, another in the cooleſt part, and a third on the table, to be uſed occaſionally in the courſe of the experiment: the frame of this laſt was made to turn back by a joint, ſo as to leave the ball and about two inches of the ſtem quite bare, that it might be more conveniently applied for aſcertaining the heat of the body, and ſeveral other purpoſes.

EXPERIMENT I.

In the firſt room the higheſt thermometer ſtood at 120°, the loweſt at 110°; in the ſecond room the heat was from 90° to 85°; the third room felt moderately warm, while the external air was below the freezing point. About three hours after breakfaſt, Dr. FORDYCE having taken off all his cloaths, except his ſhirt, in the third room, and being furniſhed with wooden ſhoes, or rather ſandals tied on with liſt, entered into the ſecond room, and ſtaid five minutes

I

minutes in a heat of 90°, when he began to ſweat gently. He then entered the firſt room, and ſtood in the part heated to 110°; in about half a minute his ſhirt became ſo wet that he was obliged to throw it aſide, and then the water poured down in ſtreams over his whole body. Having remained ten minutes in this heat of 110°, he removed to the part of the room heated to 120°; and after ſtaying there twenty minutes, he found that the thermometer placed under his tongue, and held in his hand, ſtood juſt at 100°, and that his urine was of the ſame temperature. His pulſe had gradually riſen till it made 145 pulſations in a minute. The external circulation was greatly increaſed; the veins had become very large, and an univerſal redneſs had diffuſed itſelf over the body, attended with a ſtrong feeling of heat. His reſpiration, however, was but little affected. Here Dr. FORDYCE remarks, that the moiſture of his ſkin moſt probably proceeded chiefly from the condenſation of the vapour in the room upon his body. He concluded this experiment in the ſecond room, by plunging into water heated to 100°; and, after having been wiped dry, was carried home in a chair; but the circulation did not ſubſide for two hours, after which he walked out in the open air, and ſcarcely felt the cold.

EXPERIMENT II.

In the firſt room the higheſt thermometer varied from 132° to 130°; the loweſt ſtood at 119°. Dr. FORDYCE having undreſſed in an adjoining cold chamber, went into the heat of 119°; in half a minute the water poured down in ſtreams over his whole body, ſo as to keep that part of the floor

floor where he ſtood conſtantly wet. Having remained here fifteen minutes, he went into the heat of 130°; at this time the heat of his body was 100°, and his pulſe beat 126 times in a minute. While Dr. FORDYCE ſtood in this ſituation, a Florence flaſk was brought in, by his order, filled with water heated to 100°, and a dry cloth, with which he wiped the ſurface of the flaſk quite dry; but it immediately became wet again, and ſtreams of water poured down its ſides; which continued till the heat of the water within had riſen to 122°, when Dr. FORDYCE went out of the room, after having remained fifteen minutes in an heat of 130°; juſt before he left the room his pulſe made 139 beats in a minute, but the heat under his tongue, in his hand, and of his urine, did not exceed 100°. Here Dr. FORDYCE obſerves, that as there was no evaporation, but conſtantly a condenſation of vapour on his body, no cold was generated but by the animal powers. At the concluſion of this experiment, Dr. FORDYCE went into a room where the thermometer ſtood at 43°, dreſſed himſelf there, and immediately went out into the cold air, without feeling the leaſt inconvenience; on which he remarks, that the tranſition from very great heat to cold is not ſo hurtful as might be expected, becauſe the external circulation is ſo excited, as not to be readily overcome by the cold. Dr. FORDYCE has ſince had occaſion, in making other experiments, to go frequently into a much greater heat, where the air was dry, and to ſtay there a much longer time, without being affected nearly ſo much, for which he aſſigns two reaſons;

 that

that dry air does not communicate its heat like air ſatu-rated with moiſture; and that the evaporation from the body, which takes place when the air is dry, aſſiſts its living powers in producing cold. It muſt be imme-diately perceived, that, beſides the principal object, theſe curious experiments throw great light on many other very important ſubjects of natural philoſophy.

January 23. The honourable Captain PHIPPS, Mr. BANKS, Dr. SOLANDER, and myſelf, attended Dr. FOR-DYCE to the heated chamber, which had ſerved for many of his experiments with dry air. We went in without taking off any of our cloaths. It was an oblong-ſquare room, fourteen feet by twelve in length and width, and eleven in height, heated by a round ſtove, or *cockle*, of caſt iron, which ſtood in the middle, with a tube for the ſmoke carried from it through one of the ſide walls. When we firſt entered the room, about 2 o'clock in the afternoon, the quickſilver in a thermometer which had been ſuſ-pended there ſtood above the 150th degree. By placing ſeveral thermometers in different parts of the room we afterwards found, that the heat was a little greater in ſome places than in others; but that the whole difference never exceeded 20°. We continued in the room above 20 minutes, in which time the heat had riſen about 12°, chiefly during the firſt part of our ſtay. Within an hour afterwards we went into this room again, without feeling any material difference, though the heat was con-ſiderably increaſed. Upon entering the room a third time, between five and ſix o'clock after dinner, we ob-

ſerved

ſerved the quickſilver in our only remaining thermometer at 198°(c): this great heat had ſo warped the ivory frames of our other thermometers that every one of them was broken. We now ſtaid in the room, all together, about 10 minutes; but finding that the thermometer ſunk very faſt, it was agreed, that for the future only one perſon ſhould go in at a time, and orders were given to raiſe the fire as much as poſſible. Soon afterwards Dr. SOLANDER entered the room alone, and ſaw the thermometer at 210°; but during three minutes that he ſtaid there, it ſunk to 196°. Another time, he found it almoſt five minutes before the heat was leſſened from 210° to 196°. Mr. BANKS cloſed the whole, by going in when the thermometer ſtood above 211°; he remained ſeven minutes, in which time the quickſilver had ſunk to 198°; but cold air had been let into the room, by a perſon who went in and came out again during Mr. BANKS's ſtay. The air heated to theſe high degrees felt unpleaſantly hot, but was very bearable. Our moſt uneaſy feeling was a ſenſe of ſcorching on the face and legs; our legs particularly ſuffered very much, by being expoſed more fully than any other part to the body of the ſtove, heated red-hot by the fire within. Our reſpiration was not at all affected; it became neither quick nor laborious; the only difference was a want of that refreſhing ſenſation which accompanies a full inſpiration of cool air. Our time was ſo taken up with other obſervations that we did not

(c) This thermometer ſtands, near the boiling point, about a degree too high; the ſcale is FAHRENHEIT's.

R 2 count

count our pulses by the watch: mine, to the beſt of my judgment by feeling it, beat at the rate of 100 pulſations in a minute, near the end of the firſt experiment; and Dr. SOLANDER's made 92 pulſations in a minute ſoon after we had gone out of the heated room. Mr. BANKS ſweated profuſely, but no one elſe; my ſhirt was only damp at the end of the experiment. But the moſt ſtriking effects proceeded from our power of preſerving our natural temperature. Being now in a ſituation in which our bodies bore a very different relation to the ſurrounding atmoſphere from that to which we had been accuſtomed, every moment preſented a new phænomenon. Whenever we breathed on a thermometer the quickſilver ſunk ſeveral degrees. Every expiration, particularly if made with any degree of violence, gave a very pleaſant impreſſion of coolneſs to our noſtrils, ſcorched juſt before by the hot air ruſhing againſt them when we inſpired. In the ſame manner our now cold breath agreeably cooled our fingers whenever it reached them. Upon touching my ſide, it felt cold like a corpſe; and yet the actual heat of my body, tried under my tongue, and by applying cloſely the thermometer to my ſkin, was 98°, about a degree higher than its ordinary temperature. When the heat of the air began to approach the higheſt degree which this apparatus was capable of producing, our bodies in the room prevented it from riſing any higher; and when it had been previouſly raiſed above that point, inevitably ſunk it. Every experiment furniſhed proofs of this: toward the end of the firſt, the thermo-

3 meter

meter was ſtationary: in the ſecond, it ſunk a little during the ſhort time we ſtaid in the room: in the third, it ſunk ſo faſt as to oblige us to determine that only one perſon ſhould go in at a time: and Mr. BANKS and Dr. SOLANDER each found, that his ſingle body was ſufficient to ſink the quickſilver very faſt, when the room was brought nearly to its *maximum* of heat.

Theſe experiments, therefore, prove in the cleareſt manner, that the body has a power of deſtroying heat. To ſpeak juſtly on this ſubject, we muſt call it a power of deſtroying a certain degree of heat communicated with a certain quickneſs. Therefore in eſtimating the heat which we are capable of reſiſting, it is neceſſary to take into conſideration not only what degree of heat would be communicated to our bodies, if they poſſeſſed no reſiſting power, by the heated body, before the equilibrium of heat was effected; but alſo what time that heat would take in paſſing from the heated body into our bodies. In conſequence of this compound limitation of our reſiſting power, we bear very different degrees of heat in different mediums. The ſame perſon who felt no inconvenience from air heated to 211°, could not bear quickſilver at 120°, and could juſt bear rectified ſpirit of wine at 130°; that is, quickſilver heated to 120° furniſhed, in a given time, more heat for the living powers to deſtroy, than ſpirits heated to 130°, or air to 211° (f). And

(f) Theſe numbers are the reſult of ſome experiments which were made on the firſt of February, in a room where the heat of the air was 65°. Mr. BANKS

And we had in the heated room where our experiments were made, a ftriking though familiar inftance of the fame. All the pieces of metal there, even our watch-chains, felt fo hot, that we could fcarcely bear to touch them for a moment, whilft the air, from which the metal had derived all its heat, was only unpleafant. The flownefs with which air communicates its heat was further fhewn, in a remarkable manner, by the thermometers we brought with us into the room, none of which at the end of twenty minutes, in the firft experiment, had acquired the real heat of the air by feveral degrees. It might be fuppofed, that by an action fo very different from that to which we are accuftomed, as deftroying a large quantity of heat, inftead of generating it, we muft have been greatly difordered. And indeed we experienced fome inconvenience; our hands fhook very much, and we felt a confiderable degree of languor and debility; I had alfo a noife and giddinefs in my head. But it was only a fmall part of our bodies that exerted the power of deftroying heat with fuch a violent effort as feems neceffary at firft fight. Our cloaths, contrived to guard us from cold, guarded us from the heat on the fame principles. Underneath we were furrounded with an atmo-

BANKS and I found that we could bear fpirits which had been confiderably heated and were now cooling, when the thermometer came to the 130th degree; cooling oil at 129°; cooling water at 123°; cooling quickfilver at 117°. And thefe points were pretty nicely determined; fo that though we could bear water very well at 123°, we could not bear it at 125°, an experiment in which Dr. SOLANDER joined us. And our feelings with refpect to all thefe points, feemed pretty exactly the fame.

fphere

ſphere of air, cooled on one ſide to 98°, by being in contact with our bodies, and on the other ſide heated very ſlowly, becauſe woollen is ſuch a bad conductor of heat. Accordingly I found, toward the end of the firſt experiment, that a thermometer put under my cloaths, but not in contact with my ſkin, ſunk down to 110°. On this principle it was that the animals, ſubjected by M. TILLET to the intereſting experiments related in the Memoirs of the Academy of Sciences for the year 1764, bore the oven ſo much better when they were cloathed, than when they were put in bare: the heat actually applied to the greateſt part of their bodies was conſiderably leſs in the firſt caſe than in the laſt. As animals can deſtroy only a certain quantity of heat in a given time, ſo the time they can continue the full exertion of this deſtroying power ſeems to be alſo limited; which may be one reaſon why we can bear for a certain time, and much longer than can be neceſſary to fully heat the *cuticle*, a degree of heat which will at length prove intolerable. Probably both the power of deſtroying heat, and the time for which it can be exerted, may be increaſed, like moſt other faculties of the body, by frequent exerciſe. It might be partly on this principle that, in M. TILLET's experiments, the girls who had been uſed to attend the oven bore, for ten minutes, an heat which would raiſe FAHRENHEIT's thermometer to 280°: in our experiments, however, not one of us thought he ſuffered the greateſt degree of heat that he was able to ſupport.

A principal

A principal uſe of all theſe facts is, to explode the common theories of the generation of heat in animals. No attrition, no fermentation, or whatever elſe the mechanical and chemical phyſicians have deviſed, can explain a power capable of producing or deſtroying heat, juſt as the circumſtances of the ſituation require. A power of ſuch a nature, that it can only be referred to the principle of life itſelf, and probably exerciſed only in thoſe parts of our bodies in which life ſeems peculiarly to reſide. From theſe, with which no conſiderable portion of the animal body is left unprovided, the generated heat may be readily communicated to every particle of inanimate matter that enters into our compoſition. This power of generating heat ſeems to attend life very univerſally. Not to mention other well known experiments, Mr. HUNTER found a carp preſerve a coat of fluid water round him, long after all the reſt of the water in the veſſel had been congealed by a very ſtrong freezing mixture. And as for inſects, Dr. MARTINE (g) obſerved, that his thermometer, buried in the midſt of a ſwarm of bees, roſe to 97°. It ſeems extremely probable, that vegetables, together with the many other vital powers which they poſſeſs in common with animals, have ſomething of this property of generating heat. I doubt, if the ſudden melting of ſnow which falls upon graſs, whilſt that on the adjoining gravel walk continues ſo many hours unthawed, can be adequately explained on any other ſuppoſition. Moiſt dead

(g) Eſſays Medical and Philoſophical, p. 331.

ſticks

ſticks are often found frozen quite hard, when in the ſame garden the tender growing twigs are not at all affected. And many herbaceous vegetables, of no great ſize, reſiſt every winter degrees of cold which are found ſufficient to freeze large bodies of water. It may be proper to add, that after each of the above mentioned experiments of bearing high degrees of heat, we went out immediately into the open air, without any precaution, and experienced from it no bad effect. The languor and ſhaking of our hands ſoon went off, and we have not ſince ſuffered the leaſt inconvenience.

2

Further Experiments and Observations in an heated Room. *By* Charles Blagden, *M. D. F. R. S.*

Redde, July 6, 1775.

ON the third of April, nearly the same party as before[(a)], together with Lord SEAFORTH, Sir GEORGE HOME, Mr. DUNDAS, and Dr. NOOTH, went to the heated room in which the experiments of the 23d of January were made. Dr. FORDYCE had ordered the fire to be lighted the preceding day, and kept up all night; so that every thing contained in the room, and the walls themselves, being already well warmed, we were able to push the heat to a much higher degree than before. In the course of the day several different sets of experiments were going on together; but to avoid confusion, it will be necessary to relate each series by itself, without regard to the order of time; beginning with that series which serves as a continuation of our former experiments.

Soon after our arrival, a thermometer in the room rose above the boiling point; this heat we all bore perfectly well, and without any sensible alteration in the temperature of our bodies. Many repeated trials, in successively higher degrees of heat, gave still more remarkable proofs of our resisting power. The last of these expe-

(a) See the former experiments, p. 111. of this volume.

riments

riments was made about eight o'clock in the evening, when the heat was at the greateſt: a very large thermometer, placed at a diſtance from the door of the room, but nearer to the wall than to the cockle, and defended from the immediate action of the cockle by a piece of paper hung before it, roſe one or two degrees above 260°: another thermometer, which had been ſuſpended very near the door, ſtood ſome degrees above 240°. At this time I went into the room, with the addition, to my common cloaths, of a pair of thick worſted ſtockings drawn over my ſhoes, and reaching ſome way above my knees; I alſo put on a pair of gloves, and held a cloth conſtantly between my face and the cockle: all theſe precautions were neceſſary to guard againſt the ſcorching of the red-hot iron. I remained eight minutes in this ſituation, frequently walking about to all the different parts of the room, but ſtanding ſtill moſt of the time in the cooleſt ſpot, near the loweſt thermometer. The air felt very hot, but ſtill by no means to ſuch a degree as to give pain: on the contrary, I had no doubt of being able to ſupport a much greater heat; and all the gentlemen preſent, who went into the room, were of the ſame opinion. I ſweated, but not very profuſely. For ſeven minutes my breathing continued perfectly good; but afte that time I began to feel an oppreſſion in my lungs, attended with a ſenſe of anxiety; which gradually increaſing for the ſpace of a minute, I thought it moſt prudent to put an end to the experiment, and immediately left the room. My pulſe, counted as ſoon as I came into the cool air, for the uneaſy feeling rendered me incapable of examining

examining it in the room, was found to beat at the rate of 144 pulſations in a minute, which is more than double its ordinary quickneſs. To this circumſtance the oppreſſion on my breath muſt be partly imputed, the blood being forced into my lungs quicker than it could paſs through them; and hence it may very reaſonably be conjectured, that ſhould an heat of this kind ever be puſhed ſo far as to prove fatal, it will be found to have killed by an accumulation of blood in the lungs, or ſome other immediate effect of an accelerated circulation(b); for all the experiments ſhew, that heating the air does not make it unfit for reſpiration, communicating to it no noxious quality except a power of irritating. In the courſe of this experiment, and others of the ſame kind by ſeveral of the gentlemen preſent, ſome circumſtances occurred to us which had not been remarked before. The heat, as might have been expected, felt moſt intenſe when we were in motion; and, on the ſame principle, a blaſt of the heated air from a pair of bellows was ſcarcely to be born; the ſenſation in both theſe caſes exactly reſembled that felt in our noſtrils on inſpiration. The reaſon is obvious; when the ſame air remained for any time in contact with our bodies, part of its heat was deſtroyed, and conſequently, we came to be ſurrounded with a cooler medium than the common air of the room; whereas when

(b) Since this experiment, I have obſerved the *mucus* from my lungs to be more *ſerous* than before, and to incline more to a ſaltiſh taſte, though the lungs themſelves ſeem perfectly ſound in all other reſpects; which raiſes a ſuſpicion that ſome of the ſmaller arteries ſuffered a degree of dilatation from the increaſed impulſe of the blood.

freſh

fresh portions of the air were applied to our bodies in such a quick succession, that no part of it could remain in contact a sufficient time to be cooled, we necessarily felt the full heat communicated by the stove. It was observed that our breath did not feel cool to the fingers unless they were held very near the mouth; at a distance the cooling power of the breath did not sufficiently compensate the effect of putting the air in motion, especially when we breathed with force.

A chief object of this day's experiments was to ascertain the real effect of our cloaths in enabling us to bear such high degrees of heat. With this view I took off my coat, waistcoat, and shirt, and in that situation went into the room, as soon as the thermometer had risen above the boiling point, with the precaution of holding a piece of cloth constantly between my body and the cockle, as the scorching was otherwise intolerable. The first impression of the heated air on my naked body was much more disagreeable than I had ever felt it through my cloaths; but in five or six minutes a profuse sweat broke out, which gave me instant relief, and took off all the extraordinary uneasiness: at the end of twelve minutes, when the thermometer had risen almost to 220°, I left the room, very much fatigued, but no otherwise disordered; my pulse made 136 beats in a minute. On this occasion I felt nothing of that oppression on my breath which became so material a symptom in the experiment with my cloaths when the thermometer had risen to 260°: this may be partly explained by the less quickness of my pulse, the difference being at least eight beats in a minute, and probably more, as in the experiment without my ſhirt the pulſations were counted before I had left the room; but there is a further circumſtance to be taken into conſideration, that the experiment attended with oppreſſion on the breath was made in the evening after a very plentiful meal, whereas the other was made in the forenoon, ſome hours after a moderate breakfaſt. The unuſual degree of fatigue which I felt from the experiment without my ſhirt, muſt be aſcribed in great meaſure to the more violent effort which the living powers were obliged to exert, in order to preſerve the due human temperature, when ſuch hot air came into immediate contact with my body. In the preſent caſe it appears beyond all doubt, that the living powers were very much aſſiſted by the perſpiration, that cooling evaporation which is a further proviſion of nature for enabling animals to ſupport great heats. Had we been provided with a proper balance, it would undoubtedly have rendered the experiment more complete to have taken the exact weight of my body at going into, and coming out of, the room; as from the quantity loſt ſome eſtimate might be formed of the ſhare which the perſpiration had in keeping the body cool; probably its effect was very conſiderable, but by no means ſufficient to account for the whole of the cooling, and certainly not equable enough to keep the temperature of the body to ſuch an exact pitch: For it ſhould here be remarked, that during all the experiments made this day, whenever I tried the heat of my body, the thermometer always came

3 very

very nearly to the ſame point; I could not perceive even the ſmall difference of one degree, which was obſerved in our former experiments. Should theſe conſiderations, however, be thought inſufficient to prove that evaporation was not the ſole agent in keeping the body cool, I believe that Dr. FORDYCE's experiments in moiſt air will be found to remove all doubts on this ſubject. Several of the gentlemen preſent, as well as myſelf, went into the room without ſhirts many times afterwards, when the thermometer had riſen much higher, almoſt to 260°, and found that we could bear the heat very well, though the firſt ſenſation was always more diſagreeable than with our cloaths.

In all the experiments made this day it was obſerved, that the thermometer did not ſink ſo much in conſequence of our ſtay in the room as on the 23d of January; probably becauſe a much larger maſs of matter had been heated by the longer continuance of the fire.

Our own obſervations, together with thoſe of M. TILLET in the Memoirs of the Academy of Sciences(b), had given us good reaſon to ſuſpect, that there muſt have been ſome fallacy in the experiment with a dog, made at the deſire of Dr. BOERHAAVE, and related in his Elements of Chemiſtry(c). To determine this matter more exactly, we ſubjected a bitch weighing thirty-two pounds, to the following experiment. When the thermometer had riſen to 220°, the animal was ſhut up in the heated room, incloſed in a baſket, that its feet might be defended from

(b) For the year 1764, p. 186, &c. (c) Tom. I. p. 275.

Ttt 2 the

the ſcorching of the floor, and with a piece of paper before its head and breaſt to intercept the direct heat of the cockle. In about ten minutes it began to pant and hold out its tongue, which ſymptoms continued till the end of the experiment, without ever becoming more violent than they are uſually obſerved in dogs after exerciſe in hot weather; and the animal was ſo little affected during the whole time, as to ſhew ſigns of pleaſure whenever we approached the baſket. After the experiment had continued half an hour, when the thermometer had riſen to 236°, we opened the baſket, and found the bottom of it very wet with *ſaliva*, but could perceive no particular *fœtor*. We then applied a thermometer between the thigh and flank of the animal; in about a minute the quickſilver ſunk down to 110°: but the real heat of the body was certainly leſs than this, for we could neither keep the ball of the thermometer a ſufficient time in proper contact, nor prevent the hair, which felt ſenſibly hotter than the bare ſkin, from touching every part of the inſtrument. I have ſince found, that the thermometer held in the ſame place, when the animal is perfectly cool and at reſt, will not riſe above 101°. At the end of thirty-two minutes the bitch was permitted to go out of the room; upon coming into the cold air ſhe appeared perfectly briſk and lively, not in the leaſt injured by the heat, and has now continued very well above a month. Our experiment, therefore, differs, in every eſſential circumſtance of the event, from that related by Dr. BOERHAAVE. With reſpect to this laſt it is remarkable, if the facts be properly

perly reprefented, that an intolerable ftench arofe from the dog; and that an affiftant dropped down fenfelefs upon going into the ftove.

To prove that there was no fallacy in the degree of heat fhewn by the thermometer, but that the air which we breathed was capable of producing all the well-known effects of fuch an heat on inanimate matter, we put fome eggs and a beef-fteak upon a tin frame, placed near the ftandard thermometer, and farther diftant from the cockle than from the wall of the room. In about twenty minutes the eggs were taken out, roafted quite hard; and in forty-feven minutes the fteak was not only dreffed, but almoft dry. Another beef-fteak was rather overdone in thirty-three minutes. In the evening, when the heat was ftill greater, we laid a third beef-fteak in the fame place: and as it had now been obferved, that the effect of the heated air was much increafed by putting it in motion, we blew upon the fteak with a pair of bellows, which produced a vifible change on its furface, and feemed to haften the dreffing; the greateft part of it was found pretty well done in thirteen minutes.

About the middle of the day two fimilar earthen veffels; one containing pure water, and the other an equal quantity of the fame water with a bit of wax, were put upon a piece of wood in the heated room. In one hour and an half the pure water was heated to 140° of the thermometer, whilft that with the wax had acquired an heat of 152°, part of the wax having melted and formed a film on the furface of the water, which prevented the evaporation.

evaporation. The pure water never came near the boiling point, but continued ſtationary above an hour at a much lower degree; a ſmall quantity of oil was then dropped into it, as had before been done to that with the wax; in conſequence of which, the water in both the veſſels came at length to boil very briſkly. A ſaturated ſolution of ſalt in water put into the room, was found to heat more quickly, and to an higher degree, than pure water, probably becauſe it evaporated leſs; but it could not be brought to boil till oil was added, by means of which it came toward evening into briſk ebullition, and conſequently had acquired an heat of 230°. Some rectified ſpirit of wine in a bottle ſlightly corked, which had been immerſed into this ſolution of ſalt whilſt cold, began to boil in about two hours, and ſoon afterwards was totally evaporated. Perhaps no experiments hitherto made furniſh more remarkable inſtances of the cooling effect of evaporation than theſe laſt facts; a power which appears to be much greater than hath commonly been ſuſpected. The evaporation itſelf, however, was more conſiderable in our experiments than it can be in almoſt any other ſituation, becauſe the air applied to the evaporating ſurface was uncommonly hot, and at the ſame time not more charged with moiſture than in its ordinary ſtate. A powerful aſſiſtant evaporation muſt undoubtedly prove, in keeping the living body properly cool, when expoſed to great heats; but it can act only in a *groſs* way, and by no means in ſuch a nice proportion to the momentary exigencies of the animal as would be requiſite for the exact preſervation of its temperature: that other proviſion of nature

nature which feems more immediately connected with the powers of life, is, probably, the great agent in preferving the juft balance of temperature; exerting a greater effort in proportion as the evaporation is deficient, and a lefs effort as the evaporation increafes. This idea correfponds with the general analogy of the animal occonomy, the nicer balances of which are almoft univerfally effected in that part of the body which is formed with the moft fubtile organization.

The heated room will, I hope, in time become a very ufeful inftrument in the hands of the phyfician. Hitherto the neceffary experiments have not been made to direct its application with a fufficient degree of certainty. However, we can already perceive a foundation for fome diftinctions in the ufe of this uncommon remedy. Should the object in view be to produce a profufe perfpiration, a dry heat acting on the naked body would moft effectually anfwer that purpofe. The hiftories of dropfies and fome other difeafes, fuppofed to have been cured by fuch means, are well known to every phyfician. In fome cafes alfo, a moift heat, and in others heat tranfmitted through a quantity of cloaths, might have their peculiar advantages. That the danger likely to enfue from fuch applications is lefs than has been commonly apprehended, our former experiments gave fufficient reafon to believe, and the fame was amply confirmed by thofe which make the fubject of this paper. For during the whole day, we paffed out of the heated room, after every experiment, immediately into the cold air, without

out any precaution; after expoſing our naked bodies to the heat, and ſweating moſt violently, we inſtantly went out into a cold room, and ſtaid there even ſome minutes before we began to dreſs; yet no one received the leaſt injury. I felt nothing this day of the noiſe and giddineſs in my head, which had affected me in making the former experiments; and, whether from the force of habit, or any other cauſe, the ſhaking of our hands was leſs, and we felt leſs languor, though the heat had been ſo much more intenſe.

XLVII. *A*

3

Of the Heat, &c. *of Animals and Vegetables.*

By Mr. John Hunter, *F. R. S.*

Read June 19, and Nov. 13, 1777.

IN the courſe of a variety of experiments on animals and vegetables, I have frequently obſerved that the reſult of experiments in the one has explained the œconomy of the other, and pointed out ſome principle common to both; I have therefore collected ſome experiments which relate to the heat and cold of thoſe ſubſtances. Having found variations in the degree of heat and cold in the ſame experiment, for which I could not account; I ſuſpected that this might ariſe from ſome imperfection in the conſtruction of the thermometer. I mentioned to Mr. RAMSDEN my objection to the common conſtruction of that inſtrument, and my ideas of one more perfect in its nature, and better adapted to the experiments in which I was engaged. He accordingly made me ſome very ſmall thermometers, ſix or ſeven inches long, not above $\frac{2}{12}$ths of an inch thick in the ſtem; having the external diameter of the ball very little larger than that of the

the ſtem, on which was marked the freezing point. The ſtem was embraced by a ſmall ivory ſcale ſo as to ſlide upon it eaſily, and retain any poſition. Upon the hollow ſurface of this ſcale were marked the degrees which were ſeen through the ſtem. By theſe means the ſize of the thermometer was very much reduced, and it could be applied to ſoft bodies with much more eaſe and certainty, and in many caſes in which the former ones could not be conveniently applied: I therefore repeated with it ſuch of my former experiments as were not originally ſatisfactory, and found the degrees of heat very different, not only from what I generally imagined, but alſo from what I had found in my former experiments with the thermometers of the common conſtruction.

I have obſerved in a former paper[(a)], and find it ſupported by every experiment I have made on the heat and cold of animals, that the more perfect have the greateſt power of retaining a certain degree of heat, which may be called their ſtandard heat, and allow of much leſs variation than the more imperfect animals: however, it will appear from the firſt, ſecond, and third experiments, that many, if not all of them, are not capable of keeping conſtantly to one ſtandard; but vary from their ſtandard

(a) Vide Philoſophical Tranſactions for the year 1775, vol. LXV. part II. p. 446.

heat,

heat, either by external applications, or diſeaſe. However, theſe variations are much greater below the ſtandard heat than above it; the perfect animals having a greater power of reſiſting heat than cold, ſo that they are commonly near their ultimate heat. Indeed we do not want any other proof of this variation than our own feelings: we are all ſenſible of heat and of cold, which ſenſations could not be produced without an alteration really taking place in the parts affected; which alteration in the parts could not take place, if they did not become actually warmer or colder. I have often cooled my hands to ſuch a degree, that I have put them into pump-water, immediately pumped, to warm them; therefore, my hands were really colder than the pump-water.

Real increaſe of heat muſt alter the texture or poſition of the parts, ſo as to produce the ſenſation called heat: and as this heat is diminiſhed, the texture or poſition of the parts is altered in a contrary way; which, when carried to a certain degree, becomes the cauſe of the ſenſation of cold. Now theſe effects could not take place in either caſe without a real increaſe or decreaſe of heat in the part; heat, therefore, in its different degrees, muſt be preſent. When heat is applied to the ſkin, it becomes hot, in ſome degree, according to the application; and this may be carried ſo far as actually to burn the living parts: on the contrary, in a cold atmoſphere, a man's hand ſhall become ſo cold as to loſe the ſenſation of cold altogether, and change it for that of pain. Real heat and cold can be carried ſo far, as even to alter the ſtructure of the parts upon which the actions of life depend.

As animals are ſubject to variations in their degrees of heat and cold from external applications, they are of courſe, in this reſpect, affected in ſome meaſure like inanimate matter; and therefore, as parts are elongated or recede from the common maſs, theſe effects more readily take place: for inſtance, all projecting parts and extremities, more eſpecially toes, fingers, noſe, ears, combs of fowls, particularly of the cock, are more readily cooled; and are therefore moſt ſubject to be affected by cold. Animals are not only ſubject to increaſe and decreaſe of heat, ſimilar to inanimate matter; but the tranſition from the one to the other (as far as they allow themſelves to go) is nearly as quick. However, I ſhall not confine myſelf to ſenſation alone, for it is in ſome degree ruled by habit: the habit of uniformity in the degree of the one or the other, will be the cauſe of a conſiderable increaſe of ſenſation from the ſmalleſt variation; while the habit of variation in the degree of heat and cold, will, in a proportional degree, prevent the ſenſation ariſing from either: but we ſhall be guided by actual experiment.

2 The

The parts above mentioned (*viz.* projecting parts and extremities) are ſuch as will admit of the greateſt change in their degrees of heat and cold, without materially affecting the animal. I find that they will raiſe or ſink the thermometer, in ſome degree, according to the external heat or cold applied; although not in a proportional degree to this application, as would be the caſe in inanimate matter. Nor are the living parts cooled or heated in the ſame degrees, which appears from the application of the thermometer to the ſkin; for the cuticle may be conſidered as a dead covering, capable of taking greater degrees of heat or cold, than the living parts underneath can do; and it might be ſuſpected, that the whole of the variation was in the covering. To remove this doubt I made the following experiments.

EXP. I. I ſunk the ball of my thermometer under my tongue, which lay perfectly covered by all the ſurrounding parts, kept it there ſome minutes, and found that it roſe to 97°; having continued it ſome time longer there, I found it roſe no higher. I then took ſeveral pieces of ice, about the ſize of walnuts, and put them in the ſame ſituation, allowing them to melt in part, but not wholly, that the application of cold might be better kept up, occaſionally ſpitting out the water ariſing from the ſolution: this I continued for ten minutes, and found, on

C 2 intro-

introducing my thermometer, that it fell to 77°; ſo that the mouth at this part had loſt 20° of heat. It gradually roſe to 97° again; but the thermometer in this experiment did not ſink ſo low as it would have done in the hand, if a piece of ice had been held in it ſo long. Perhaps one reaſon may be aſſigned: the ſurface under the tongue being ſurrounded with warm parts, renders it next to an impoſſibility to cool it to any greater degree: but I ſuſpect ſtill another reaſon, *viz.* parts which have been in a habit of conſiderably varying in this reſpect, as the hand, will allow of greater latitude, being as it were inſenſibly drawn into cold, nor ſo ſuſceptible of it, as has been already obſerved.

As a further proof, that the more perfect animals are capable of varying their heat, in ſome degree, according to the external heat applied, I ſhall adduce the following experiments made on the human ſubject.

The mouth being a part ſo frequently in contact with the external atmoſphere in the action of breathing, whatever is put into it will be ſuppoſed to be influenced by that atmoſphere; this will always render an experiment made in the mouth, relative to heat and cold, in ſome degree doubtful. I imagined that the urethra would anſwer better, becauſe it is an internal cavity, and can be only influenced by heat and cold applied to the external

external ſkin of the parts. I imagined alſo, that whatever effects heat or cold might have, when applied, would ſooner take place in the urethra than in any other part of the body, as it is a projecting part; and therefore, if living animal matter was in any degree ſubject to the common laws of matter in this reſpect, the urethra would be readily affected: for this purpoſe I got a perſon, who allowed me to make ſuch experiments as I thought neceſſary.

EXP. II. I introduced the ball of my thermometer into the urethra about an inch; after it had remained there a minute, the quickſilver roſe only to 92°; at two inches, it roſe to 93°; at four inches, the quickſilver roſe to 94°; and when the ball had got as far as the bulb of the urethra, where it is ſurrounded by warm parts, the quickſilver roſe to 97°

EXP. III. Theſe parts being immerſed in water heated only to 65° for one minute, and the thermometer introduced about an inch and a half into the urethra, the quickſilver roſe to 79°: this was repeated ſeveral times with the ſame ſucceſs. To find if there was any difference in the quickneſs of the tranſition of heat and cold in living and dead parts, and alſo if the latitudes to which each would go were alſo different, I made the following experiments. As this (*viz.* the urethra) ſtill

4 appeared

appeared to me to be the very beſt part of any animal body for experiments of this kind, I had recourſe to it; and as all comparative experiments ſhould be as ſimilar to one another as poſſible, excepting in thoſe points where the difference (if there is any) makes the eſſential part of the experiment, I procured a dead penis.

EXP. IV. The heat of the penis of a living perſon, an inch and a half in the urethra, was 92° exactly. I firſt heated the dead one to the ſame degree, and then had the living one immerſed in water at 50°, at the ſame time immerſing the dead one in the ſame water; when, introducing the thermometer at different times, I obſerved their comparative quickneſs in cooling from 92°. The dead one cooled faſter; but only by two or three degrees. The living came down to 58°, and the dead to 55°. After having continued the thermometer there ſome time longer, it fell no lower. I repeated the ſame experiment ſeveral times, with the ſame ſucceſs; although ſometimes there was a ſmall difference in the degrees of heat from thoſe of others, the heat of the water alſo differing; but the difference in the reſult was nearly in proportion, in all the three different trials, therefore the ſame concluſions are to be drawn from them. In theſe laſt experiments we find very little difference between the cooling of a part of a dead body, and that of the living; but we cannot ſuppoſe that this can take

place

place through the whole body, as in this caſe a living man ſhould always be of the ſame degree of heat with the atmoſphere in which he lives. The man not chooſing to be cooled lower than 53° or 54°, put it out of my power to ſee if the powers of generating heat were exerted in a higher degree, when the heat was brought ſo low as to threaten deſtruction; but from ſome experiments on mice, which will be related hereafter, it will appear, that the animal powers are called upon to exert themſelves in this, when neceſſary.

From the experiments related I found, that parts of an animal were capable of becoming much colder than the common or natural heat: I therefore made farther experiments, with a view to ſee whether the ſame parts were capable of becoming much hotter than the ſtandard heat of animals. The experiments were made in the ſame manner as the former, only the water was now hotter than the natural heat of the animal.

EXP. V. The natural heat of the parts being 92°, they were now immerſed in water heated to 113° for two minutes, and the thermometer being introduced as before, the quickſilver roſe to 100°½. This experiment I alſo repeated ſeveral times, but could not raiſe the heat of the penis beyond 100°½: this was probably owing to the perſon not being able at this time to bear the application of water warmer than 113°. As theſe were only ſingle expe-

experiments, I chose to make a comparative one with the dead part.

EXP. VI. Both the living and dead part being immersed in water, gradually made warmer and warmer from 100° to 118°, and continued in this heat for some minutes, the dead part raised the thermometer to 114°, while the living could not raise it higher than 102°¼. It was observed, by the person on whom the experiment was made, that, after the parts had been in the water about a minute, the water did not feel hot; but, on its being agitated, it felt so hot that he could hardly bear it. Upon applying the thermometer to the sides of the living gland, the quicksilver immediately fell from 118° to about 104°, while it did not fall above a degree when put close to the dead; so that the living gland produced a cold space of water around it[(b)].

EXP. VII. The heat of the rectum in the same man was 98°½ exactly.

In the second, third, fourth, fifth, and sixth experiments, we had an internal cavity, which is both very vascular and sensible, evidently influenced by external heat and cold, though only applied to the skin of the part;

(*b*) This might furnish an useful hint respecting bathing in water, whether colder or warmer than the heat of the body: for if intended to be either colder or hotter, it will soon be of the same temperature with that of the body; therefore in a large bath, the patient should move from place to place: and in a small one, there should be a constant succession of water of the intended heat.

while,

while, in the ſeventh experiment, another part of the ſame body, where external heat and cold can make little or no impreſſion, was of the ſtandard heat. Although we ſhall find hereafter, from experiment, that the rectum is not the warmeſt part of an animal; yet, in order to determine how far the heat could be increaſed by ſtimulating the conſtitution to a degree ſufficient to quicken the pulſe, I repeated the ſeventh experiment after the man had eaten a hearty ſupper, and drank a bottle of wine, which increaſed the pulſe from 73° to 87°, and yet the thermometer only roſe to 98°½.

Having formerly made experiments upon dormice in the ſleeping ſeaſon, with a view to ſee if there was any alteration in the animal œconomy at that time, I find amongſt theſe experiments the following which appear to be to our preſent purpoſe: but, that I might be more certain of the accuracy of my former experiments, I repeated them with my new thermometer.

EXP. VIII. In a room, in which the air was at between 50° and 60° of temperature, a ſmall opening was made in the belly of a dormouſe, of a ſufficient ſize to admit the ball of my thermometer, which, being introduced into the belly at about the middle of that cavity, roſe to 80°, and no higher.

EXP. IX. The moufe was put into a cold atmofphere of 15° above 0, and left there for fifteen minutes; after which, the thermometer being introduced a fecond time, it rofe to 85°.

EXP. X. The moufe was again put into a cold atmofphere for fifteen minutes more; and the thermometer being then introduced, the quickfilver rofe to 72° only, but gradually came up to 83°, 84°, and 85°.

EXP. XI. It was put a third time into the cold atmofphere, and allowed to ftay there for thirty minutes; the lower part of the moufe was at the bottom of the difh, and almoft frozen; the whole of the animal was a little numbed, and a good deal weakened. When the thermometer was introduced, it varied according to the different parts of the belly; in the pelvis, near the parts moft expofed to the cold, it was as low as 62°; in the middle, among the inteftines, about 70°; but near the diaphragm it rofe to 80°, 82°, 84°, and 85°; fo that in the middle of the body the heat had decreafed 10°. Finding a variation in different parts of the fame cavity in the fame animal, I repeated the fame experiments upon another dormoufe.

EXP. XII. I took a healthy dormoufe, which had been afleep in a room in which there was a fire (the atmofphere at 64°): I put the thermometer into its belly, nearly at the

middle,

middle, between the thorax and pubis, and the quick-ſilver roſe to 74° or 75°; when I turned the ball towards the diaphragm, it roſe to 80°; and when I applied it to the liver, it roſe to 81°½.

EXP. XIII. The mouſe was put into an atmoſphere at 20°, and left there half an hour; when taken out, it was very lively, much more ſo than when put in. I intro-duced the thermometer into the lower part of the belly, and it roſe to 91°; and upon turning it up to the liver, to 93°.

EXP. XIV. The animal was put back into the cold at-moſphere at 30° for an hour, when the thermometer was again introduced into the belly; at the liver it roſe to 93°; in the pelvis, to 92°: it was ſtill very lively.

EXP. XV. It was again put back into the cold atmo-ſphere at 19°, and left there an hour; the thermometer at the diaphragm was 87°; in the pelvis, 83°; but the animal was now leſs lively.

EXP. XVI. It was put into its cage, and two hours after the thermometer, placed at the diaphragm, was at 93°.

From theſe experiments we have actual heat increaſed and decreaſed by the application of external cold; and likewiſe the heat varied according to the powers of life, as well in the ſame parts, as alſo in the different parts, of the ſame animal; for at firſt the natural heat of the

D 2 animal

animal was much below the common ſtandard, and, by the application of cold, and the powers of reſiſtance to the cold being thus increaſed, the heat was conſiderably augmented; but when the animal was weakened by thoſe exertions, it fell off with reſpect to the power of producing heat, and this in proportion to the diſtance from the heart.

Why the heat of this animal ſhould be ſo low as 80° in an atmoſphere of between 50° and 60°, is not eaſily accounted for, except upon the principle of ſleep. But I ſhould very much ſuſpect, that the ſimple principle of ſleep is out of the queſtion, as ſleep is an effect that takes place in all degrees of heat and cold. In thoſe animals where the voluntary actions are ſuſpended, it appears to be an effect ariſing from a certain degree of cold acting as a ſedative, under which the animal faculties are proportionably weakened, but ſtill retain the power of carrying on all the functions of life under ſuch circumſtances; but beyond this degree cold ſeems to act as a ſtimulant, and the animal powers are rouſed to action for ſelf-preſervation. It is more than probable, that moſt animals are under this predicament; and that every order has its degree of cold, in which the voluntary actions can be ſuſpended.

When man is aſleep, he is colder than when awake; and I find, in general, that the difference is about one de-

2 gree

gree and a half, ſometimes leſs. But this difference in the degree of cold between ſleeping and waking is not a cauſe of ſleep, but an effect; for many diſeaſes produce a much greater degree of cold in the animal, without giving the leaſt tendency to ſleep; therefore the inactivity of animals from cold is different from ſleep. Beſides, all the operations of perfect life are going on in the time of natural ſleep, at leaſt in the perfect animals, ſuch as digeſtion, ſenſations, &c.; but none of theſe operations are performed in the latter tribe.

To ſee how far the reſult of theſe experiments upon dormice was peculiar to them, I repeated the ſame expements upon common mice. I procured two; one ſtrong and vigorous, the other weakened by faſting.

EXP. XVII. The common atmoſphere being at 60°, I introduced the thermometer into the abdomen of the ſtrong mouſe: the ball being at the diaphragm, the quickſilver was raiſed to 99°, but at the pelvis only to 96°$\frac{3}{4}$.

Here there was a real difference of about 9° in two animals of the ſame ſize, in ſome degree of the ſame genus, and at the ſame ſeaſon of the year, and the atmoſphere of nearly the ſame temperature.

EXP. XVIII. The ſame mouſe was put into a cold atmoſphere of 13°, for an hour, and then the thermometer

was

was introduced as before; the quickſilver at the diaphragm was raiſed to 83°, in the pelvis only to 78°.

Here the real heat of the animal was diminiſhed 16° at the diaphragm, and 18° in the pelvis.

EXP. XIX. In order to determine whether an animal that is weakened, has the ſame powers, with reſpect to preſerving heat and cold, as one that is vigorous and ſtrong, I introduced the ball of the thermometer into the belly of the weak mouſe; the ball being at the diaphragm, the quickſilver roſe to 97°; in the pelvis to 95°: the mouſe being put into the cold atmoſphere as the other, and the thermometer again introduced, the quickſilver ſtood at 79° at the diaphragm, and at 74° in the pelvis.

In this experiment the heat at the diaphragm was diminiſhed 18°, in the pelvis 21°.

Here was a diminution of heat in the ſecond greater than in the firſt, we may ſuppoſe proportional to the decreaſed power of the animal ariſing from want of food.

To determine how far different parts of other animals than thoſe mentioned were of different degrees of heat; I made the following experiments upon a healthy dog.

EXP. XX. The ball of the thermometer was introduced two inches within the rectum, the quickſilver roſe to 100°½ exactly. The cheſt of the dog was opened, and a wound a wound made into the right ventricle of the heart, and the ball immediately introduced; the quickfilver rofe to 101° exactly. A wound was next made fome way into the fubftance of the liver; and the ball being introduced, the quickfilver rofe to $100°\frac{3}{4}$. It was next introduced into the cavity of the ftomach, where it ftood exactly at 101°. All thefe experiments were made in a few minutes.

EXP. XXI. The fame experiments were made upon oxen; the quickfilver rofe exactly to $99°\frac{1}{2}$.

EXP. XXII. The fame were alfo made upon a rabbit, and the quickfilver rofe to $99°\frac{1}{2}$.

From the experiments on mice, and thofe upon the dog, it plainly appears, that every part of an animal is not of the fame degree of heat; and hence we may reafonably infer, that the heat of the vital parts of man is greater than what it is found to be either in the mouth, the rectum, or the urethra.

To determine how far my idea, that animals could have their heat varied in proportion to their imperfections, is juft, I made the following experiments upon fowls, which I confider to be one remove below what are commonly called quadrupeds.

EXP. XXIII. I introduced the ball of the thermometer fucceffively into the *inteftinum rectum* of feveral hens, and

and found that the quickſilver roſe as high as 103°, 103°$\frac{1}{2}$, and in one of them to 104°.

EXP. XXIV. I made the ſame experiments on ſeveral cocks, and the reſult was the ſame.

EXP. XXV. To determine if the heat of the hen was increaſed when ſhe was prepared for incubation, I repeated the twenty-third experiment upon ſeveral ſitting or clocking hens; in one the quickſilver roſe to 104°; in the others, to 103$\frac{1}{2}$, 103°, as in the twenty-third experiment.

EXP. XXVI. Under the hen, who raiſed the quickſilver to 104°, I placed the ball of the thermometer, and found the heat there as great as in the rectum.

EXP. XXVII. I took ſome of the eggs from under the ſame hen, where the chick was about three parts formed, broke a hole in the ſhell, &c. and introduced the ball of the thermometer, and found that the quickſilver roſe to 99°$\frac{1}{2}$. In ſome that were addled, I found their heat not ſo high by two degrees; ſo that the life in the living egg aſſiſted in ſome degree to ſupport its own heat.

It may be aſked, whether thoſe three or four degrees of heat, which are found in the fowl more than in the quadruped, are for the purpoſe of incubation? We found that the heat of the eggs, which was cauſed and ſupported by this heat, was not above the ſtandard of the quadrupeds;

and

and that it muſt probably have been leſs, if the heat of the hen had not been ſo great.

Finding from the above experiments, that fowls were ſome degrees warmer than that claſs commonly called quadrupeds (although certainly not ſo perfect animals) I choſe to continue the experiments upon the ſame principles, and made the following upon thoſe of a ſtill inferior order. The next remove from the fowl are thoſe commonly called amphibious.

EXP. XXVIII. I took a healthy viper, and introduced the thermometer into its ſtomach and anus; the quickſilver roſe from 58° (the heat of the atmoſphere in which it was) to 68°; ſo that in a common atmoſphere it is 10° warmer.

EXP. XXIX. The viper was put into a pan, and the pan into a cold mixture of about 10°; after being there about ten minutes, its heat was reduced to 37°. It was allowed to ſtay ten minutes longer, the mixture being at 13°, and its heat was reduced to 35°. It was allowed to ſtay ten minutes more, the mixture at 20°, its heat at 31°, and it did not become lower; its tail was beginning to freeze; and it was now very weak. It may be remarked, that it became cold much ſlower than many of the following animals.

 The

The frog being, in its ſtructure, more ſimilar to the viper than to either fowl or fiſh, I made the following experiments on that animal.

EXP. XXX. I introduced the ball of the thermometer into its ſtomach, and the quickſilver ſtood at 44°. I then put it into a cold mixture, and the quickſilver ſunk to 31°; the animal appeared almoſt dead, but recovered very ſoon: beyond this point it was not poſſible to leſſen the heat, without deſtroying the animal. But its decreaſe of heat was quicker than in the viper, although the mixture was nearly the ſame.

The next order of animals were fiſh.

EXP. XXXI. I aſcertained the heat of water in a pond, where there were carp, and found it $65^{\circ}\frac{1}{2}$. I then took a carp out of the ſame water, and introduced the thermometer into the ſtomach; the quickſilver roſe to 69°; ſo that the difference between the water and the fiſh was only $3^{\circ}\frac{1}{2}$.

EXP. XXXII. In an eel, the heat in the ſtomach, which at firſt was at 37°, ſunk, after it had been ſome time in the cold mixture, to 31°. The animal at that time appeared dead, but was alive the next day.

EXP. XXXIII. In a ſnail, whoſe heat was at 44°, it ſunk, after it had been put into the cold mixture, to 31°, and then the animal froze.

4 EXP.

EXP. XXXIV. Several leaches having been put into a bottle, and that bottle immerfed in the cold mixture, the ball of the thermometer being placed in the middle of them, the quickfilver funk to 31°; and by continuing the immerfion for a fufficient time to deftroy life, the quickfilver rofe to 32°, and then the leaches froze. In all thefe experiments none of the animals returned to life when they became thawed.

Finding that thefe imperfect claffes of animals are capable of varying their heat to that ftandard which can freeze the folids or fluids when dead, and not much farther before death enfues, I wifhed to determine to what degree of heat the animal could be brought.

EXP. XXXV. A healthy viper was put into an atmofphere of 108°, and allowed to ftay feven minutes, when the heat of the animal in the ftomach and anus was found to be 92°$\frac{1}{2}$, beyond which it would not rife in the above heat. The fame experiment was made upon frogs with nearly the fame fuccefs.

EXP. XXXVI. An eel very weak, its heat at 44°, which was nearly that of the atmofphere, was put into water at 65°, for fifteen minutes; and, upon examination, it was of the fame degree of heat with the water.

EXP. XXXVII. A tench, whofe heat was 41°, was put into water at 65°, and left there ten minutes; the

E 2 ball

ball of the thermometer being introduced both into the ſtomach and rectum, the quickſilver roſe to 55°. Theſe experiments were repeated with nearly the ſame ſucceſs.

To determine whether life had any power of reſiſting heat and cold in theſe claſſes of animals, I made comparative trials between living and dead ones.

EXP. XXXVIII. I took a living and a dead tench, and a living and a dead eel, and put them into warm water; they all received heat equally faſt; and when they were put into the cold, both the living and the dead received it equally.

I long ſuſpected, that the principle of life was not wholly confined to animals, or animal ſubſtance endowed with viſible organization and ſpontaneous motion; but I conceived, that the ſame principle exiſted in animal ſubſtances, devoid of apparent organization and motion, where the power of preſervation ſimply was required.

I was led to this notion twenty years ago, when I was making drawings of the growth of the chick in the proceſs of incubation. I then obſerved, that whenever an egg was hatched, the yolk (which is not diminiſhed in the time of incubation) was always perfectly ſweet to the very laſt; and that part of the albumen, which is not expended on the growth of the animal, ſome days before hatching,

2 was

was alſo perfectly ſweet; although both were kept in a heat of 103°, in the hen's egg for three weeks, and in the duck's for four; but I obſerved, that if an egg was not hatched, that egg became putrid in nearly the ſame time with any other dead animal matter.

To determine how far eggs would ſtand other teſts of a living principle, I made the following experiments.

EXP. XXXIX. I put an egg into cold at about 0, and froze it, then allowed it to thaw; from this proceſs I conceived, that the preſerving powers of the egg muſt be loſt. I then put this egg into the cold mixture, and with it one newly laid; and the difference in freezing was ſeven minutes and a half, the freſh one taking ſo much longer time in freezing.

EXP. XL. A new laid egg was put into a cold atmoſphere, fluctuating between 17° and 15°; it took above half an hour to freeze; but, when thawed and put into an atmoſphere at 25°, it froze in half the time. This experiment was repeated ſeveral times, with nearly the ſame ſucceſs.

To determine the comparative heat between a living and a dead egg, and alſo to determine whether a living egg be ſubject to the ſame laws with the more imperfect animals, I made the following experiments.

EXP.

EXP. XLI. A freſh egg, and one which had been frozen and thawed, were put into the cold mixture at 15°; the thawed one ſoon came to 32°, and began to ſwell and congeal; the freſh one ſunk to 29°$\frac{1}{3}$, and in twenty-five minutes after the dead one, it roſe to 32°, and began to ſwell and freeze.

The reſult of this experiment upon the freſh egg was ſimilar to the above experiments upon the frog, eel, ſnail, &c. where life allowed the heat to be diminiſhed 2° or 3° below the freezing point, and then reſiſted all further decreaſe; but the powers of life were expended by this exertion, and then the parts froze like any other dead animal matter.

From theſe experiments in general it muſt appear, that a freſh egg has the power of reſiſting heat, cold, and putrefaction, equal to many of the more imperfect animals; and it is more than poſſible, that this power ariſes from the ſame principle in both.

From ſome of theſe experiments it appears, that the more imperfect animals are capable of having their heat and cold varied very conſiderably, not according to the extent of the heat or cold of the ſurrounding medium in which they can live, but according to the degree of cold which is capable of altering the parts in a dead ſtate, below which the living power will not go far: for

for whenever the ſurrounding cold brings them to that degree, the power of generating heat takes place till life is gone, then the animal freezes, and is immediately capable of admitting any degree of cold.

From theſe circumſtances of thoſe imperfect animals (upon which I made my experiments) varying their heat ſo readily, we may conclude, that heat is not ſo very eſſential to life in them as in the more perfect; although it be eſſential to many of the operations, or what may be called the ſecondary actions of life, ſuch as digeſting food[(b)], and the propagation of their ſpecies, which requires the greateſt power an animal can exert, more eſpecially the laſt; and, as moſt of the more perfect of theſe imperfect animals are commonly employed in the firſt, we may ſuppoſe their heat to be ſuch as this action of life requires, although in them it be never eſſentially neceſſary to be ſo high as to produce propagation[(c)].

Therefore

(b) How far this idea holds good with fiſh I am not certain.

(c) How far the animal heat is lowered in the more perfect animals, when theſe ſecondary actions are not neceſſary, as in the bat, hedge-hog, bear, &c. I have not been able to determine, not having opportunities of examining theſe animals in their involuntary ſtate. Dormice are in a mixed ſtate between the voluntary and involuntary, and we find the heat diminiſhed when the actions are not vigorous; and from a general review of this whole ſubject it would appear, that a certain degree of heat in the animal is neceſſary for digeſtion, and that neceſſary heat will be according to the nature of the animal. A frog will digeſt food when its heat is at 60°, but not when at 35° or 40°; and it is

very

Therefore, whenever theſe imperfect animals are in a cold ſo low as to weaken their powers, and diſable them from performing the firſt of theſe ſecondary actions, they become in ſome meaſure involuntary, and remain in a torpid ſtate during the degree of cold which will always occur in ſome part of the winter in ſuch countries as they inhabit; and the food of ſuch animals in general not being produced in the cold ſeaſon, affords another reaſon for their torpidity.

From the circumſtances of their heat being allowed to ſink to the freezing point, or ſomewhat lower, and then becoming ſtationary; and of the animal not being able to ſupport life in a much greater degree of cold for a conſiderable time, we ſee a reaſon why thoſe animals always endeavour to procure ſuch places of abode in the winter as ſeldom arrive at that point. Thus we have toads burrowing, frogs living under large ſtones, ſnails protected under the ſhelter of ſtones and in holes, fiſh hav-

very probable that, when the heat of the bear, hedge-hog, dormouſe, bat, &c. is reduced to 70°, 75°, or 80°, they loſe their power of digeſtion; or rather that the body, in ſuch a degree of cold, has no call upon the ſtomach. That animals, in a certain degree of heat, muſt always have food, is further illuſtrated by the inſtance of bees. The conſtruction of a bee is very ſimilar to a fly, a waſp, &c. A fly and a waſp can allow their heat to diminiſh as in the fiſh, ſnake, &c. without loſing life, but a bee cannot; therefore a bee is obliged to keep up its heat as high as what we may call its digeſtive heat, but not its propagating; for which purpoſe they provide againſt ſuch cold as would deprive them even of their digeſtive heat, if they had not food to preſerve it.

ing

ing recourſe to deep water, all which places are generally above the freezing point in our hardeſt froſts: however, our froſts are ſometimes ſo ſevere as to kill many whoſe habitations are not very ſecure.

When the froſt is more intenſe and of longer ſtanding than common, or in countries where the winters are always ſevere, there is generally ſnow, and the water freezes: the advantage ariſing from theſe two circumſtances are great; the ſnow ſerving as a blanket to the earth, and the ice to the water[(e)].

(e) Snow and ice are perhaps the worſt conductors of heat of any ſubſtance yet known. In the firſt place, they never allow their own heat to riſe above the freezing point, ſo that no heat can paſs through ice or ſnow when at 32°, by which means they become an abſolute barrier to all heat that is at or above that degree; ſo that the heat of the earth, or whatever ſubſtance they cover, is retained: but they are conductors of heat below 32°. Perhaps that power decreaſes in proportion as the heat decreaſes under that part.

In the winter 1776 a froſt came on, the ſurface of the ground was frozen; but a conſiderable fall of ſnow alſo came on, and continued ſeveral weeks; the atmoſphere at this time was often at 15°, but it was not allowed to affect the ſurface of the earth conſiderably, ſo that the ſurface of the ground thawed, and the earth retained the heat of 34°, in which beans and peas grow.

The ſame thing took place in water, in a pond where the water was frozen on the ſurface to a conſiderable thickneſs; a large quantity of ſnow fell and covered the ice; the heat of the water was preſerved and thawed the ice, and the ſnow at its under ſurface was found mixed with the water.

The heat of the water under the ſnow was at 35°, in which the fiſh lived very well.

It would be worthy of the attention of the philoſopher, to inveſtigate the cauſe of the heat of the earth, upon what principle it is preſerved, &c.

As all the experiments I ever made upon the freezing of animals, with a view to ſee if it were poſſible to reſtore the actions of life when thawed, were made upon whole ones, and as I never ſaw life return by thawing[(f)]; I wiſhed to ſee how far parts were ſimilar to the whole in this reſpect; eſpecially as we have it aſſerted, and with ſome authority, that parts of a man may be frozen, and afterwards recover: for this purpoſe I made the following experiments upon an animal of the ſame order as ourſelves.

In January 1777, I mixed ſalt and ice till the cold was about 0; on the ſide of the veſſel was a hole, through which I introduced the ear of a rabbit. To carry off the heat as faſt as poſſible, it was held between two flat pieces of iron that went farther into the mixture. That part of the ear projecting into the veſſel became ſtiff, and when cut did not bleed; and the part cut off by a pair of ſciſſars, flew from between the blades like a hard chip.

The ear remained in the mixture nearly an hour: when taken out it ſoon thawed, and began to bleed; it became very flaccid, ſo as to double upon itſelf, having loſt its natural elaſticity. When out of the mixture nearly an hour, it became warm, and this warmth in-

(f) Vide Phil. Tranſ. for the year 1775, vol. LXV. part. II. p. 446.

creaſed

creaſed to a conſiderable degree; while the other ear continued in its uſual cold, and alſo began to thicken. The day following the frozen ear was ſtill warm; and two days after it ſtill retained its heat and thickneſs, which continued for many days after.

About a week after this, the mixture being the ſame as the former, I introduced both ears of the ſame rabbit through the hole, and froze them both: the ſound one, however, froze firſt, probably from its being conſiderably colder at the beginning. When withdrawn, they ſoon thawed, and ſoon both became warm, and the freſh ear thickened as the other had done before.

Feb. 23, 1777, I repeated theſe experiments. I froze the ear of a white rabbit till it became as hard as a board. It was longer in thawing than in the former experiment, and much longer before it became warm; however, in about two hours it became a little warm, and the day following it was very warm and thickened.

In the ſpring 1776, I obſerved that the cocks I had in the country had their combs ſmooth with an even edge, and not ſo broad as formerly, appearing as if near one half of them had been cut off. Having inquired into the cauſe of this, my ſervant told me, that it had been common in that winter during the hard froſt. He obſerved, that they had become in part dead, and at laſt dropped off:

F 2 alſo,

alſo, that the comb of another cock had dropped intirely off, which I did not ſee, as by accident he burnt himſelf to death. I naturally imputed this effect to thoſe combs having been frozen in the time of the ſevere froſt; and having, conſequently, loſt the life of that part by this operation. I endeavoured to try the ſolidity of this reaſoning by experiment.

I attempted to freeze the comb of a very large young cock (which was of a conſiderable breadth) but could only freeze the ſerrated edges (which proceſſes were full half an inch long); the comb itſelf being very thick and warm reſiſted the cold. The frozen parts became white and hard; and, when I cut off a little bit, it did not bleed, nor did the animal ſhew any ſigns of pain. I next introduced into the cold mixture one of his wattles, which was very broad and thin; it froze very readily: upon thawing both the comb and wattle, they became warm, but were of a purple colour, having loſt that tranſparency which the other parts of the comb and the other wattle had. The wound in the comb now bled freely.

Both comb and wattle recovered perfectly in about a month. The natural colour returned firſt neareſt to the ſound parts, increaſing gradually till the whole was become perfectly ſound.

I There

There was a very material difference in the effect between those fowls, the serrated edges of whose combs I suspected to have been frozen in the winter of 176$\frac{5}{6}$, for they must have dropped off. The only way in which I can account for this difference is, that in those fowls the parts were kept so long frozen, that the unfrozen or active parts had time to inflame, and had brought about a separation of the frozen parts, treating them exactly as dead, similar to a mortified part; and that before they thawed, the separation was so far compleated as to deprive them of farther support.

As it is confidently asserted, that fish are often frozen and come to life again, and as I had never succeeded in any of my experiments of this kind upon whole fish; I made some partial experiments upon this class of animals, being led to it by having found a material difference in my experiments upon whole individuals and only parts of the more perfect order of animals.

I froze the tail of a tench (as high as the anus) which became as hard as a board; when it thawed, that part was whiter than common; and when it moved, the whole tail moved as one piece, and the termination of the frozen part appeared like the joint on which it moved.

On the same day I froze the tails of two gold fish till they became as solid as a piece of wood. They were put into

into cold water to thaw: they appeared at firſt, for ſome days, to be very well; but that part of the tail which had been frozen had not the natural colour, and the fin of the tail became ragged. About three weeks after a furr came all over the frozen part; the tail became lighter, ſo that the fiſh was ſuſpended in the water perpendicularly, and they had almoſt loſt the power of motion; at laſt they died. The water in which they were kept was New River water, ſhifted every day, and about ten gallons in quantity.

I made ſimilar experiments upon an order of animals ſtill inferior, *viz.* common earth worms.

I firſt froze the whole of an earth worm as a ſtandard; when thawed it was perfectly dead.

I then froze the anterior half of another earth worm; but the whole died.

I next froze the poſterior half of an earth worm; the anterior half lived, and ſeparated itſelf from the dead part.

As I had formerly in making my experiments upon animals, relative to heat and cold, made ſimilar ones on vegetables, and had generally found a great ſimilarity between them in theſe reſpects, I was led to purſue the ſubject upon the ſame plan; but I was ſtill farther induced to continue my experiments upon vegetables, as I ima-

I imagined I ſaw a material difference between them in their power of ſupporting cold.

From obſervations and the foregoing experiments it plainly appears, that the living principle will not allow the heat of ſuch animals to ſink much lower than the freezing point, although the ſurrounding atmoſphere be much colder, and that in ſuch a ſtate they cannot ſupport life long; but it may be obſerved, that moſt vegetables of every country can ſuſtain the cold of their climate. In very cold regions, as in the more Northern parts of America, where the thermometer is often 50° below 0, where peoples feet are known to freeze and their noſes to drop off if great care be not taken, yet the ſpruce-fir, birch, juniper, &c. are not affected.

Yet that vegetables can be affected by cold, daily experience evinces; for the vegetables of every country are affected if the ſeaſon be more than ordinarily cold for that country, and ſome more than others; for in the cold climates abovementioned, the life of the vegetable is often obliged to give way to the cold of the country: a tree ſhall die by the cold, then freeze and ſplit into a great number of pieces, and in ſo doing produce conſiderable noiſe, giving loud cracks which are often heard at a great diſtance.

In this country the ſame thing ſometimes happens to exotics from warmer climates: a remarkable inſtance of this

this kind happened this winter in his Majeſty's garden at Kew. The *Erica arborea* or Tree-heath, a native of Spain and Portugal, which had kept its health extremely well againſt a garden-wall for four or five years, though covered with a mat, was killed by the cold, and then being frozen ſplit into innumerable pieces(g). But the queſtion is, is every tree dead that is frozen? I can only ſay, that in all the experiments I ever made upon trees and ſhrubs, whether in the growing or active ſtate, or in the paſſive, that whole or part which was frozen, was dead when thawed.

The winter 177$\frac{5}{6}$ afforded a very favourable opportunity for making experiments relative to cold, which I carefully availed myſelf of. However, previous to that winter, I had made many experiments upon vegetables reſpecting their temperature comparatively with that of the atmoſphere, and when they were in their different ſtates of activity: I therefore examined them in different ſeaſons, with a

(g) This muſt be owing to the ſap in the tree freezing, and occupying a larger ſpace when frozen than in a fluid ſtate, ſimilar to water; and that there is a ſufficient quantity of ſap in a tree newly killed is proved by the vaſt quantity which flows out upon wounding a tree. But what appeared moſt remarkable to me was, that in a walnut-tree, on which I made many of my experiments, I obſerved that more ſap iſſued out in the winter than in the ſummer. In the ſummer, a hole being bored, ſcarcely any came out; but in the winter it flowed out abundantly.

view

view to ſee what power vegetables have. I ſhall relate theſe experiments in the order in which they were made.

They were begun in the ſpring, the actions of life upon which growth depends being then upon the increaſe; and they were continued till thoſe actions were upon the decline, and alſo when all actions were at an end, but whilſt the paſſive powers of life were ſtill retained.

The firſt were made on a walnut tree, nine feet high in the ſtem, and ſeven feet in circumference in the middle.

A hole was bored into it on the North ſide, five feet above the ſurface of the ground, eleven inches deep towards the centre of the tree, but obliquely upwards, to allow any ſap, which might ooze through the wounded ſurface, to run out.

I then fitted to this part a box about eight inches wide and five deep, and faſtened it to the tree: the bottom of the box opened like a door with a hinge. I ſtuffed the box with wool, excepting the middle, oppoſite to the hole in the tree: for this part I had a plug of wool to ſtuff in, which, when the door was ſhut, incloſed the whole. The intention of this was to keep off as much as poſſible all immediate external influence either of heat or cold.

 The

The ſame thermometer with which I made my former experiments, ſeven inches and a half long, was ſunk into a long feather of a peacock's tail, with a ſlit upon one ſide to ſhow the degrees; by this means the ball of the thermometer could be introduced into the bottom of the hole.

EXP. I. March 29th, I began my experiments at ſix in the morning, the atmoſphere at $57^{\circ}\frac{1}{2}$, the thermometer in the tree at 55°; when it was withdrawn the quickſilver ſunk to 53°, but ſoon roſe to $57^{\circ}\frac{1}{2}$[(b)].

This experiment was repeated three times with the ſame ſucceſs. Here the tree was cooler than the atmoſphere; when one ſhould rather have expected to have found it warmer, ſince it could not be ſuppoſed to have as yet loſt its former day's heat.

EXP. II. April 4th, half paſt five in the evening, the tree at 56°, the atmoſphere at 62°; the tree therefore ſtill cooler than the atmoſphere.

EXP. III. April 5th, wind in the North, a coldiſh day, ſix o'clock in the evening, the thermometer in the tree was at 55°, the atmoſphere at 47°; the tree warmer than the atmoſphere.

(*b*) The ſinking of the quickſilver upon being withdrawn I imputed to the evaporating of the moiſture of the fluid upon the ball.

EXP.

EXP. IV. April 7th, a cold day, wind in the North, cloudy, at three o'clock in the afternoon, the thermometer in the tree was at 42°, the atmoſphere at 42° alſo.

EXP. V. April 9th, a cold day, with ſnow, hail, and wind, in the North-eaſt; at ſix in the evening the thermometer in the tree at 45°, the atmoſphere at 39°.

Here the tree was warmer than the atmoſphere, juſt as might have been expected. If theſe experiments prove any thing, it is that there is no ſtandard; and probably theſe variations aroſe from ſome circumſtance which had no immediate connection with the internal powers of the tree; but it may alſo be ſuppoſed to have ariſen from a power in the tree to produce or diminiſh heat, as ſome of them were in oppoſition to the atmoſphere.

After having endeavoured to find out the comparative heat between vegetables and the atmoſphere, when the vegetables were in action; I next made my experiments upon them when they were in the paſſive life.

As the difference was very little when in their moſt active ſtate, I could expect but very little when the powers of the plant were at reſt.

From experiment upon the more imperfect claſſes of animals it plainly appears, that although they do not reſiſt the effects of extreme cold till they are brought to the freezing point, they then appear to have the power of refifting it, and of not allowing their cold to be brought much lower.

To fee how far vegetables are fimilar to thofe animals in this refpect, I made feveral experiments: I however fufpected them not to be fimilar, becaufe fuch animals will die in a cold in which vegetables do live; I therefore fuppofed that there is fome other principle.

I did not confine thefe experiments to the walnut tree, but made fimilar ones on feveral trees of different kinds, as pines, yews, poplars, &c. to fee what was the difference in different kinds of trees. The difference proved not to be great, not above a degree or two: however, this difference, although fmall, fhews a principle in life, all other things being equal; for as the fame experiments were made on a dead tree, which ftood with its roots in the ground, fimilar to the living ones, they became more conclufive.

In October I began the experiments upon the walnut tree, when its powers of action were upon the decline, and when it was going into its paffive life.

EXP. VI. October 18th, at half paft fix in the morning, the atmofphere at $51°\frac{1}{2}$, the thermometer in the tree was at $55°\frac{1}{2}$; but, on withdrawing and expofing i for a few minutes in the common atmofphere, it fell to $50°\frac{1}{2}$.

EXP.

EXP. VII. October 21ſt, ſeven o'clock in the morning, the atmoſphere at 41°, the tree at 47°.

EXP. VIII. October 21ſt, in the evening at five o'clock, the atmoſphere at 51°½, the tree at 57°.

EXP. IX. October 22d, at ſeven in the morning, the atmoſphere at 42°, the tree at 48°

EXP. X. October 22d, one o'clock after noon, the atmoſphere at 51°, the tree at 53°.

EXP. XI. October 23d, in the evening of a wet day, the atmoſphere at 46°, the tree at 48°.

EXP. XII. October 28, a dry day, the atmoſphere at 45°, the tree at 46°.

EXP. XIII. October 29th, a fine day, the atmoſphere at 45°, the tree at 49°.

EXP. XIV. November 2d, wind Eaſt, the atmoſphere at 43°, the tree at 43°.

EXP. XV. November 5th, wet day, the atmoſphere at 43°, the tree at 45°.

EXP. XVI. Nov. 10th, atmoſphere at 49°, the tree at 55°.

EXP. XVII. November 18th, atmoſphere at 42°, the tree at 44°.

EXP. XVIII. November 20th, fine day, the atmoſphere at 40°, the tree at 42°.

EXP.

EXP. XIX. December 2d, the atmoſphere at 54°, the tree at 54°.

In all theſe experiments, which were made at very different times in the day, *viz.* in the morning, at noon, and in the evening, the tree was in ſome degree warmer than the atmoſphere, excepting in one, when their temperatures were equal. For the ſake of brevity I have drawn up my other experiments (which were made on different trees) into four tables, as they were made at four different degrees of heat of the atmoſphere, including thoſe made in the time of the very hard froſt in the winter of $177\frac{5}{6}$. They were as follows.

1ſt.

Atmoſphere.	Names.	Height. Ft. In.	Diameter. Ft. In.	Heat. °
29 deg.	Carol. poplar,	2	2	$29\frac{1}{2}$
	Engl. poplar,	4	$2\frac{1}{4}$	$29\frac{1}{2}$
	Orien. plane,	3	$1\frac{1}{4}$	30
	Occid. plane,	3.6	2	30
	Carol. plane,	1	$1\frac{3}{4}$	30
	Birch,	3.6	$2\frac{1}{2}$	$29\frac{1}{2}$
	Scotch fir,	3.6	4	$28\frac{1}{2}$
	Cedar libanon,	2.2	$4\frac{1}{2}$	$28\frac{1}{2}$
	Arbutus,	2.6	$3\frac{1}{3}$	30
	Arbor vitæ,	2.8	$3\frac{1}{2}$	29
	Diffid. cyprus,	3	$2\frac{1}{2}$	30
	Lacker varniſh,	3.6	2	30
	Walnut tree,	5	2.4	31

The

The old hole in the walnut tree being full of ſap was frozen up, but a freſh one was made.

2d.

Atmoſphere.	Names.	Height. Ft. In.	Diameter. In.	Heat. °
27 deg.	Spruce fir,	4	$2\frac{1}{2}$	32
	Scotch fir,	1.$5\frac{1}{2}$	$1\frac{1}{2}$	28
	Silver fir,	3.11	$2\frac{1}{2}$	30
	Weymouth fir,	4.6	$2\frac{1}{2}$	30
	Yew,	3.7	3	30
	Holly,	2.6	2	30
	Plumb tree,	4.6	3	$31\frac{1}{2}$
	Dead cedar,	3.11	3	29
	Ground under ſnow,	3 deep	—	34

3d.

Atmoſphere.	Names.	Heat.
24 deg.	Spruce fir,	23°
	Scotch fir,	23
	Silver fir,	23
	Weymouth fir,	23
	Yew,	22
	Holly,	23
	Dead cedar,	24

The ſame trees we mentioned when the thermometer was at 29°, in new holes made at the ſame height, and left ſome time pegged up till the heat produced by the gimlet was gone off; but in which, as they were moiſt

I from

from the ſap, the heat could be very little, eſpecially as the gimlet was not in the leaſt heated by the operation.

4th.

16 deg.	Car. poplar,	17°
	Eng. poplar,	17
	Ori. plane,	17
	Occ. plane,	17
	Carol. plane,	17
	Birch,	17
	Scotch fir,	$16\frac{1}{2}$

It will be neceſſary to obſerve, that the ſap of the walnut tree, which flowed out in great quantity, froze at 32°. I did not try to freeze the ſap of the others.

Now, ſince the ſap of a tree, when taken out, freezes at 32°; alſo, ſince the ſap of the tree, when taken out of its proper canals, freezes when the heat of the tree is at 31°; and ſince the heat of the tree can be ſo low as 17° without freezing; by what power are the juices of the tree, when in their proper canals, kept fluid in ſuch a cold? Is it the principle of vegetation? Or is the ſap incloſed in ſuch a way as that the proceſs of freezing cannot take place, which we find to be the caſe when water is confined in globular veſſels? If ſo, its confinement muſt be very different from the confinement of the moiſture in dead vegetables; but the circumſtance of vege-

vegetables dying with the cold, and then freezing, appears to anſwer the laſt queſtion. Theſe, however, are queſtions which at preſent I ſhall not endeavour to ſolve.

I have made ſeveral experiments upon the ſeeds of vegetables ſimilar to thoſe on the eggs of animals; but, as inſerting them would draw out this paper to too great a length, I will reſerve them for another.

Editor's Comments on Papers 4–12

Le Milieu Intérieur

II

Frederic L. Holmes, whose doctoral dissertation is entitled "Claude Bernard's Concept of the Milieu Intérieur," states that,

> The concept of the milieu intérieur ranks, with the ancient idea of a nutritive fluid and the twentieth century idea of a vehicle for the integrative action of hormones, among the fundamental conceptions of the biological role of the blood. But physiologists have come to identify Bernard's milieu intérieur too easily as a preliminary version of Walter Cannon's definition of homeostasis, so that much of the conceptual richness and historical complexity of Bernard's own thought about the subject have dropped from view. Bernard contemplated and discussed his idea of the internal environment for over 20 years, adding gradually new facets of meaning until it came to epitomize for him his whole approach to experimental physiology and medicine, as well as his understanding of the organization of living organisms themselves.

These are certainly wholly valid statements as the papers that follow will show, but they also determine with equal clarity that le milieu intérieur is the true base upon which the concept of homeostasis was developed.

In this section, the writings of Charles Robin, Claude Bernard, Eduard Pflüger, Leon Fredericq, and Charles Richet are presented. Charles Robin is included because apparently he used the term before Bernard did. In 1853, Robin published a book, *Treatise of Anatomical and Physiological Chemistry, Normal and Pathologic.* On page 14, he discusses the relationship of the living organism and its environment. He then refers to "the humors (this milieu de l'intérieur)" but he goes no further with the concept. A year later, in 1854, Bernard presented his first lecture as the occupant of the newly created chair of physiology of the University of Paris. This lecture was promptly published in *Le Moniteur des Hopitaux.* In this opening lecture, Bernard discusses broad, general principles. He defines the field of physiology and then states that "There are two general conditions necessary for the understanding of the mechanism of life: 1. the environment (le milieu); 2. the organism." This environment is the external one. In his second lecture he speaks of the internal environment without actually using the expression "le milieu intérieur." He says, ". . . mais si l'on arrive à l'individu, où existent un grand nombre de molécules, il est impossible qu'elles soient toutes en rapport avec l'extérieur; il faut donc, pour qu'elles soient en rapport avec le milieu, un artifice. . . . Cet artifice, c'est la circulation; le sang est ce milieu." So the blood is the internal environment, or at least an important component of it and, at least in this early lecture, "intérieur" was merely implied.

In later writings, as the reader will see, "intérieur" was added and, more importantly, the concept was steadily developed to embrace physiological mechanisms responsible for the maintenance of the constancy of the internal environment which is, after all, the very definition of homeostasis. Robin did nothing to develop this idea. Yet, according to Holmes, Robin later claimed that he had been overlooked as the originator of the concept. Clearly, he should be given credit for the first use of the term, but little for its development.

Charles Philippe Robin (1821–1885) was a prominent histologist of the Faculty of Medicine. He organized biology into four fields, one of which was entitled "The science which studies the influence of the environment, or, if one prefers, of external agents, on living organisms." He helped found the Society of Biology, as did Claude Bernard. They were thus contemporaries, both noted biologists, and Bernard may well have first heard Robin use the expression "milieu de l'intérieur."

Claude Bernard (1813–1878) studied French, Latin, history, and elementary mathematics. During this period, corresponding to our high school, he had no contact with science. He had little idea what he wanted to do. To make a living, he took a job in a pharmacy. Somehow he came to fancy himself an author, a playwright, and even enjoyed limited success in this endeavor. Flushed with success in a small town, he went to Paris to win fame and fortune, but there the best thing he won was the sound advice to abandon any thought of making a career in the theater. He took that advice, gave up the theater, began the study of medicine, and came under the influence of the leading French physiologist, François Magendie. He graduated from medical school in 1843 and turned immediately to physiology. Beyond doubt, although his contributions were many, his greatest was the development, introduction, and advocation of the experimental method. A life of thought and work on this subject is expressed magnificently in his *Introduction to the Study of Experimental Medicine* published in 1865.

After reading the writings of Bernard, one may well wonder why publications of Pflüger, Fredericq, and Richet are included. The reason is simple. Walter Cannon, who coined the term "homeostasis" mentions this trio as being important in the development of the concept. Their key utterances are included so that the reader can judge for himself.

Eduard Pflüger (1829–1910) was born in Hanau, Germany. He studied under J. P. Muller and Du Bois Reymond in Bonn. In 1868, he founded the *Archiv fur die gesamte Physiologie,* and ten years later, the new Institute of Physiology at Bonn.

His range in physiology was broad. He worked and published in the fields of metabolism, embryology, and respiration. In embryology, he carried out experiments with cross-fertilization of the eggs of different species of frogs; in respiration, he established that the respiratory center (seat of respiration) lies not in the blood or the lungs, but rather in the nervous system. All of this certainly establishes him as a solid investigator, as dedicated to the experimental method as Bernard, but the article cited by Cannon, "Die teleologische Mechanik der lebendigen Natur," published a year before Bernard's death would seem to predate Bernard by years, even centuries, for it is a bewildering mixture of philosophy, theology, and pure speculation with but a sprinkling of what could today be called physiology.

The statement in this article that Cannon quotes is, "The cause of every need of a living being is also the cause of the satisfaction of that need." This is what Pflüger calls the "law of teleologic causality." He then goes on to cite numerous examples. For example, in his most noted field, respiration, he states, "The respiratory motions are primarily regulated by need and demand." No argument there, but there is also no discussion of mechanism, merely the conclusion that if there is a greater need for oxygen, breathing increases to satisfy that need. Teleology, so proudly championed by Pflüger in this very long and rambling article, is the despair of every teacher of physiology because it is so dearly depended upon by so many inadequately prepared students. One is hard pressed to believe that Cannon really read this article, or read it in its entirety. In the following pages, only long excerpts are presented since few readers could be expected to wade through the entire 46 pages.

Leon Fredericq (1852–1935) was a very distinguished Belgian physiologist. He was a modern experimentalist who made important contributions in many fields, but particularly in circulation and respiration. He was among the first to recognize the role of the atrial-ventricular node; he studied the relationship between respiration and blood pressure oscillations. He established much of our basic understanding of the regulation of breathing by the nervous system using sophisticated and pioneering techniques. He was also well-known for his work on the mechanism of temperature regulation, on the protein content of blood, the mechanism of coagulation, and the respiratory exchange of gases. Fredericq worked with a variety of animals and was greatly interested in comparative physiology. He attained world stature and in 1952, the centennial of his birth was commemorated at the University of Liege.[1]

Leon Fredericq published extensively but only a short note is included here because it is the one from which Cannon quoted. Fredericq clearly understood the importance of regulatory mechanisms that served to maintain the internal environment despite variations in the external environment. His work is obviously of primary importance in the development of the concept of homeostasis.

Charles Robert Richet (1850–1935) was born in Paris, the son of a well-known surgeon. He obtained his M.D. in 1877. In 1881, he was appointed co-director of the Revue Scientifique and in 1877, he became professor of physiology in the faculty of medicine of the University of Paris. He worked on the physiology of respiration, on epilepsy, the treatment of tuberculosis and, most importantly, immunity. Interest-

---

[1]Leon Fredericq, Un Pionnier de la Physiologie. Sciences et Lettres, Liege, 1953.

ingly, Richet was not only an outstanding scientist, but he also won fame as a pioneer of aviation, an earnest worker on behalf of peace, and as a poet, novelist, and playwright.

The apparently tireless Richet published extensively including a multiple volume *Dictionnaire de Physiologie.* In discussing "Defense (Fonctions de)," he mentions self-regulation. He states, "Life is a perpetual self-regulation, an adaptation to changing external conditions." Further, ". . . the living being is stable . . . because it is modifiable." In other words, by virtue of a dynamic self-regulating mechanism, stability, constancy, is maintained.

In summary, then, this section presents publications that carry the concept from the birth of the idea of "le milieu intérieur" to the recognition of the mechanism responsible for the maintenance of the constancy of this environment. Throughout, the term "le milieu intérieur" has been translated as "the internal environment," rather than "the internal milieu" as is used by some authors.

4

Treatise of Anatomical and Physiological Chemistry, Normal and Pathologic

CHARLES ROBIN and
F. VERDEIL

Article II

Comparative Domains of Anatomy, Physiology, and Chemistry

8. Anatomy has as its subject everything that constitutes organization (economy). The parts, be they solid or liquid, contribute no less to its constitution; they are no less indispensable one than another for realization of the phenomena studied in physiology. It is certainly not in the consistency, in the state of fluidity or solidity of the objects studied, that lies the distinction between anatomy and physiology on the one hand, between anatomy and chemistry on the other.

Physiology is the branch of biology which has as its subject organized beings in the dynamic state, in action, in a state of activity, and as the object or goal the knowledge of the acts they carry out. The most general is nutrition which thus characterizes life. As this knowledge can be reduced to the notion of a certain number of general facts or laws, it is sometimes said that the object of physiology is the knowledge of the laws of organic activity or of life. As life has often been personified to make of it an entity which has become an agent of the organic or vital phenomena, it is advisable to reject, at least temporarily, the expression of "laws of life, of vitality."

In anatomy it is the agent that is studied, in physiology, it is the act: that is what separates the two sciences. No clear-cut and serious study is possible as long as one does not have a precise idea of the difference existing between the static or resting state, and the dynamic or activity of subjects in general and of organized beings in particular.

In chemistry it is the raw matter, its conditions of molecular activity, also the general and special phenomena of this activity that are pursued; in anatomy it is the disposition of parts or principles constituting directly organized matter, and this matter with its different successive modes of arrangement that are studied.

The *subject* and the *object* of the studies are therefore very different in the two cases and we say the subject and the object because they must always both be considered since, in some exceptional circumstances, the subject of study can be the same but the goal different.

Thus, the otolith, which is described in anatomy, is formed of crystals of $CaCO_3$, studied also in chemistry, but the examination of that structure is made with an objective so clearly anatomical that there is not a moment of doubt as to the natural separation of the two orders of research. That case which is exceptional for the whole of anatomy presents itself rather frequently in one of its subdivisions: the one which treats the direct principles. Hence we recognize that, by having neglected in our definitions the subject and the object of each science of anatomy and chemistry in particular, we often apply to one that which belongs to the other, and vice versa.

A. Logical Need to Include in Anatomy Everything which Belongs to its Attributes

9. Would you like, from the practical point of view, to have a criterion allowing one to distinguish immediately what is included in the field of anatomy and what belongs to physiology? Try to reconstruct in your mind a living being using the parts of the organism described in the treatises of anatomy and imagine it in action at a given moment. You will see at once that for this activity to be possible, the liquids which fill the vessels, moisten the synovials and the arachnoid, fill the glandular spaces and the excretory ducts are necessary. There is a whole class of parts which contribute to the constitution of the organism and without which there is only a solid frame, a sort of skeleton, unthinkable in action. The existence of acts by the organism cannot be conceived as long as one does not consider the humors, but to say therefore that their study belongs to physiology would be childish reasoning; for evidently solids are no less indispensable, and as important physiological phenomena take place therein as in the liquids.

We are, however, obliged to underline such objections since they are among those made to us most frequently in discussions by authoritative names. Notice then that it is impossible to conceive of a living organism without an environment from which it draws and into which it discharges; one is the agent, the other provides the conditions of activity. The agent in its turn is subdivided into various orders of parts, which are also indispensable to each other: on the one hand are the solids which are primarily active and on the other hand the humors which keep them capable of acting, humors which are the conditions of action having the same relationship to solids as the internal environment has to the total organism and lastly which establish the link between the interior and the exterior, between the general environment and the organized being. Should the general environment disappear or be altered, the agent ceases to act; should the humors (this *milieu de l'intérieur*) change, all activity stops in the solids as if they disappeared themselves, as if they had been destroyed.

There is after all no need to dwell at great length on this point which has already been treated elsewhere in articles which are only the introduction to the present article which will itself be continued by a consideration of general anatomy.

B. General Reasons for the Lack of Precision in Anatomy, Physiology, and Chemistry

10. As long as it is thought that each static or anatomic disposition is not intimately connected with a corresponding dynamic or physiological state, the treatises of anatomy, physiology and chemistry will fight over the study of a certain number of parts of organized beings. But, each gets away only with shreds and the result of the conflict is that no good description of the constitution of any humor, even of the blood, exists.

If this concept is overlooked it is no longer recognized that everything which is part of the body has to be studied in anatomy before the study of corresponding actions is approached and it becomes difficult to combine anatomic descriptions with the study of fundamental physiological phenomena. It is then that we hear that general anatomy is more and more only a branch of physiology, a description which fully characterizes the aberration that we just pointed out. To say that for the physiologist general anatomy is as necessary to know as any other branch of anatomy is indisputable, but to say that general anatomy blends more and more into physiology is first to establish a shocking opposition between words with very plain meaning, secondly it is to return to the state of confusion between these two sciences from which it is so hard to escape and which has so much damaged the progress of both: which happens every time a clear distinction is not made between the study of the conditions of activity and the study of the acts themselves.

Thereon lies the whole difficulty with which anatomists, physiologists, and chemists have to grapple when they are dealing with the study of humors and first principles.

5

General Physiology Course of the Faculty of Sciences Opening Lecture: Introduction of the Method

CLAUDE BERNARD

The creation of a chair of physiology, we shall not say of *general* physiology, because we do not fully understand the meaning of the word, even after M. Claude Bernard's definition, but at least of physiology, if not of comparative physiology is an event important enough to be of interest to us in order to know which direction this new course shall take, what kind of spirit will reign. The most varied suppositions have been put forward, the greatest hopes expressed. True, the background, the excellent judgment, the eminence of M. Cl. Bernard were for us sufficient guarantee that the learned professor would not, in his new position, vary from the principles which have guided him so far, from the methods or rather from the method which will bring forth progress in the science, but which alone can make of physiology as well as of medicine a real science. A large and brilliant audience composed chiefly of hospital interns and among which M. Rayer had not hesitated to take his place, shared evidently the feeling that we just expressed; it is therefore with a lively interest that this audience listened to the professor; from the first words on, the audience expressed its great satisfaction. M. Cl. Bernard has remained what he was and the method which will guide him is the method which has always guided him: the experimental method.

We owe to a kind and attentive listener, a hospital intern, the editing of that first Lecture in which M. Cl. Bernard set forth his scientific philosophy in the clearest manner and if at times the form leaves something to be desired, our readers will note with great satisfaction on what ground this philosophy is founded; they will draw from it useful inspiration, be it for the direction of their studies, or for the path they must follow if they want to undertake original investigations with success.

M. Cl. Bernard has been rewarded by warm approval for his perseverance on the road of science and he has been able to convince himself that the only real success to which a scientist can aspire today is that obtained by appealing to human reason and not by indulging in fantasies which are neither French nor intelligible.

Here is the text of the first Lecture of M. Cl. Bernard:

Gentlemen, physiology such as understood today encompasses the study of all manifestations of life in living beings. With regard to its object, it is as old as all the sciences which deal with organized beings. One understands, however, that physiology whose aim is the interpretation of life mechanisms had to come after all those sciences because it needs them; it was therefore preceded by anatomy or the study of tissues and by physico-chemical sciences or the study of the ambient environment. In view of this intimate and necessary union between physiology and anatomy, physics and chemistry, one understands that physiology had to be appropriated by the anatomists, the chemists, or the physicists according to which one of these sciences prevailed at the moment. History shows us that physiology was helped by the discoveries of the anatomists, the chemists, and the physicists who have carried physiology along with them, thus physiology was seen to come under the influence of these doctrines. I shall not try to trace the history of these variations and to demonstrate how physiology detached itself from the other sciences and became independent. Suffice it to say that today this separation is an accomplished fact.

Physiology must be an independent science; it has its own method of experimentation. It is our task here to lay down this new method applied to life phenomena; it is to this method that we owe the rapid progress made in life science in the last few years. We shall see what rules are to be followed for the use of the method which contains the future of physiology, but first I shall tell you what must be the domain of general physiology, the object of this course.

Gentlemen, is general physiology the entire or universal study of life manifestations *in toto*, in their particularities and generalities, is it the study of living atoms? Following that lead one is lost in an infinity of details, the real aim disappears from sight. Should we say that this science is the most general possible expression of phenomena of living beings or call it with Burdach the unifying point of all generalities? Should we rack our brains to know life's essence? include living and nonliving beings? Lastly shall we have to pursue the vital forces under three points of view, idealism, dynanism, mechanism?

It is enough to reflect on life phenomena and on their knowledge obtained so far to see how wrong these general ideas are. I say that these ideas are wrong because, confusing here biological and physical sciences with mathematics and moral sciences, we tried to apply to them one and the same method. We must go up from facts to laws and not go down from laws to facts since the laws are not constituted.

After having reflected on it, in the strict and truly scientific spirit of our age, it seemed to me that general physiology could be more simply defined: "I shall say that it is a methodic exposition of our experimentally acquired knowledge of the whole of life phenomena."

Our whole knowledge must stem from experience; like physicists and chemists we proceed by experimentation. While in courses of comparative physiology, we aim to determine particular functions in certain species, in general physiology we are not concerned with species. There are two general conditions necessary for the understanding of the mechanism of life: 1. the environment; 2. the organism.

Life is neither in one nor in the other; it is in the fusion of the two. This is why physiology could not do without anatomical, physical, and chemical sciences. These two conditions, environment and organism, are indispensable: suppress the environment or the organism, life ceases since it exists only by their fusion. Life itself is localized in the living organism, in the living molecule. This aptitude for life which is shown by vital acts is a special force unique to living beings; this is what we have to accept, not define; let us be content to see the effects of this force. If it be said that life is a property of matter resulting from thc arrangement of certain molecules and that the disturbance of these same molecules is death, it does not alter the fact that the properties of matter are not cleared up by definitions; we have to admit them and judge them by their effects; we have first to determine these properties, then to examine what manifestations or reactions are given by these vital molecules so as to arrive at an understanding of life phenomena in their totality.

. . .

In the foregoing, I have only wanted to make a simple comparison, since in the organic and the inorganic kingdoms the phenomena are of a different nature. In nutrition there is an unceasing and rapid movement which is not found in inorganic beings in a continuous state. While a tissue feeds incessantly on substances drawn from its environment, at the same time it "unfeeds", it loses from its substance in its continuous functioning. Feeding brings disassimilation; these two things are in direct relation one with another. The more a fiber loses as a result of its functioning, the more it has to absorb to redress the balance. This exchange is the more rapid, the more active the manifestations of life. Experiments done with precision on higher animals have shown that in a subject, in the space of 24 hours, approximatively 1/10 of its weight of new substance enters and that the subject gets rid of about an equal weight.

Not only does this movement exist, but there is still another mediator which acts through the fluids taken into the intestinal cavities and which bathes the substances to facilitate their absorption. Thus, in each individual, each molecule is renewed by a tenth of its weight. Tissues or organic molecules possess their properties only in function of the relation between intake and loss. The muscle fiber has the property to contract; surrounded by blood, at each contraction it loses some of its substance and immediately takes up new substance from the surrounding blood. If the muscle fiber were isolated from the organism, we would see that this renewal is no longer possible; if we ask how much time is necessary for a fiber separated from the body to remain active, we see that this time differs according to the levels on the animal scale. In lower animals, the mollusks and even in reptiles, we see that the properties of the muscle fiber are preserved for a long time after death because their nutritional phenomena proceed more slowly.

Therefore, for the maintenance of the organism, an unceasing movement is necessary which implies a rapid renewal. The ambient environment wherein the molecules draw the elements which must reconstitute them is the blood in higher animals; therefore the blood is mainly destined to rebuild the solid matter, muscle fibers, nerves, etc. In a word, if we envisage an isolated organic living molecule with access to alimentary matter which has to be absorbed by it for the maintenance of its function, we understand that this molecule can live in that environment but if we come to the individual where a great number of molecules exist they cannot possibly all be in contact with the exterior; to be in contact with their environment they need a device. This device is the circulation; the blood is this environment.

In a complicated organism we always find that the living molecules are in contact with the blood; therefore what is blood? It is a fluid which contains all the substances which surround the individual and which must nourish him. It contains oxygen, it contains substances dissolved by digestion and which are going to get in touch with all the molecules. Therefore blood must be regarded as an *environment* containing all substances necessary for the maintenance of life and which places substances in contact with all living elements. So we have here a property which is the first, the most important of all and which has to be acknowledged first: nutrition. It is a property or affinity which is organic instead of chemical. In the same way as certain substances attract or absorb molecules identical with themselves by affinity, so do organic molecules.

6

Lectures on the Physiological Properties and the Pathological Alterations of the Liquids of the Organism

First Lecture

A. CLAUDE BERNARD

Dec. 9, 1857

Summary: On the experimental method. Experimentation and its perfections. On the experimental critique.

Gentlemen:

We shall have to examine this year the physiological properties and the pathological changes of the different liquids of the organism. Following my custom, before going into the subject matter, I shall devote the first session of the course to generalities on some points of the experimental method applied to the study of life phenomena.

You know in what respect the teaching of the College de France differs from the teaching of the Faculties: here it cannot be our aim to bring before you solely the notions already established in science about the subjects that we treat. We must endeavour to enlarge the field of our knowledge, be it in making discoveries, be it in approaching especially the obscure and unsettled questions in order to elucidate or verify the facts connected therewith. In a word, we have here to present, not simple, expository lectures in which the mind of the listener always remains somewhat passive before established scientific results, but, on the contrary, lectures of research and investigation in which the mind of the listener joining the mind of the professor pursues with him the solution of a problem which preoccupies both.

In these conditions, the generalities of an opening lecture are always a very useful introduction because they have the advantage, in putting us immediately at a common point of view, of allowing us to follow and to appreciate, with the same philosophical approach, all the questions of details which shall crop up in the course of our research.

Today, the biological sciences are fast finding their way. The experimental method is as definitely established as in the other sciences. To our century belongs the glory of this achievement and the name of my illustrious mentor, my predecessor in this chair will remain attached to this definitive advent of the experimental method in the physiological sciences.

However, several reasons prevent this method from providing the physiologist all the services that he should expect from it. First, the scope of the method is still very limited: it is applied to very complex phenomena and what especially complicates it is that it is very often used by men who are in no way trained in it.

What should preoccupy the physiologist today is no longer the introduction of experimentation into scientific usage; that is an accomplished fact. It should be his aim to apply the method properly and to determine its rules. In this lecture, I shall talk to you of the present refinements of experimentation in physiology.

But because I have sometimes read or heard wrong or too narrow definitions of the experimental method, I would like to say a few words to you about the manner in which this method must, in my opinion, be understood.

The experimental method is in the last resort only logic applied to the coordination of the phenomena of nature in order to discover its laws. In that respect, it has general principles which are common to all sciences. But, as we shall see, it is necessary in applying this method to the science of living beings to take into account some very important special indications.

In every case, the experimental method must be applied to a series of findings on facts given in nature (true observation) or on evoked facts (experimentation). Then these observations must be marshalled logically in order to serve as verification of a preconceived hypothesis or idea. I say that we must seek the verification and not the proof of our idea, because only in the first case does the experimenter find himself in a favorable frame of mind to see clearly when he is prepared in advance to accept all results of the experiment, be they favorable or contradictory to the hypothesis which served him as starting point, or even if they were unrelated to it. If, on the contrary, his only preoccupation is to seek suitable arguments to justify his opinion or to refute someone else's, his mind being exclusively fixed on the facts which he wishes to see realized, he finds himself predisposed, as we mentioned elsewhere, to be dominated by a fixed idea which causes him to exaggerate all that relates to the object he pursues while neglecting all the rest. But apart from the fact that such a procedure is incapable of leading to an exact appreciation of the facts, it has also the grave inconvenience of robbing him who uses it of the happy chance, frequent in sciences as little advanced as physiology, of making unexpected discoveries while pursuing something else.

In my opinion, experimentation must have as its object not only the verification of ideas founded on facts previously acquired, but at the same time, so as to be complete and fruitful, it must seek to gain new ideas which will spontaneously arise from the unexpected facts which the conduct of experiments always brings up. The verification alone of a preconceived idea leads in general only to the confirmation or extension of a known theory while the appearance of an unexpected fact constitutes discovery par excellence because from it emerges a new notion which in its turn will engender new experimental verification if a fact which serves as basis for this hypothesis is not available beforehand.

If, for example, a toxic substance about whose origin and nature he had no information were given to a physiologist, it would be impossible for him to have a rational starting point in order to conceive a probable hypothesis. Then, this physiologist would make a first experiment, haphazardly so to say (exploratory experiment) so that the result, whatever it be, suggests the first lead in establishing the hypothesis which will require new experiments for its verification and so on.

In short, without entering into developments which have been discussed elsewhere, I wish only to remind you that in experimentation, particularly when the sciences are not very advanced as is the case for physiology, the unexpected is always more fruitful than the expected aspect because the observation of natural phenomena is more instructive than the idea we make of it for ourselves. Consequently, the best thing to do in all cases is to look first at what there is; the question of knowing then if our forecasts were invalidated or confirmed as entirely secondary from the point of view of the truth we are seeking. But I hasten, after this digression, to come to experimentation of which I have more specifically to speak to you.

Experimentation is the art of provoking the appearance of phenomena by appropriate means, in conditions chosen and defined by the proposed goal. Experimentation must modify its methods and even sometimes its point of view according to the nature of the sciences to which it is applied. I want to prove to you today that the conditions of experimentation have to be envisaged differently acording to whether one experiments on living beings or on inanimate bodies. In my opinion, the whole accuracy of physiological experimentation and the logic of the experimental critique rests on this fundamental consideration.

Everybody understands the importance of perfecting the art of experimentation and this thought is presently the special preoccupation of physiologists and doctors. Everywhere weight and measures are introduced in the evaluation of life phenomena. Everyone appreciates the importance of a rigorous experimentation because, as long as it has not been achieved, it will remain impossible to compare facts, to deduce their laws and thereon to build physiological science.

It is the awareness of this need for exactness which causes, in all appearing papers, each experimenter to be more precise than his forerunners. Everyday new processes or more perfect instruments are invented destined to measure phenomena which, until then, had more or less escaped the observation of the scrutinizers of nature.

Gentlemen, I am not going to enumerate to you, here, all the ways and means of research that the physiologist and the physician borrow from physics and chemistry. It suffices to recognize in that respect the realization of a vast process which tends to grow everyday. This process consists in the acquisition of a mass of very precise instruments and of investigating methods of all kinds which are applied with rigor to the determination and measuring of phenomena under observation. All this must constitute, in fact, the first indispensable condition for the achievement of an exact experiment.

But to obtain a good experimental result, it is not enough to have good instruments, one must also be able to use them competently. For that, it is necessary to place oneself always in identical, comparable experimental conditions.

To implement this second condition of the experiment, physiologists do one thing which seems very simple; they imitate the physicists and the chemists in the use of the instruments they borrow from them. With the aid of the barometer, the thermometer, etc., they can put themselves in determined conditions of pressure, temperature, etc., then as the weight of different animals differs, they reduce to a common unit, the kilogram, all the physiological results obtained. Thus it is the procedure generally adopted today to make animals comparable, and in well done papers appearing today on respiration, digestion, or secretions, for example, each phenomenon is evaluated in function of kilogram animal.

In this progressive perfection of the experimental art, a natural and entirely logical scientific evolution took place: experimentation was first introduced and perfected in the physico-chemical sciences where the complexity of phenomena is less. Later, after a long series of fruitless attempts, this experimentation was finally and definitively introduced into the much more complex biological sciences. Since then, physiologists rightly take care to emulate their elders in the experimental career, the physicists and chemists, whose instruments and methods they borrowed. Thanks to these efforts, one must recognize that today experimentation is sufficiently perfected to yield results of great delicacy, obtained in experimental conditions beyond reproach from the physical, chemical, mechanical, or instrumental point of view.

But here comes the important question in my opinion: Is it sufficient for a good physiological experiment to be irreproachable from the external physico-chemical or the purely instrumental point of view?

Certainly not, for these external conditions which are so important to the physicist and the chemist are of relatively slight importance to the physiologist. The internal vital conditions of the animal under experiment, which are normally neglected by the physicist, have to take priority in all physiological experiments. This is after all quite understandable when one considers the distinctive fundamental character which separates living beings from inanimate bodies.

In fact an inanimate body has by itself no spontaneity; all modifications it can undergo come only from external circumstances and it is evident that, by taking these into account, one is certain to have all experimental conditions necessary for the conception of the experiment.

By contrast, in living beings, there is a spontaneous organic evolution which, although it needs the environment to manifest itself, is nevertheless independent therefrom in its course. This is proved by the fact that one sees a living being born, developing, sickening, and dying without the conditions of the external world changing for the observer or vice versa. The child and the aged, the healthy man and the sick, are they not subject to the same atmospheric pressure? Don't they breathe the same air? Are they not warmed by the same sun and chilled by the same winter?

This kind of independence shown by the organism in the external environment comes from the fact that, in the living being, the tissues are, in reality, removed from direct external influences and protected by a true internal environment (*milieu intérieur*) mostly constituted by fluids circulating in the body. This independence moreover increases as the living being rises in the scale of organization, that is, it is endowed with an internal environment protecting it more completely. In vegetables and lower animals these conditions of independence diminish in intensity and create a more direct relationship between the organism and the environment. In cold-blooded vertebrates, we see the environment having still a larger influence on the form of the phenomena, but in man and warmblooded animals, the independence of the external environment and of the internal environment is such as to make it possible to consider these beings as living in their own organic environment. We have not been able so far to penetrate with our instruments into this internal environment of living beings, but its influence is very great. We shall designate this spontaneous, vital activity for the moment by the name of organic or physiological conditions.

Therefore, I say that when it is a question of devising an experiment on a living being, it will not suffice to keep the external physico-chemical and instrumental conditions constant, as the chemist or physicist does; in addition, and above all, one must standardize the organic or physiological internal conditions of the living being on which the observation is carried out. There are, as may be seen, two very distinct orders of consideration, and this is what makes physiological experiments very much more difficult and complex from experiments of pure physics or chemistry.

7

Lectures on the Physiological Properties and the Pathological Alterations of the Liquids of the Organism

Third Lecture

CLAUDE BERNARD

Dec. 16, 1857

Summary: Blood. Its general function as organic environment. Effects of its withdrawal. Influence of blood on the tissues and of tissues on the blood. Study of the properties of blood. Temperature. Heat producing properties in warm-blooded and in cold-blooded animals. Heat production. Comparative temperatures of venous and arterial blood. Experiments. Critique of previous experiments, causes of error.

Gentlemen:

We cannot imagine the existence of any organic fluid without considering blood. The role assigned to it, so important for the maintenance of the organic phenomena of life, marks it as prime object of the series of studies on which we are about to embark.

Observers were always perfectly aware of the importance of blood; it has been called the nutrient fluid, thus being lumped together, under this general description, with chyle and lymph. Thus, the general fluid which bathes the different tissues and is enclosed in a closed system, without communication with the outside, was conceived as a physiological unit.

In higher animals and in man where we have to consider the nutrient fluid, we distinguish, according to their color, blood, chyle, and lymph. Blood would be the nutrient fluid colored red, chyle the nutrient fluid colored white, and lymph the nutrient fluid which is colorless. Although there is nothing absolute in these distinctions we shall keep them since their usage is long established, except that we shall later see the more precise meaning that we must give them.

What characterizes all these fluids of which we have been talking and what should bring them together is the fact that they are in motion within the organism. Blood, driven into every part of the body, is returned to the center only to be pumped out again into the whole organism. The result is that blood is in constant contact with all the molecules of our tissues, while, at the same time, it is in indirect relationship with the outside world from which it takes matter for repair and into which it puts products which have become unsuitable for life, with the aid of a number of mechanisms.

Thus envisaged in a general way, blood constitutes a true organic environment, intermediate between the external environment wherein lives the whole individual and the living molecules which could not with impunity be put in direct contact with this external environment. Blood, therefore, contains all the ingredients necessary for life, ingredients which it absorbs from outside by means of certain organic tools. Thereupon it acts as vehicle of all the influences which, coming from outside, act on the tissue fibers: oxygen, nutrients, temperature, etc.

Through the respiratory system, blood is in contact with air and therefrom absorbs oxygen which is then carried through the whole organism. Through the system of alimentary absorption, blood takes up from the outside all the liquid materials which are then transported into the organism to serve as nourishment for the tissues. On the other hand, all the organic decomposition products which result from the accomplishment of the nutritional function are poured into the blood and circulate with it to be secreted in the exterior, be it in gaseous form through the skin or the lungs, or in liquid form through the kidneys.

As I mentioned, blood is therefore a true environment into which all the tissues reject their decomposition products and where they find, for the achievement of their functions, invariable conditions of temperature, humidity, and oxygenation, as well as nitrogenous matter, carbohydrates, and salts without which organs cannot thrive. In this nourishment of the organs, however, we ought to be aware of the fact that the tissues are active and act upon the blood so as to appropriate, according to their nature, the various materials from which they are built. It would not be accurate to look upon the blood as containing all the immediate elements of the organs and only depositing them by some sort of selection at such and such a tissue, as Bordeu seemed to have assumed when he said that blood is flowing flesh and forms the tissues. This idea which sprang perhaps from the fact that the blood fibrin becomes somewhat carnified during coagulation is no longer tenable today since we know that the muscle fiber does not consist at all of a substance analogous to blood fibrin.

In a word, blood does not form the organs, and the proof is that the organs exist in the embryo, formed before the appearance of blood; the organs are only nourished with the elements of the blood which they appropriate, each in its own way.

Blood has always been regarded as a fluid indispensable to life, by the simple reason that its loss leads immediately to death.

This fact can be demonstrated by a very simple experiment which we shall certainly not miss showing you. Using a rather small animal and introducing a catheter through the jugular vein into the right ventricle, if we now, with a sufficiently large syringe, withdraw a considerable part of his blood, the animal at once drops as if he were dead. We may however bring him back to life if, without waiting for this state to be prolonged, we push in the piston of the syringe to thus return to him the blood which had been taken from him.

On the need—always recognized—of the existence of blood for the maintenance of the phenomena of life was based a very old idea, on the strength of which it was believed possible to change blood of inferior quality and to renew it. Thus was born the operation of transfusion. This operation, you see, was based on a wrong idea: it was thought that the starting point of the phenomena of nutrition was in the blood, that this fluid somehow became organized. Yet what ages in the organism is not the blood, but the organs. The blood is constantly renewed; it is constantly young and if it does not always have the same properties, it is, above all, due to the conditions in which it is formed, to the state of integrity of the functions which, directly or indirectly, govern its elaboration. It would not be possible to rejuvenate an old man by submitting him to the alimentary regime of a child. In the circle of mutual influences which maintain life, each must be given its distinctive part: the tissues find in the blood the materials of their nutrition, but they must know how to extract them; to them belongs the active role: the blood is entirely passive, it yields what is taken, yet there must be organs to do the taking.

Although our ideas of the role of the blood in the nutritional phenomena do not sufficiently account for the role of the tissues and for the developmental force by virtue of which they are maintained by a mechanism identical with that of their formation, it is not without reason that the cause of a host of diseases has been laid to the changes of the blood. From that idea to the idea of relieving the organism of a deleterious fluid was only one step; bleeding was called upon to fulfill this indication. Who was the physician who first practiced it? We don't know him. Anyway, once this operation had passed into the practice, the theoretical ideas were able to change; each new doctrine could get hold of the result and justify it for different reasons.

Gentlemen, here again we find the influence of the erroneous idea which arose about the role of the blood in the intimate phenomena of nutrition. It is obvious that one can have the idea to replace bad blood, but it should be possible to replace it by a better blood. Yet in the living being the relationship between the organs and the fluid within which they live is such that, to justify this withdrawal of a deleterious fluid, the cause of its alteration would always have to come from outside and not be due to a fault or a perversion of the organs, as is the case so often. The blood is made for the organs, it is true; but I could not repeat it to you too many times, it is also made by the organs. To take notice of only one of the terms of this double influence means to expose oneself to all the consequences of an erroneous theory. If the organs are sick, the blood will be sick; and withdraw it as much as one might, one will never regenerate it.

We shall see later how it is indispensable to keep both these elements of the question in mind and how their solidarity accounts for otherwise inexplicable phenomena; how the chemical acts which occur in the blood are regulated or suspended by the influence of the nervous system, an influence capable of modifying or preventing the physical conditions of their production. Thus you have seen how the influence of the nervous system increased, decreased, or stopped the production of sugar in the organism by acting on the mechanism which makes this production possible.

Perhaps we shall have to occupy ourselves later with the physiological phenomena which accompany the operations of bloodletting and transfusion, as well as with the physiological characteristics and the pathological modifications of the pulse, an object of special studies to which the ancient physicians attached the greatest importance.

The studies on the blood are excessively numerous; they date from every epoch and were undertaken from every point of view. I do not have to recite here the history of circulation; this mechanical aspect of the phenomena was the first to be examined.

The chemical studies came later, towards the middle of the last century; the blood globules were described as especially characteristic for this fluid by Leeuwenhoeck, Malpighi, etc. More recently, iron was discovered to be a constituent element of the blood. Towards the beginning of this century it was recognized that this metal is specifically localized in the red cells. Meanwhile methods of analysis have been developed to study the blood in the physiological state and in various pathological states. These methods are nowadays commonplaces in science; they have been and still are used daily by highly skilled people.

Despite the multiplicity of these investigations, we must admit that hematology is still a very obscure subject owing to our inability to find the law of all the variations obtained in the analyses which were made. Without mentioning the refinement which chemical procedures or analyses, still very deficient even in the opinion of their inventors, can and certainly will undergo, we think that the principal causes of these discrepancies are in the ignorance of the conditions in which the blood was collected. Yet, as we shall see, these conditions which are excessively manifold can bring about, in the nature of the blood, the most remarkable differences, such that in the analytical study of blood just as in the study of all other fluids of the organism, we think we have to determine first of all the organic conditions of the formation of this fluid so as to eventually relate all the observations to one and the same physiological condition. This is the position which we had to take, convinced that it was impossible that the study of the organic fluids could acquire a certain precision and have a real usefulness for science unless it was done in a determined organic condition.

It is only after having acquired some data on the law of physiological changes that we can hope to find the organic conditions to which these chemical changes react. We shall, therefore, look for what are, in the physiological state and under different normal conditions, the mechanical phenomena of blood circulation, the chemical modifications of blood, their influence on the nourishment of organs and on the formation of other fluids, etc.

The order that we shall follow in this study is practically unimportant, but it is useful to indicate that it is always with the view of clarifying the formation of other fluids of the organism that we shall examine the properties of blood.

We shall treat successively:

1. The temperature of blood in a general manner so as to be able later to examine the same property in its relationship with the formation of other secretions.

2. The conditions of pressure to which blood is exposed in the circulatory system, studied in a general way and in their relation to secretions.

3. The color of blood in general and in particular in the secretory organs.

4. The various physico-chemical properties of the blood, such as coagulability, content of water, salts, gas, always examined first in a general way and later in their relation with the formation of other fluids of the organism.

In connection with each one of these studies on blood, we shall take special note of the influence of the nervous system and of the mechanism whereby it can modify the organic fluids physically and chemically.

Lastly, we shall examine the general causes of alterations suffered by the blood and what influences these causes can have on other fluids.

Blood presents a series of physical properties which are related to the uses it is destined to perform. Presently we will consider these properties in the living animal; later we shall deal with the changes occurring in the blood withdrawn from the body.

Among the physical properties there is one which is of the utmost importance: we mean the temperature of the blood. We shall begin the study of the blood by the examination of this property, observed mainly in warm-blooded animals, because it is mainly to man that the results of our research must relate.

A fact which aroused the attention of experimenters from way back is the relation observed between the temperature of animals and the energy of the vital phenomena of which they are the site. While cold-blooded animals follow the variations of the ambient temperature, we see the higher and warm-blooded animals enjoy a temperature of their own, up to a certain point independent of that of the environment and related especially to the energy of vital functions.

As a rule, the blood temperature is higher than that of the environment; in man and mammals it ordinarily varies between 38 and 41°C and these rather feeble fluctuations depend upon the performance of the functions; one sees the temperature drop in the fasting animal during rest and sleep while it rises during wakefulness, under the influence of movement and during digestion.

The result of these observations was expressed and generalized by saying that in higher animals the organism offered a certain resistance to temperature variations. It is true, in fact, that in an environment warmer than the blood, the higher animals maintain themselves at a temperature below the outside air. Magendie and I saw that if an animal was placed in an incubator at 40°C and the temperature of the incubator was gradually raised to 100°C, for a while the animal resisted this increase of temperature by keeping its blood at 40°C or 45°C; but after some time it overheated, the blood successively reached 41, 42, 43, 44°C; at last, at 45°C, death occurs without fail.

The temperature of blood is therefore one of the most essential attributes of this fluid since in higher animals it can fluctuate only in very restricted limits.

What we saw happen to the animal placed in too hot an environment will take place, although in wider limits, where we expose it to cooling. In a cooled environment the temperature of an animal drops. When the animal's blood has dropped to 25 or 30°C, if it is left in that cold environment, it is going to perish. But it can be brought back to life if, at that moment, its temperature is slowly and gradually raised. This property of animals to produce heat is of capital importance; it is all the more necessary to life the higher the animals are in the zoological scale.

It can be said that, in higher animals, tissues do not really feel the effects of temperature of the environment because they are steeped in another environment, a liquid internal environment which is the blood wherein the organs live like the embryo in the fluids which surround him. This internal environment is very important and it can be seen that, with regard to heat-producing phenomena, the animal carries in itself an environment which has its own temperature, 38 to 40°C. Therefore, it is here that one should look for the mechanism whereby an animal can maintain a constant temperature in spite of such large variations of the external temperature. Thus it has always been recognized that research upon animal heat had to be directed towards the blood. It had been seen, by placing a dead animal in an incubator, that organic tissues were poor heat conductors, consequently the temperature of the body had to be provided and spread by the fluids.

When physiological theories got hold of this question, the theory of Lavoisier on combustion triumphed in science. Condensing into a concept, which remains one of the finest efforts of the human mind, all the phenomena involving oxygen, Lavoisier has related the cause of animal heat to an oxidation, to a true combustion. And as it was in the lung that oxygen was absorbed and CO_2 was released, Lavoisier at first assumed that the heat which maintains the animal temperature was produced in the lung. Therefore he assigned a cause to the phenomena of heat production and furthermore he localized its site of action in the lung.

Since then, many experiments were made which were put forward as confirming Lavoisier's ideas; perhaps even today there are still authors who assume that the lung is the seat of animal heat. As a necessary consequence of these views it was also assumed that arterial blood was warmer than venous blood. Let us therefore examine this first question formulated thus: Is arterial blood warmer than venous blood? Is the lung the seat of the phenomena of heat production which maintains animal temperature at an approximately constant degree? Gentlemen, we shall not discuss these questions by reasonings; we shall set forth historically the experiments undertaken to solve them and we shall conclude according to facts.

Among the experiments cited, a great number confirm Lavoisier's views, but let us see what these experiments are. In a great number of these experiments the temperature of the blood was taken directly in an artery or in a vein, and a temperature difference was found in favor of arterial blood. But, in examining these experiments, one sees that they are very different and not at all comparable: some have experimented in vessels of a certain caliber, others in the right or left ventricle, sometimes by opening, sometimes without opening the chest; lastly others took the temperature by placing the thermometer in the jet of blood which burst through the open vessel. For these reasons these experiments were not identical.

But here we meet with another cause of error which in physiology appears at every step and which we have often and at length stressed. The unanimity which without reserve and in all its implication greeted the law formulated by Lavoisier could lead you to believe that the experiments which attempted to verify it were all confirmatory. Far from it, and if in a great number of experiments we find the arterial blood warmer than the venous blood, there are others where, on the contrary, venous blood was found warmer than arterial blood. Well, confirmatory experiments of the Lavoisier theory were taken account of in the evaluation of the facts while contradictory experiments had been left in the shade as insignificant or doubtful. Now, Gentlemen, this is the time to remind you of what I told you in my first lesson. It is an essentially vicious critique which consists in excluding so-called negative facts in the name of other so-called positive facts. There are no negative facts, there are only poorly understood or incompletely observed facts; one must, by new experiments, complete the observation or rectify the evaluation but one must not reject any fact, one must admit them all since they all have their cause and the evaluation consists in finding an interpretation which includes them all in assigning to them value and relationship. This is what we shall try to do after we shall have given to you: 1. the historical table of all the experiments in which arterial blood was found warmer than venous blood; 2. the chronological table of authors who found the opposite, e.g., venous blood warmer than arterial blood.

As a comparison and parallel to the preceding table [pages 96–97], we are going to list the authors who found results opposite to Lavoisier's ideas, e.g., venous blood warmer than arterial blood.

Coleman (1791) found venous blood at first warmer by one degree Fahrenheit but later, arterial blood was found warmer than venous blood. (These undetailed experiments were undoubtedly done on the heart.)

Astley Cooper arrived at results similar to those of Coleman.

Thakerah (1817) obtained similar results.

Mayer found in the horse the blood of the jugular vein colder by one to two degrees than the blood of the carotid; but he never found any difference between the two kinds of blood in the heart of the freshly killed animal.

Autenrieth found the blood of the right ventricle warmer by 0.55 to 1.10 degree than the blood of the left ventricle in recently expired animals in which artificial respiration was maintained. But afterwards, when the blood started to clot, he says, it proved to be warmer in the left ventricle.

Berger (1833) found 40.90 degrees in the left cavities and 40.40 degrees in the right cavities of the heart; the difference was 0.50 degree in favor of the right cavities. He experimented on a sheep but did not indicate his method.

Collard de Martigny and *Malgaigne* (1832) found the blood warmer by one degree in the right heart, the chest of the animal having been partly opened.

Magendie and *Claude Bernard* (1844) observed in horses that the blood of the right heart was warmer than the blood of the left heart. In the standing, living animal a long thermometer was pushed into the heart cavities through the jugular vein and through the right carotid opened in the neck as far down as possible. The temperature difference in favor of the right heart was greater when the animal had just run.

Claude Bernard (1849) reported to the Biology Society the first results of his experiments which prove that blood becomes warmer in passing through the liver, so that the blood of the hepatic veins is warmer than the blood of the ventral aorta and of the portal vein.

Hering (1850) introduced a thermometer into the heart cavities of a calf. He

Table I
Experiments in Which Arterial Blood Was Found Warmer Than Venous Blood

Authors	Arterial blood (°C)	Venous blood (°C)	Experimented vessels
Haller (1760)	37.20	36.10	Undesignated
Crawford (1799)	38.80	37.50	Carotid artery, jugular vein
Krimer (1823)	38.18	37.50	Temporal artery, jugular vein
Scudamore (1826)	37.70	36.60	Carotid artery, jugular vein
	36.10	35.50	Temporal artery, brachial vein
Saissy (1815)	38.30	38.00	Right heart, left heart
	36.50	36.00	Right heart, left heart
	38.00	37.50	Right heart, left heart
	31.40	31.00	Right heart, left heart
Davy (1815)	40.00	39.10	Carotid, jugular
	40.50	40.00	Carotid, jugular
	40.50	40.00	Carotid, jugular
	40.50	39.70	Carotid, jugular
	40.50	40.00	Carotid, jugular
	40.00	39.10	Carotid, jugular
	40.20	39.70	Carotid, jugular
	40.00	39.40	Carotid, jugular
	38:60	37.70	Carotid, jugular
	38.30	38.30	Carotid, jugular
	41.10	40.80	Right heart, left heart
Nasse (1843)	41.10	40.50	Right heart, left heart
	41.10	40.80	Right heart, left heart
	41.10	40.80	In the ventricles
	41.10	40.80	Pulmonary veins
Becquerel and Breschet (1837)	41.10	40.80	Leg artery and vein
	41.10	40.80	Leg artery and vein
	41.10	40.80	Carotid artery and leg vein
	38.90	38.00	Jugular vein, carotid artery
	38.90	38.00	Carotid artery, leg artery
	38.90	38.00	Jugular vein and leg vein

[1]The cloaca 0.10° to 0.60° warmer than the left ventricle. This result was obtained 8 times out of 12 experiments. In two others the temperature was equal and in the other the blood of the left ventricle was 0.10° warmer than the cloaca.

Difference (°C)	Animals	Experimental method
1.10	Undesignated	Not indicated
1.30	Sheep	Thermometer in vessels
0.68	Man	Thermometer in bloodstream
1.10	Sheep	Thermometer in vessels
0.60	Man	Thermometer in bloodstream
0.50	Woodchuck	Puncture of the auricle, open thorax
0.50	Hedgehog	Puncture of the auricle, open thorax
0.50	Squirrel	Puncture of the auricle, open thorax
0.40	Bat	Puncture of the auricle, open thorax
1.10	Lamb	Thermometer in vessels
0.50	Lamb	Thermometer in vessels
0.50	Lamb	Thermometer in vessels
0.80	Lamb	Thermometer in vessels
0.50	Lamb	Thermometer in vessels
1.10	Sheep	Thermometer in vessels
0.50	Sheep	Thermometer in vessels
0.60	Sheep	Thermometer in vessels
0.90	Ox	Thermometer in vessels
0.00	Ox	Thermometer in vessels
0.30	Lamb	Thermometer in the heart; open chest; dead animal; temperature in rectum 40°
0.60	Lamb	Thermometer in the heart; open chest; dead animal; temperature in rectum 40.5°
0.30	Lamb	Thermometer in the heart; open chest; dead animal; temperature in rectum 48.5°
0.30	Chicken	Thermometer in heart cavities; open chest; living animal[1]
0.30	Chicken	Open vessels; thermometer in bloodstream[2]
1.12	Dog	Thermoelectric apparatus
0.84	Dog	Thermoelectric apparatus
0.84	Dog	Thermoelectric apparatus
0.90	Dog	Thermoelectric apparatus
0.15	Dog	Thermoelectric apparatus
0.30	Dog	Thermoelectric apparatus

[2]Blood of the pulmonary veins 0.56° to 1.56° colder than the cloaca. By the same method, Nasse finds the right ventricle 0.96° to 1.80° colder than the cloaca while he admits that the left ventricle has a temperature nearly equal to the cloaca. The general conclusion of Nasse is that the left ventricle is 1° warmer than the right.

found 38.77° for arterial blood and 39.30° for the blood of the right cavities; a difference in favor of the right cavities of 0.53°.

G. Liebig (1854) published an extensive and well done paper in which he comes to the same conclusion: that venous blood is warmer than arterial blood.

Fick (1855) published a paper containing numerous and detailed experiments leading to the same conclusions.

Confronted with the two classes of facts that we just reported, should we adopt solely the first and declare arterial blood warmer than venous blood? Or should we take account only of the latter and look at arterial blood as less warm than venous blood? Should we take an average of the results and say that arterial and venous blood have the same temperature?

No, Gentlemen, we must admit absolutely everything and our conclusion will be like the facts, that is, that there are cases in which the temperature of arterial blood is higher than the temperature of venous blood and other cases in which the temperature of venous blood is higher than the temperature of arterial blood: hence, we admit in principle that the conditions of these experiments is different. After all, it is always thus in physiology and elsewhere: we can only achieve concordant results by studying the conditions in which they were obtained.

Let us search first, in the experiments referred to in support of the conclusions offered if there are not some likely to furnish conclusions less general than those which were drawn from them and closer to the observed facts. In fact we find that when arterial and venous blood was examined comparatively in the limbs, the result was constantly identical: arterial blood was always warmer than venous blood. The blood coming to the limb has unanimously been found warmer than the blood leaving it.

Now, if instead of observing the temperature of blood in the vessels of the limbs, we take it from the right and the left ventricles, we find an opposite result every time the experiment has been properly done. (We shall explain presently the conditions which make this test satisfactory or useless.)

We see already from these two categories of experiments that the venous blood is warmer or less warm than the arterial blood according to the regions where it is examined; the response to the experiments thus formulated would not exclude any case. There stood the question at the time we thought it necessary to take up these experiments again.

However, before going any further, I must tell you that among the experiments cited some have to be explained because they are deficient in the experimental method used. Thus, the right and the left hearts having walls of unequal thickness, if one opens the chest walls to take the temperature of the ventricles, the right heart will cool much faster than the left heart, hence a very evident cause of error. Experiments done on dead animals or animals with interrupted circulation are equally bad: to have physiologically valid experiments, the animal must be alive, circulation must take place in the vessels under observation, the chest must not be opened. Besides, we shall soon stress the precautions to be taken in these experiments when we talk about those we have designed to examine these questions with the collaboration of M. Walferdin.

The study of animal temperature had, therefore, to be reopened *in toto*; the experiments already done had to be repeated to ascertain the conditions in which they were planned and to make sense of the apparently contradictory results that had to be interpreted in order to understand why an opposite temperature was obtained, when the comparative examination of the two kinds of blood was carried out in the heart and in the periphery.

One understands, up to a certain point, the cooling of blood returning from the limbs. The arterial blood, coming to these parts which have a large surface in relation to their volume, with a higher temperature than that of the ambient environment is in the most favorable conditions for a significant loss of heat. This can explain the lower temperature of the venous blood in the peripheral parts.

But how could the venous blood, colder than the arterial blood when examined in the veins of the limbs, be warmer when its temperature is taken in the right heart?

This venous blood must necessarily have been warmed in returning to the heart. This is effectively what takes place: we shall see where and how.

When we examine what happens in the different venous trunks, proceeding from the periphery to the center, we see that the blood of the superior vena cava was not heated in its centripetal course but it is otherwise for the blood of the inferior vena cava: the latter is heated and the excess of heat that it carries gives to the total amount of blood arriving at the right heart a higher temperature than that of the blood of the left ventricle.

The blood cools therefore in the peripheral organs; but, on its return, it acquires an excess of heat by receiving, at one point of its journey, a warmer blood which not only makes good its losses but even gives it a temperature higher than the one it had when it left the left heart. This point, where the venous blood is warmed by mixing with a warmer blood, is situated in the inferior vena cava, at the level where the cava meets the hepatic veins which bring to it the blood coming from the abdominal visceral organs, blood which is venous twice over since, after having passed through the capillaries of the general circulation, it was taken up by the portal vein, to nourish the capillary circulation of the liver. It is the blood of vegetative life which brings heat to the blood of animal life through the hepatic veins. In operating on the aorta and the vena cava above the liver, one finds that aortic blood possesses a lower temperature than the blood of the vena cava. We shall establish experimentally that it is indeed there that the increase of temperature takes place.

You see, therefore, where our critical method has taken us, which must be that of every wise observer in the presence of contradictory results: he must reject nothing, destroy nothing but look for the reason of discrepancies between the observed results: there is the solution to the question.

It is after all remarkable, as another unique feature of the human mind, that the beautiful theory of Lavoisier in its exclusive localization in the lungs was no longer accepted while nevertheless, people persisted in holding on to the facts consistent with this first localization. Lagrange, W. Edwards, Magnus had already done experiments and shown that if the lung was a seat of lively combustion, it would not resist the heat developed by the combination of the gaseous elements which meet there. If, therefore, one would assume a combustion, it could not be a local and active combustion but a slow and general combustion. In the lung there is mainly exchange between the oxygen of the air and the CO_2 of the venous blood. Then the oxygen transported by the arterial blood to all tissues produces there those oxidations which constitute the slow combustions.

Granting the preceding considerations, it is difficult to understand how one would persist in assuming, to sustain Lavoisier's theory, that arterial blood in the heart has to be warmer than venous blood. It is just the opposite that should be assumed.

In the next session, we shall give you the experiments which relate to this subject and which show that venous blood, cooled in the extremities, is warmed up in the inferior vena cava, at the expense of abdominal venous blood, so as to become warmer at this point than the arterial blood of the aorta.

8

An Introduction to the Study of Experimental Medicine

CLAUDE BERNARD

PART TWO

EXPERIMENTATION WITH LIVING BEINGS

CHAPTÊR I

EXPERIMENTAL CONSIDERATIONS COMMON TO LIVING THINGS AND INORGANIC BODIES

I. The Spontaneity of Living Beings Is no Obstacle to the Use of Experimentation

The spontaneity enjoyed by beings endowed with life has been one of the principal objections urged against the use of experimentation in biological studies. Every living being indeed appears to us provided with a kind of inner force, which presides over manifestations of life more and more independent of general cosmic influence in proportion as the being rises higher in the scale of organization. In the higher animals and in man, for instance, this vital force seems to result in withdrawing the living being from general physico-chemical influences and thus making the experimental approach very difficult.

Inorganic bodies offer no parallel; whatever their nature, they are all devoid of spontaneity. As the manifestation of their properties is therefore absolutely bound up in the physico-chemical conditions surrounding them and forming their environment, it follows that the experimenter can reach them and alter them at will.

On the other hand, all the phenomena of a living body are in such reciprocal harmony one with another that it seems impossible to separate any part without at once disturbing the whole organism. Especially in higher animals, their more acute sensitiveness brings with it still more notable reactions and disturbances.

Many physicians and speculative physiologists, with certain anatomists and naturalists, employ these various arguments to attack experimentation on living beings. They assume a vital force in opposition to physico-chemical forces, dominating all the phenomena of life, subjecting them to entirely separate laws, and making the organism an organized whole which the experimenter may not touch without destroying the quality of life itself. They even go so far as to say that inorganic bodies and living bodies differ radically from this point of view, so that experimentation is applicable to the former and not to the latter. Cuvier, who shares this opinion and thinks that physiology should be a science of observation and of deductive anatomy, expresses himself thus: "All parts of a living body are interrelated; they can act only in so far as they act all together; trying to separate one from the whole means transferring it to the realm of dead substances; it means entirely changing its essence."

If the above objections were well founded, we should either have to recognize that determinism is impossible in the phenomena of life, and this would be simply denying biological science; or else we should have to acknowledge that vital force must be studied by special methods, and that the science of life must rest on different principles from the science of inorganic bodies. These ideas, which were current in other times, are now gradually disappearing; but it is essential to extirpate their very last spawn, because the so-called vitalistic ideas still remaining in certain minds are really an obstacle to the progress of experimental science.

I propose, therefore, to prove that the science of vital phenomena must have the same foundations as the science of the phenomena of inorganic bodies, and that there is no difference in this respect between the principles of biological science and those of physico-chemical science. Indeed, as we have already said, the goal which the experimental method sets itself is everywhere the same; it consists in connecting natural phenomena with their necessary conditions or with their immediate causes. In biology, since these conditions are known, physiologists can guide the manifestation of vital phenomena as physicists guide the natural phenomena, the laws of which they have discovered; but in doing so, experimenters do not act on life.

Yet there is absolute determinism in all the sciences, because every phenomenon being necessarily linked with physico-chemical conditions, men of science can alter them to master the phenomenon, i.e., to prevent or to promote its appearing. As to this, there is absolutely no question in the case of inorganic bodies. I mean to prove that it is the same with living bodies, and that for them also determinism exists.

II. Manifestation of Properties of Living Bodies Is Connected with the Existence of Certain Physico-Chemical Phenomena Which Regulate Their Appearance

The manifestation of properties of inorganic bodies is connected with surrounding conditions of temperature and moisture by means of which the experimenter can directly govern mineral phenomena. Living bodies at first sight do not seem capable of being thus influenced by neighboring physico-chemical conditions; but that is merely a delusion depending on the animal having and maintaining within himself the conditions of warmth and moisture necessary to the appearance of vital phenomena. The result is that an inert body, obedient to cosmic conditions, is linked with all their variations, while a living body on the contrary remains independent and free in its manifestations; it seems animated by an inner force that rules all its acts and liberates it from the influence of surrounding physico-chemical variations and disturbances. This quite different aspect of the manifestations of living bodies as compared with the behavior of inorganic bodies has led the psysiologists, called vitalists, to attribute to the former a vital force ceaselessly at war with physico-chemical forces and neutralizing their destructive action on the living organism. According to this view, the manifestations of life are determined by spontaneous action of this special vital force, instead of being, like the manifestations of inorganic bodies, the necessary results of conditions or of the physico-chemical influences of a surrounding environment. But if we consider it, we shall soon see that the spontaneity of living bodies is simply an appearance and the result of a certain mechanism in completely determined environments; so that it will be easy, after all, to prove that the behavior of living bodies, as well as the behavior of inorganic bodies, is dominated by a necessary determinism linking them with conditions of a purely physico-chemical order.

Let us note, first of all, that this kind of independence of living beings in the cosmic environment appears only in complex higher animals. Inferior beings, such as the infusoria, reduced to an elementary organism, have no real independence. These creatures exhibit the vital properties with which they are endowed, only under the influence of external moisture, light or warmth, and as soon as one or more of these conditions happens to fail, the vital manifestation ceases, because the parallel physico-chemical phenomenon has stopped. In vegetables the manifestation of vital phenomena is linked in the same way with conditions of warmth, moisture and light in the surrounding environment. It is the same again with cold-blooded animals; the phenomena of life are benumbed or stimulated according to the same conditions. Now the influences producing or retarding vital manifestations in living beings are exactly the same as those which produce, accelerate or retard manifestations of physico-chemical phenomena in inorganic bodies, so that instead of following the example of the vitalists in seeing a kind of opposition or incompatibility between the conditions of vital manifestations and the conditions of physico-chemical manifestations, we must note, on the contrary, in these two orders of phenomena a complete parallelism and a direct and necessary relation. Only in warm-blooded animals do the conditions of the organism and those of the surrounding environment seem to be independent; in these animals indeed the manifestation of vital phenomena no longer suffers the alternations and variations that the cosmic conditions display; and an inner force seems to join combat with these influences and in spite of them to maintain the vital forces in equilibrium. But fundamentally it is nothing of the sort; and the semblance depends simply on the fact that, by the more complete protective mechanism which we shall have occasion to study, the warm-blooded animal's internal environment comes less easily into equilibrium with the external cosmic environment. External influences, therefore, bring about changes and disturbances in the intensity of organic functions only in so far as the protective system of the organism's internal environment becomes insufficient in given conditions.

III. Physiological Phenomena in the Higher Animals Take Place in Perfected Internal Organic Environments Endowed with Constant Physico-Chemical Properties

Thoroughly to understand the application of experimentation to living beings, it is of the first importance to reach a definite judgment on the ideas which we are now explaining. When we examine a higher, i.e., a complex living organism, and see it fulfill its different functions in the general cosmic environment common to all the phenomena of nature, it seems to a certain extent independent of this environment. But this appearance results simply from our deluding ourselves about the simplicity of vital phenomena. The external phenomena which we perceive in the living being are fundamentally very complex; they are the resultant of a host of intimate properties of organic units whose manifestations are linked together with the physico-chemical conditions of the internal environment in which they are immersed. In our explanations we suppress this inner environment and see only the outer environment before our eyes. But the real explanation of vital phenomena rests on study and knowledge of the extremely tenuous and delicate particles which form the organic units of the body. This idea, long ago set forth in biology by great physiologists, seems more and more true in proportion as the science of the organization of living beings makes progress. We must, moreover, learn that the *intimate particles* of an organism exhibit their vital activity only through a necessary physico-chemical relation with immediate environments which we must also study and know. Otherwise, if we limit ourselves to the survey of total phenomena visible from without, we may falsely believe that a force in living beings violates the physico-chemical laws of the general cosmic environment, just as an untaught man might believe that some special force in a machine, rising in the air or running along the ground, violated the laws of gravitation. Now, a living organism is nothing but a wonderful machine endowed with the most marvellous properties and set going by means of the most complex and delicate mechanism. There are no forces opposed and struggling one with another; in nature there can be only order and disorder, harmony or discord.

In experimentation on inorganic bodies, we need take account of only one environment, the external cosmic environment; while in the higher living animals, at least two environments must be considered, the external or extra-organic environment and the internal or intra-organic environment. In my course on physiology at the Faculty of Sciences, I explain each year these ideas on organic environment,—new ideas which I regard as fundamental in general physiology; they are also necessarily fundamental in general pathology, and the same thoughts will guide us in adapting experimentation to living beings. For, as I have said elsewhere, the great difficulties that we meet in experimentally determining vital phenomena and in applying suitable means to altering them are caused by the complexity involved in the existence of an internal organic environment.[1]

Physicists and chemists experimenting on inert bodies need consider only the external environment; by means of the thermometer, barometer and other instruments used in recording and measuring the properties of the external environment, they can always set themselves in equivalent conditions. For physiologists these instruments no longer suffice; and yet the internal environment is just the place where they should use them. Indeed, the internal environment of living beings is always in direct relation with the normal or pathological vital manifestations of organic units. In proportion as we ascend the scale of living beings, the organism grows more complex, the organic units become more delicate and require a more perfected internal environment. The circulating liquids, the blood serum and the intra-organic fluids all constitute the internal environment.

In living beings the internal environment, which is a true product of the organism, preserves the necessary relations of exchange and equilibrium with the external cosmic environment; but in proportion as the organism grows more perfect, the organic environment becomes specialized and more and more isolated, as it were, from the surrounding environment. In vegetables and in cold-blooded animals, as we have said, this isolation is less complete than in warm-blooded animals; in the latter the blood serum maintains an almost fixed and constant temperature and composition. But these differing conditions do not constitute differences of nature in different living beings; they are merely improvements in the isolating and protecting mechanisms of their environment. Vital manifestations in animals vary only because the physico-chemical conditions of their internal environments vary; thus a mammal, whose blood has been chilled either by natural hibernation or by certain lesions of the nervous system, closely resembles a really cold-blooded animal in the properties of its tissues.

To sum up, from what has been said we can gain an idea of the enormous complexity of vital phenomena and of the almost insuperable difficulties which their accurate determination opposes to physiologists forced to carry on experimentation in the internal or organic

[1] Claude Bernard, *Leçons sur la physiologie et la pathologie du système nerveux. Leçon d'ouverture*, Dec. 17, 1856. Paris, 1858, Vol. I.—*Cours de pathologie expérimentale.* (*The Medical Times*, 1860.)

environments. These obstacles, however, cannot terrify us if we are thoroughly convinced that we are on the right road. Absolute determinism exists indeed in every vital phenomenon; hence biological science exists also; and consequently the studies to which we are devoting ourselves will not all be useless. General physiology is the basic biological science toward which all others converge. Its problem is to determine the elementary condition of vital phenomena. Pathology and therapeutics also rest on this common foundation. By normal activity of its organic units, life exhibits a state of health; by abnormal manifestation of the same units, diseases are characterized; and finally through the organic environment modified by means of certain toxic or medicinal substances, therapeutics enables us to act on the organic units. To succeed in solving these various problems, we must, as it were, analyze the organism, as we take apart a machine to review and study all its works; that is to say, before succeeding in experimenting on smaller units we must first experiment on the machinery and on the organs. We must, therefore, have recourse to analytic study of the successive phenomena of life, and must make use of the same experimental method which physicists and chemists employ in analyzing the phenomena of inorganic bodies. The difficulties which result from the complexity of the phenomena of living bodies arise solely in applying experimentation; for fundamentally the object and principles of the method are always exactly the same.

IV. The Aim of Experimentation Is the Same in Study of Phenomena of Living Bodies as in Study of Phenomena of Inorganic Bodies

If the physicist and the physiologist differ in this, that one busies himself with phenomena taking place in inorganic matter, and the other with phenomena occurring in living matter, still they do not differ in the object which they mean to attain. Indeed, they both set themselves a common object, viz., getting back to the immediate cause of the phenomena which they are studying.

Now, what we call the immediate cause of a phenomenon is nothing but the physical and material condition in which it exists or appears. The object of the experimental method or the limit of every scientific research is therefore the same for living bodies as for inorganic bodies; it consists in finding the relations which connect any phenomenon with its immediate cause, or putting it differently, it consists in defining the conditions necessary to the appearance of the phenomenon. Indeed, when an experimenter succeeds in learning the necessary conditions of a phenomenon, he is, in some sense, its master; he can predict its course and its appearance, he can promote or prevent it at will. An experimenter's object, then, is reached; through science, he has extended his power over a natural phenomenon.

We shall therefore define physiology thus: the science whose object it is to study the phenomena of living beings and to *determine* the material conditions in which they appear. Only by the analytic or experimental method can we attain the determination of the conditions of phenomena, in living bodies as well as in inorganic bodies; for we reason in identically the same way in experimenting in all the sciences.

For physiological experimenters, neither spiritualism nor materialism can exist. These words belong to a philosophy which has grown old; they will fall into disuse through the progress of science. We shall never know either spirit or matter; and if this were the proper place I should easily show that on one side, as on the other, we quickly fall into scientific negations. The conclusion is that all such considerations are idle and useless. It is our sole concern to study phenomena, to learn their material conditions and manifestations, and to determine the laws of those manifestations.

First causes are outside the realm of science; they forever escape us in the sciences of living as well as in those of inorganic bodies. The experimental method necessarily turns aside from the chimerical search for a vital principle; vital force exists no more than mineral force exists, or, if you like, one exists quite as much as the other. The word, force, is merely an abstraction which we use for linguistic convenience. For mechanics, force is the relation of a movement to its cause. For physicists, chemists and physiologists, it is fundamentally the same. As the essence of things must always remain unknown, we can learn only relations, and phenomena are merely the results of relations. The properties of living bodies are revealed only through reciprocal organic relations. A salivary gland, for instance, exists only because it is in relation with the digestive system, and because its histological units are in certain relations one with another and with the blood. Destroy these relations by isolating the units of the organism, one from another in thought, and the salivary gland simply ceases to be.

A scientific law gives us the numerical relation of an effect to its cause, and that is the goal at which science stops. When we have the law of a phenomenon, we not only know absolutely the conditions determining its existence, but we also have the relations applying to all its variations, so that we can predict modifications of the phenomenon in any given circumstances.

As a corollary to the above we must add that neither physiologists nor physicians need imagine it their task to seek the cause of life or the essence of disease. That would be entirely wasting one's time in pursuing a phantom. The words, life, death, health, disease, have no objective reality. We must imitate the physicists in this matter and say, as Newton said of gravitation: "Bodies fall with an accelerated motion whose law we know: that is a fact, that is reality. But the first cause which makes these bodies fall is utterly unknown. To picture the phenomenon to our minds, we may say that the bodies fall as if there were a force of attraction toward the centre of the earth, *quasi esset attractio.* But the force of attraction does not exist, we do not see it; it is merely a word used to abbreviate speech." When a physiologist calls in vital force or life, he does not see it; he merely pronounces a word; only the vital phenomenon exists, with its material conditions; that is the one thing that he can study and know.

To sum up, the object of science is everywhere the same: to learn the material conditions of phenomena. But though this goal is the same in the physico-chemical and in biological sciences, it is much harder to reach in the latter because of the mobility and complexity of the phenomena which we meet.

V. The Necessary Conditions of Natural Phenomena Are Absolutely Determined in Living Bodies as Well as in Inorganic Bodies

We must acknowledge as an experimental axiom that in living beings as well as in inorganic bodies the necessary conditions of every phenomenon are absolutely determined. That is to say, in other terms, that when once the conditions of a phenomenon are known and fulfilled, the phenomenon must always and necessarily be reproduced at the will of the experimenter. Negation of this proposition would be nothing less than negation of science itself. Indeed, as science is simply the determinate and the determinable, we must perforce accept as an axiom that, in identical conditions, all phenomena are identical and that, as soon as conditions are no longer the same, the phenomena cease to be identical. This principle is absolute in the phenomena of inorganic bodies as well as in those of living beings, and the influence of life, whatever view of it we take, can nowise alter it. As we have said, what we call vital force is a first cause analogous to all other first causes, in this sense, that it is utterly unknown. It matters little whether or not we admit that this force differs essentially from the forces presiding over manifestations of the phenomena of inorganic bodies, the vital phenomena which it governs must still be determinable; for the force would otherwise be blind and lawless, and that is impossible. The conclusion is that the phenomena of life have their special law because there is rigorous determinism in the various circumstances constituting conditions necessary to their existence or to their manifestations; and that is the same thing. Now in the phenomena of living bodies as in those of inorganic bodies, it is only through experimentation, as I have already often repeated, that we can attain knowledge of the conditions which govern these phenomena and so enable us to master them.

Everything so far said may seem elementary to men cultivating the physico-chemical sciences. But among naturalists and especially among physicians, we find men who, in the name of what they call vitalism, express most erroneous ideas on the subject which concerns us. They believe that study of the phenomena of living matter can have no relation to study of the phenomena of inorganic matter. They look on life as a mysterious supernatural influence which acts arbitrarily by freeing itself wholly from determinism, and they brand as materialists all who attempt to reconcile vital phenomena with definite organic and physico-chemical conditions. These false ideas are not easy to uproot when once established in the mind; only the progress of science can dispel them. But vitalistic ideas, taken in the sense which we have just indicated, are just a kind of medical superstition,—a belief in the supernatural. Now, in medicine, belief in occult causes, whether it is called vitalism or is otherwise named, encourages ignorance and gives birth to a sort of unintentional quackery; that is to say, the belief in an inborn, indefinable science. Confidence in absolute determinism in the phenomena of life leads, on the contrary, to real science, and gives the modesty which comes from the consciousness of our little learning and the difficulty of science. This feeling incites us, in turn, to work toward knowledge; and to this feeling alone, science in the end owes all its progress.

I should agree with the vitalists if they would simply recognize that living beings exhibit phenomena peculiar to themselves and unknown in inorganic nature. I admit, indeed, that manifestations of life cannot be wholly elucidated by the physico-chemical phenomena known in inorganic nature. I shall later explain my view of the part played in biology by physico-chemical sciences; I will here simply say that if vital phenomena differ from those of inorganic bodies in complexity and appearance, this difference obtains only by virtue of determined or determinable conditions proper to themselves. So if the sciences of life must differ from all others in explanation and in special laws, they are not set apart by scientific method. Biology must borrow the experimental method of physico-chemical sciences, but keep its special phenomena and its own laws.

In living bodies, as in inorganic bodies, laws are immutable, and the phenomena governed by these laws are bound to the conditions on which they exist, by a necessary and absolute determinism. I use the word determinism here as more appropriate than the word fatalism, which sometimes serves to express the same idea. Determinism in the conditions of vital phenomena should be one of the axioms of experimenting physicians. If they are thoroughly imbued with the truth of this principle, they will exclude all supernatural intervention from their explanations; they will have unshaken faith in the idea that fixed laws govern biological science; and at the same time they will have a reliable criterion for judging the often variable and contradictory appearance of vital phenomena. Indeed, starting with the principle that immutable laws exist, experimenters will be convinced that phenomena can never be mutually contradictory, if they are observed in the same conditions; and if they show variations, they will know that this is necessarily so because of the intervention or interference of other conditions which alter or mask phenomena. There will be occasion thenceforth to try to learn the conditions of these variations, for there can be no effect without a cause. Determinism thus becomes the foundation of all scientific progress and criticism. If we find disconcerting or even contradictory results in performing an experiment, we must never acknowledge exceptions or contradictions as real. That would be unscientific. We must simply and necessarily decide that conditions in the phenomena are different, whether or not we can explain them at the time.

I assert that the word exception is unscientific; and as soon as laws are known, no exception indeed can exist, and this expression, like so many others, merely enables us to speak of things whose causation we do not know. Every day we hear physicians use the words: ordinarily, more often, generally, or else express themselves numerically by saying, for instance: nine times out of ten, things happen in this way. I have heard old practitioners say that the words "always" and "never" should be crossed out of medicine. I condemn neither these restrictions nor the use of these locutions if they are used as empirical approximations about the appearances of phenomena when we are still more or less ignorant of the exact conditions in which they exist. But certain physicians seem to reason as if exceptions were necessary; they seem to believe that a vital force exists which can arbitrarily prevent things from always happening alike; so that exceptions would result directly from the action of mysterious vital force. Now this cannot be the case; what we now call an exception is a phenomenon, one or more of whose conditions are unknown; if the conditions of the phenomena of which we speak were known and determined, there would be no further exceptions, medicine would be as free from them as is any other science. For instance, we might formerly say that sometimes the itch was cured and sometimes not; but now that we attack the cause of this disease, we cure it always. Formerly it might be said that a lesion of the nerves brought on paralysis, now of feeling, and again of motion; but now we know that cutting the anterior spinal nerve paralyzes motion only. Motor paralysis occurs consistently and always, because its condition has been accurately determined by experimenters.

The certainty with which phenomena are determined should also be, as we have said, the foundation of experimental criticism, whether applied to one's self or to others. A phenomenon, indeed, always appears in the same way if conditions are similar; the phenomenon never fails if the conditions are present, just as it does fail to appear if the conditions are absent. Thus an experimenter who has made an experiment, in conditions which he believes were determined, may happen not to get the same results in a new series of investigations as in his first observation; in repeating the experiment, with fresh precautions, it may happen again that, instead of his first result, he may encounter a wholly different one. In such a situation, what is to be done? Should we acknowledge that the facts are indeterminable? Certainly not, since that cannot be. We must simply acknowledge that experimental conditions, which we believed to be known, are not known. We must more closely study, search out and define the experimental conditions, for the facts cannot be contradictory one to another; they can only be indeterminate. Facts never exclude one another, they are simply explained by differences in the conditions in which they are born. So an experimenter can never deny a fact that he has seen and observed, merely because he cannot rediscover it. In the third part of this introduction, we shall cite instances in which the principles of experimental criticism which we have just suggested, are put in practice.

VI. To Have Determinism for Phenomena, in Biological as in Physico-Chemical Sciences, We Must Reduce the Phenomena to Experimental Conditions as Definite and Simple as Possible

As a natural phenomenon is only the expression of ratios and relations and connections, at least two bodies are necessary to its appearance. So we must always consider, first, a body which reacts or which manifests the phenomenon; second, another body which acts and plays the part of environment in relation to the first. It is impossible to imagine a body wholly isolated in nature; it would no longer be real, because there would be no relation to manifest its existence.

In phenomenal relations, as nature presents them to us, more or less complexity always prevails. In this respect mineral phenomena are much less complex than vital phenomena; this is why the sciences dealing with inorganic bodies have succeeded in establishing themselves more quickly. In living bodies, the complexity of phenomena is immense, and what is more, the mobility accompanying vital characteristics makes them much harder to grasp and to define.

The properties of living matter can be learned only through their relation to the properties of inorganic matter; it follows that the biological sciences must have as their necessary foundation the physico-chemical sciences from which they borrow their means of analysis and their methods of investigation. Such are the necessary reasons for the secondary and backward evolution of the sciences concerned with the phenomena of life. But though the complexity of vital phenomena creates great obstacles, we must not be appalled, for, as we have already said, unless we deny the possibility of biological science, the principles of science are everywhere the same. So we may be sure that we are on the right road and that in time we shall reach the scientific result that we are seeking, that is to say, determinism in the phenomena of living beings.

We can reach knowledge of definite elementary conditions of phenomena only by one road, viz., by experimental analysis. Analysis dissociates all the complex phenomena successively into more and more simple phenomena, until they are reduced, if possible, to just two elementary conditions. Experimental science, in fact, considers in a phenomenon only the definite conditions necessary to produce it. Physicists try to picture these conditions to themselves, more or less ideally in mechanics or mathematical physics. Chemists successively analyze complex matters; and in thus reaching either elements or definite substances (individual compounds or chemical species), they attain the elementary or irreducible conditions of phenomena. In the same way, biologists should analyze complex organisms and reduce the phenomena of life to conditions that cannot be analyzed in the present state of science.

Experimental physiology and medicine have no other goal. When faced by complex questions, physiologists and physicians, as well as physicists and chemists, should divide the total problem into simpler and simpler and more and more clearly defined partial problems. They will thus reduce phenomena to their simplest possible material conditions and make application of the experimental method easier and more certain. All the analytic sciences divide problems, in order to experiment better. By following this path, physicists and chemists have succeeded in reducing what seemed the most complex phenomena to simple properties connected with well-defined mineral species. By following the same analytic path, physiologists should succeed in reducing all the vital manifestations of a complex organism to the play of certain organs, and the action of these organs to the properties of well-defined tissues or organic units. Anatomico-physiological experimental analysis, which dates from Galen, has just this meaning, and histology, in pursuing the same problem to-day, is naturally coming closer and closer to the goal.

Though we can succeed in separating living tissues into chemical elements or bodies, still these elementary chemical bodies are not elements for physiologists. In this respect biologists are more like physicists than chemists, for they seek to determine the properties of bodies and are much less preoccupied with their elementary composition. In the present state of the science, it would be impossible to establish any relation between the vital properties of bodies and their chemical composition; because tissues and organs endowed with the most diverse properties are at times indistinguishable from the point of view of their elementary chemical composition. Chemistry is most useful to physiologists in giving them means of separating and studying individual compounds, true organic products which play important parts in the phenomena of life.

Organic individual compounds, though well defined in their properties, are still not active elements in physiological phenomena; like mineral matter, they are, as it were, only passive elements in the organism. For physiologists, the truly active elements are what we call anatomical or histological units. Like the organic individual compounds, these are not chemically simple; but physiologically considered, they are as simplified as possible in that their vital properties are the simplest that we know,—vital properties which vanish when we happen to destroy this elementary organized part. However, all ideas of ours about these elements are limited by the present state of our knowledge; for there can be no question that these histological units, in the condition of cells and fibres, are still complex. That is why certain naturalists refuse to give them the names of elements and propose to call them elementary organisms. This appellation is in fact more appropriate; we can perfectly well picture to ourselves a complex organism made up of a quantity of distinct elementary organisms, uniting, joining and grouping together in various ways, to give birth first to the different tissues of the body, then to its various organs; anatomical mechanisms are themselves only assemblages of organs which present endlessly varied combinations in living beings. When we come to analyze the complex manifestations of any organism, we should therefore separate the complex phenomena and reduce them to a certain number of simple properties belonging to elementary organisms; then synthetically reconstruct the total organism in thought, by reuniting and ordering the elementary organisms, considered at first separately, then in their reciprocal relations.

When physicians, chemists or physiologists, by successive experimental analyses, succeed in determining the irreducible element of a phenomenon in the present state of their science, the scientific problem is simplified, but its nature is not changed thereby; and men of science are no nearer to absolute knowledge of the essence of things. Nevertheless, they have gained what it is truly important to obtain, to wit, knowledge of the necessary conditions of the phenomenon and determination of the definite relation existing between a body manifesting its properties and the immediate cause of this manifestation. The object of analysis, in biological as in physico-chemical science, is, after all, to determine and, as far as possible, to isolate the conditions governing the occurrence of each phenomenon. We can act on the phenomena of nature only by reproducing the natural conditions in which they exist; and we act the more easily on these conditions in proportion as they have first been better analyzed and reduced to a greater state of simplicity. Real science exists, then, only from the moment when a phenomenon is accurately defined as to its nature and rigorously determined in relation to its material conditions, that is, when its law is known. Before that, we have only groping and empiricism.

VII. In Living Bodies, Just as in Inorganic Bodies, the Existence of Phenomena Is Always Doubly Conditioned

The most superficial examination of what goes on around us shows that all natural phenomena result from the reaction of bodies one against another. There always come under consideration the *body,* in which the phenomenon takes place, and the outward circumstance or the environment which determines or invites the body to exhibit its properties. The conjunction of these conditions is essential to the appearance of the phenomenon. If we suppress the environment, the phenomenon disappears, just as if the body had been taken away. The phenomena of life, as well as those of inorganic bodies, are thus doubly conditioned. On the one hand, we have the organism in which vital phenomena come to pass; on the other hand, the cosmic environment in which living bodies, like inorganic bodies, find the conditions essential to the appearance of their phenomena. The conditions necessary to life are found neither in the organism nor in the outer environment, but in both at once. Indeed, if we suppress or disturb the organism, life ceases, even though the environment remains intact; if, on the other hand, we take away or vitiate the environment, life just as completely disappears, even though the organism has not been destroyed.

Thus phenomena appear as results of contact or relation of a body with its environment. Indeed, if we absolutely isolate a body in our thought, we annihilate it in so doing; and if, on the contrary, we multiply its relations with the outer world, we multiply its properties.

Phenomena, then, are definite relations of bodies; we always conceive these relations as resulting from forces outside of matter, because we cannot absolutely localize them in a single body. For physicists, universal attraction is only an abstract idea; manifestation of this force requires the presence of two bodies; if only one body is present, we can no longer conceive of attraction. For example, electricity results from the action of copper and zinc in certain chemical conditions; but if we suppress the interrelation of bodies, electricity,—an abstraction without existence in itself,—ceases to appear. In the same way, life results from contact of the organism with its environment; we can no more understand it through the organism alone than through the environment alone. It is therefore a similar abstraction, that is to say, a force which appears as if it were outside of matter.

But however the mind conceives the forces of nature, that cannot alter an experimenter's conduct in any respect. For him the problem reduces itself solely to determining the material conditions in which a phenomenon appears. These conditions once known, he can then master the phenomenon; by supplying or not supplying them, he can make the phenomenon appear or disappear at will. Thus physicists and chemists exert their power over inorganic bodies; thus physiologists gain empire over vital phenomena. Living bodies, however, seem at first sight to elude the experimenter's action. We see the higher organisms uniformly exhibit their vital phenomena, in spite of variations in the surrounding cosmic environment, and from another angle we see life extinguished in an organism after a certain length of time without being able to find reasons in the external environment for this extinction. But, as we have already said, there is an illusion here, resulting from incomplete and superficial analysis of the conditions of vital phenomena. Ancient science was able to conceive only the outer environment; but to establish the science of experimental biology, we must also conceive an inner environment. I believe I was the first to express this idea clearly and to insist on it, the better to explain the application of experimentation to living beings. Since the outer environment, on the other hand, infiltrates into the inner environment, knowing the latter teaches us the former's every influence. Only by passing into the inner, can the influence of the outer environment reach us, whence it follows that knowing the outer environment cannot teach us the actions born in, and proper to, the inner environment. The general cosmic environment is common to living and to inorganic bodies; but the inner environment created by an organism is special to each living being. Now, here is the true physiological environment; this it is which physiologists and physicians should study and know, for by its means they can act on the histological units which are the only effective agents in vital phenomena. Nevertheless, though so deeply seated, these units are in communication with the outer world; they still live in the conditions of the outer environment perfected and regulated by the play of the organism. The organism is merely a living machine so constructed that, on the one hand, the outer environment is in free communication with the inner organic environment, and, on the other hand, the organic units have protective functions, to place in reserve the materials of life and uninterruptedly to maintain the humidity, warmth and other conditions essential to vital activity. Sickness and death are merely a dislocation or disturbance of the mechanism which regulates the contact of vital stimulants with organic units. In a word, vital phenomena are the result of contact between the organic units of the body with the *inner physiological environment;* this is the pivot of all experimental medicine. Physiologists and physicians gain mastery over the phenomena of life by learning which conditions, in this inner environment, are normal and which abnormal, for the appearance of vital activity in the organic units; for apart from complexity of conditions, phenomena exhibiting life, like physico-chemical phenomena, result from contact between an active body and the environment in which it acts.

VIII. In Biological as in Physico-Chemical Science, Determinism Is Possible, Because Matter in Living as in Inorganic Bodies Can Possess No Spontaneity

To sum up, the study of life includes two things: (1) Study of the properties of organized units; (2) study of the organic environment, i.e., study of the conditions which this environment must fulfill to permit the appearance of vital activities. Physiology, pathology and therapeutics rest on this double knowledge; apart from this, neither medical science nor any truly scientific or effectual therapeutics exists.

In living organisms it is convenient to distinguish between three kinds of definite bodies: first, chemical elements; second, organic and inorganic individual compounds; third, organized anatomical units. Of about 70 elements known to chemistry to-day, only 16 are found in that most complex of organisms, the organism of man. But these 16 elements combine with one another to form the various liquid, solid and gaseous substances of the organism. Oxygen and nitrogen, however, are merely dissolved in the organic fluids; and in living beings, seem to act as elements. The inorganic individual compounds (earthy salts, phosphates, chlorides, sulphates, etc.) are essential constituents in the composition of living bodies, but are taken ready-made directly from the outer world. Organic individual compounds are also constituents of living bodies, but by no means borrowed from the outer world; they are made by the vegetable or animal organism; among such substances are starch, sugar, fat, albumen, etc., etc. When extracted from the body, they preserve their properties because they are not alive; they are organic products, but not organized. Anatomical units stand alone as organized living parts. These parts are irritable and, under the influence of various stimulants, exhibit properties exclusively characteristic of living beings. They live and nourish themselves, and their nourishment creates and preserves their properties, which means that they cannot be cut off from the organism without more or less rapidly losing their vitality.

Though very different from one another in respect to their functions in the organism, these three classes of bodies all show physico-chemical reactions under the influence of the outer stimuli,—warmth, light, electricity; but living parts also have the power of being irritable, i.e., reacting under the influence of certain stimuli in a way specially characteristic of living tissues, such as muscular contraction, nervous transmission, glandular secretion, etc. But whatever the variety presented by the three classes of phenomena, whether the reaction be physico-chemical or vital, it is never in any way spontaneous. The phenomenon always results from the influence exerted on the reacting body by a physico-chemical stimulant outside itself.

Every definite substance, whether inorganic, organic or organized, is autonomous; that is to say, it has characteristic properties and exhibits independent action. Nevertheless, each one of these bodies is inert, that is, it is incapable of putting itself into action; to do this, it must always enter into relation with another body, from which it receives a stimulus. Thus every mineral body in the cosmic environment is stable; it changes its state only when the circumstances in which it is placed are rather seriously changed, either naturally or through experimental interference. In any organic environment, the substances created by animals and vegetables are much more changeable and less stable, but still they are inert and exhibit their properties only as they are influenced by agents outside themselves. Finally, anatomical units themselves, which are the most changeable and unstable of substances, are still inert, that is, they never break into vital activity unless some foreign influence invites them. A muscle-fibre, for instance, has the vital property peculiar to itself of contracting, but this living fibre is inert in the sense that if nothing changes in its environmental or its inner conditions, it cannot bring its functions into play, and it will not contract. For the muscular fibre to contract, a change must necessarily be produced in it, by its coming into relation with a stimulation from without, which may come either from the blood or from a nerve. We may say as much of all the histological units, nerve units, blood units, etc. Different living units thus play the part of stimuli, one in relation to another; and the functional manifestations of an organism are merely the expression of their harmonious reciprocal relations. The histological units react either separately or one against another by means of vital properties which are themselves in necessary connection with surrounding physico-chemical conditions; and this relation is so intimate that we may say the intensity of physico-chemical phenomena taking place in an organism may be used to measure the intensity of its vital phenomena. Therefore, as has already been said, we must not set up an antagonism between vital phenomena and physico-chemical phenomena, but, on the contrary, we must note the complete and necessary parallelism between the two classes of phenomena. To sum up, living matter is no more able than inorganic matter to get into activity or movement by itself. Every change in matter implies intervention of a new relation, i.e., an outside condition or influence. The rôle of men of science is to try to define and determine the material conditions producing the appearance of each phenomenon. These conditions once known, experimenters master the phenomenon in this sense, that they can give movement to matter, or take it away, at pleasure.

What we have just said is equally true for the phenomena of living bodies and the phenomena of inorganic bodies. Only in the case of the complex higher organisms, physiologists and physicians must study the stimuli of vital phenomena, not in the relations of the whole organism with the general cosmic environment, but rather in the organic conditions of the inner environment. Considered in the general cosmic environment, the functions of man and of the higher animals seem to us, indeed, free and independent of the physico-chemical conditions of the environment, because its actual stimuli are found in an inner, organic, liquid environment. What we see from the outside is merely the result of physico-chemical stimuli from the inner environment; that is where physiologists must build up the real determinism of vital functions.

Living machines are therefore created and constructed in such a way that, in perfecting themselves, they become freer and freer in the general cosmic environment. But the most absolute determinism still obtains, none the less, in the inner environment which is separated more and more from the outer cosmic environment, by reason of the same organic development. A living machine keeps up its movement because the inner mechanism of the organism, by acts and forces ceaselessly renewed, repairs the losses involved in the exercise of its functions. Machines created by the intelligence of man, though infinitely coarser, are built in just this fashion. A steam engine's activity is independent of outer physico-chemical conditions, since the machine goes on working through cold, heat, dryness and moisture. But physicists going down into the inner environment of the machine, find that this independence is only apparent, and that the movement of its every inner gear is determined by physical conditions whose law they know. As for physiologists, if they can go down into the inner environment of a living machine, they find likewise absolute determinism that must become the real foundation of the science of living bodies.

IX. The Limits of Our Knowledge Are the Same in the Phenomena of Living Bodies and in the Phenomena of Inorganic Bodies

The nature of our mind leads us to seek the essence or the *why* of things. Thus we aim beyond the goal that it is given us to reach; for experience soon teaches us that we cannot get beyond the *how*, i.e., beyond the immediate cause or the necessary conditions of phenomena. In this respect the limits of our knowledge are the same in biological as in physico-chemical sciences.

When, by successive analyses, we find the immediate cause determining the circumstances in which a phenomenon presents itself, we reach a scientific goal beyond which we cannot pass. When we know that water, with all its properties, results from combining oxygen and hydrogen in certain proportions, we know everything we can know about it; and that corresponds to the *how* and not to the *why* of things. We know how water can be made; but why does the combination of one volume of oxygen with two volumes of hydrogen produce water? We have no idea. In medicine it is equally absurd to concern one's self with the question "why." Yet physicians ask it often. It was probably to make fun of this tendency, which results from lack of the sense of limits to our learning, that Molière put the following answer into the mouth of his candidate for the medical degree. Asked why opium puts people to sleep, he answered: *"Quia est in eo virtus dormitiva, cujus est natura sensus assoupire."* This answer seems ludicrous and absurd; yet no other answer could be made. In the same way, if we wished to answer the question: "Why does hydrogen, in combining with oxygen, produce water?" we should have to answer: "Because hydrogen has the quality of being able to beget water." Only the question "why," then, is really absurd, because it necessarily involves a naïve or ridiculous answer. So we had better recognize that we do not know; and that the limits of our knowledge are precisely here.

In physiology, if we prove, for instance, that carbon monoxide is deadly when uniting more firmly than oxygen with the hemoglobin, we know all that we can know about the cause of death. Experience teaches us that a part of the mechanism of life is lacking; oxygen can no longer enter the organism, because it cannot displace the carbon monoxide in its union with the hemoglobin. But why has carbon monoxide more affinity than oxygen for this substance? Why is entrance of oxygen into the organism necessary to life? Here is the limit of our knowledge in our present state of learning; and even assuming that we succeed in further advancing our experimental analysis, we shall reach a blind cause at which we shall be forced to stop, without finding the primal reason for things.

Let us add that, when the relative determinism of a phenomenon is established, our scientific goal is reached. Experimental analysis of the conditions of the phenomenon, when pushed still further, gives us fresh information, but really teaches us nothing about the nature of the phenomenon originally determined. The conditions necessary to a phenomenon teach us nothing about its nature. When we know that physical and chemical contact between the blood and the cerebral nerve cells is necessary to the production of intellectual phenomena, that points to conditions, but it cannot teach us anything about the primary nature of intelligence. Similarly, when we know that friction and that chemical action produce electricity, we are still ignorant of the primary nature of electricity.

We must therefore, in my opinion, stop differentiating the phenomena of living bodies from those of inorganic bodies, by a distinction based on our own ability to know the nature of the former and our inability to know that of the latter. The truth is that the nature or very essence of phenomena, whether vital or mineral, will always remain unknown. The essence of the simplest mineral phenomenon is as completely unknown to chemists and physicists to-day as is the essence of intellectual phenomena or of any other vital phenomenon to physiologists. That, moreover, is easy to apprehend; knowledge of the inmost nature or the absolute, in the simplest phenomenon, would demand knowledge of the whole universe; for every phenomenon of the universe is evidently a sort of radiation from that universe to whose harmony it contributes. In living bodies absolute truth would be still harder to attain; because, besides implying knowledge of the universe outside a living body, it would also demand complete knowledge of the organism which, as we have long been saying, is a little world (microcosm) in the great universe (macrocosm). Absolute knowledge could, therefore, leave nothing outside itself; and only on condition of knowing everything could man be granted its attainment. Man behaves as if he were destined to reach this absolute knowledge; and the incessant *why* which he puts to nature proves it. Indeed, this hope, constantly disappointed, constantly reborn, sustains and always will sustain successive generations in the passionate search for truth.

Our feelings lead us at first to believe that absolute truth must lie within our realm; but study takes from us, little by little, these chimerical conceits. Science has just the privilege of teaching us what we do not know, by replacing feeling with reason and experience and clearly showing us the present boundaries of our knowledge. But by a marvellous compensation, science, in humbling our pride, proportionately increases our power. Men of science who carry experimental analysis to the point of relatively determining a phenomenon doubtless see clearly their own ignorance of the phenomenon in its primary cause; but they have become its master; the instrument at work is unknown, but they can use it. This is true of all experimental sciences in which we can reach only relative or partial truths and know phenomena only in their necessary conditions. But this knowledge is enough to broaden our power over nature. Though we do not know the essence of phenomena, we can produce or prevent their appearance, because we can regulate their physico-chemical conditions. We do not know the essence of fire, of electricity, of light, and still we regulate their phenomena to our own advantage. We know absolutely nothing of the essence even of life; but we shall nevertheless regulate vital phenomena as soon as we know enough of their necessary conditions. Only in living bodies these conditions are much more complex and more difficult to grasp than in inorganic bodies; that is the whole difference.

To sum up, if our feeling constantly puts the question *why*, our reason shows us that only the question *how* is within our range; for the moment, then, only the question *how* concerns men of science and experimenters. If we cannot know *why* opium and its alkaloids put us to sleep, we can learn the mechanism of sleep and know *how* opium or its ingredients puts us to sleep; for sleep takes place only because an active substance enters into contact with certain organic substances which it changes. Learning these changes will give us the means of producing or preventing sleep, and we shall be able to act on the phenomenon and regulate it at pleasure.

In the knowledge that we acquire, we should distinguish between two sets of notions: the first corresponds to the *cause* of phenomena, the second to the *means* of producing them. By the cause of a phenomenon we mean the constant and definite condition necessary to existence; we call this the relative determinism or the *how* of things, i.e., the immediate or determining cause. The means of obtaining phenomena are the varied processes by whose aid we may succeed in putting in action the single determining cause which produces the phenomenon. The necessary cause in the formation of water is the combination of two volumes of hydrogen with one of oxygen; this is the single cause which always determines the phenomenon. We cannot conceive of water apart from this essential condition. Subordinate conditions or processes in the formation of water may be extremely varied; only all these processes reach the same result, viz., combination of oxygen and hydrogen in invariable proportions. Let us take another example. I assume that we wish to transform starch into glucose; we have any number of means or processes for doing this, but fundamentally there will always be the identical cause, and a single determinism will beget the phenomenon. This cause is fixation of one more unit of water in the substance, to bring about its transformation. Only we may produce this hydration in any number of conditions and by any number of methods: by means of acidulated water, of heat, of animal or vegetable enzymes; but all these processes finally come to a single condition, hydrolysis of the starch. The determinism, i.e., the cause of the phenomenon, is therefore single, though the means for making it appear may be multiple and apparently very various. It is most important to establish this distinction especially in medicine, where the greatest confusion reigns, precisely because physicians recognize a multitude of causes for the same disease. To convince ourselves of what I am urging we have only to open a treatise on pathology. By no means all the circumstances enumerated are causes; at most they are means or processes by which a disease can be produced. But the real and effective cause of a disease must be *constant* and *determined,* that is unique; anything else would be a denial of science in medicine. It is true that determining causes are much harder to recognize and define in the phenomena of living beings; but they exist nevertheless, in spite of the seeming diversity of means employed. Thus in certain toxic phenomena we see different poisons lead to one cause and to a single determinism for the death of histological units, for example, the coagulation of muscular substance. In the same way, varied circumstances producing the same disease must all correspond to a single and determined pathogenic action. In a word, determinism which insists on identity of effect bound up with identity of cause is an axiom of science which can no more be transgressed in the sciences of life than in the sciences of inorganic matter.

X. In the Sciences of Living Bodies, as in Those of Inorganic Bodies, Experimenters Create Nothing; They Simply Obey the Laws of Nature

We know the phenomena of nature only through their relations with the causes which produce them. Now the *law* of phenomena is nothing else than this relation numerically established, in such a way as to let us foresee the ratio of cause to effect in any given case. This ratio, established by observation, enables astronomers to predict celestial phenomena; this same ratio, established by observation and experiment, again enables physicists, chemists, physiologists, not only to predict the phenomena of nature, but even to modify them at pleasure and to a certainty, provided they do not swerve from the ratio which experience has pointed out, i.e., the law. In other terms, we can guide natural phenomena only by submitting to laws that govern them.

Observers can only observe natural phenomena; experimenters can only modify them; it is not given them to create or to destroy them utterly, because they cannot change natural law. We have often repeated that experimenters act, not on phenomena themselves, but on the physico-chemical conditions necessary to their appearance. Phenomena are just the actual expression of the ratio of these conditions; hence, when conditions are similar, the ratio is constant and the phenomenon identical, and when conditions change, there is another ratio, and a different phenomenon. In a word, to make a new phenomenon appear, experimenters merely bring new conditions to pass, but they create nothing, either in the way of force or of matter. At the end of the last century science proclaimed a great truth, to wit, that with respect to matter, nothing is lost, neither is anything created in nature; the bodies whose qualities ceaselessly vary under our eyes are all only transmutations of aggregations of matter of equal weight. In recent times science has proclaimed a second truth which it is still seeking to prove and which in some sense is truly complementary to the first, to wit, that with respect to forces nothing is lost and nothing created in nature; it follows that all the infinitely varied forms of phenomena in the universe are only equivalent transformation of forces, one into another. I reserve for treatment elsewhere the question whether differences separate the forces of living bodies from those of inert bodies; let it suffice for the moment to say that the two preceding truths are universal, and that they embrace the phenomena of living bodies as well as those of inert bodies.

All phenomena, to whatever order they belong, exist implicitly in the changeless laws of nature; and they show themselves only when their necessary conditions are actualized. The bodies and beings on the surface of our earth express the harmonious relation of the cosmic conditions of our planet and our atmosphere with the beings and phenomena whose existence they permit. Other cosmic conditions would necessarily make another world appear in which all the phenomena would occur which found in it their necessary conditions, and from which would disappear all that could not develop in it. But no matter what infinite varieties of phenomena we conceive on the earth, by placing ourselves in thought in all the cosmic conditions that our imagination can bring to birth, we are still forced to admit that this would all take place according to the laws of physics, chemistry and physiology, which have existed without our knowledge from all eternity; and that whatever happens, nothing is created by way either of force or of matter; that only different relations will be produced and through them creation of new beings and phenomena.

When a chemist makes a new body appear in nature, he cannot flatter himself with having created the laws which brought it to birth; he produced only the conditions which the creative law demanded for its manifestation. The case of organic bodies is the same. Chemists and physiologists, in their experiments, can make new beings appear only by obeying the laws of nature which they cannot alter in any way.

It is not given to man to alter the cosmic phenomena of the whole universe nor even those of the earth; but the advances of science enable him to alter the phenomena within his reach. Thus man has already gained a power over mineral nature which is brilliantly revealed in the applications of modern science, still at its dawn. The result of experimental science applied to living bodies must also be to alter vital phenomena, by acting solely on the conditions of these phenomena. But here our difficulties are greatly increased by the delicacy of the conditions of vital phenomena and the complexity and interrelation of all the parts grouped together to form an organized being. This is why man can probably never act as easily on animal or vegetable, as on mineral, species. His power over living beings will remain more limited, especially where they form higher, i.e., more complicated organisms. Nevertheless, the difficulties obstructing the power of physiologists do not pertain to the nature of vital phenomena, but merely to their complexity. Physiologists will first begin by getting at phenomena of vegetables and of animals in easier relations with the outer cosmic environment. It appears, at first sight, as if man and the higher animals must escape from its power to change, because they seem freed from the direct influence of the outer environment. But we know that vital phenomena in man, as in the animals nearest him, are connected with the physico-chemical conditions of an inner organic environment. This inner environment we must first seek to know, because this must become the real field of action for physiology and experimental medicine.

9

Lessons on the Phenomena of Life Common to Animals and Vegetables Second Lecture: The Three Forms of Life*

CLAUDE BERNARD

We have said that life cannot be explained, as it had been believed, by the existence of an internal principle of action acting independently of physico-chemical forces, and, above all, contrary to them. Life is a conflict. Its manifestations result from the intervention of two factors:

1. *Preestablished laws* that regulate the phenomena in their succession, their concert, and their harmony;

2. Definite *physico-chemical conditions* necessary for the appearance of phenomena.

We have no influence on these laws; they are the result of what can be called the *anterior state*; they derive by atavism from organisms which the living being continues and repeats, and they can be followed back to the very origin of living beings. This is why certain philosophers and physiologists have believed they could say that life is only a *memory*; for myself, I have written that the seed seems to retain the memory of the organism from which it proceeds.

Only the conditions of vital manifestations are accessible to us. Knowledge of the external conditions that determine the appearance of vital phenomena suffice, as we have already said, for the purposes of physiological science, since it gives us the means of influencing and of mastering these phenomena.

For us, in a word, life results from a conflict, from a close and harmonious relation between the external conditions and the pre-established constitution of the organism. It is not by warfare against the cosmic conditions that the organism develops and maintains itself, but on the contrary, by an adaptation, an accord with them.

---

*Courtesy of Prof. Hebbel Hoff.

Thus the living animal does not constitute an exception to the grand natural harmony which makes things adapt to one another; it does not break any accord; it is neither in contradiction nor in strife with the general cosmic forces; indeed, far from that, it participates in the universal concert of things, and the life of the animals, for example, is only a fragment of the total life of the universe.

The nature of the relations between the living being and the ambient cosmic conditions permits us to consider three forms of life according to whether it is in a completely narrow dependence upon external conditions, or within a lesser degree of dependence, or in a relative independence. These three forms of life are:

1. *Latent life*; non-manifest life.
2. *Oscillating life*; life with variable manifestations, and dependent upon the external environment.
3. *Constant life*; life with free manifestations, and independent of the external environment.

I. Latent Life

Latent life, according to us, is exhibited by beings whose organism has fallen into the state of *chemical indifference*.

Tiedemann, as we have seen above, believed that life is derived from an internal principle of action which prevented the organism from ever falling into the state of chemical indifference, so that the course of its vital manifestations could never be arrested or interrupted.

Observation and experiment do not permit the adoption of this proposition. We see beings which in some way live only potentially, without any manifestations characteristic of life. These beings are encountered in both the animal kingdom and the plant kingdom.

Active or manifest life, however attenuated it might be, is characterized by relationships between the living being and the environment; relations of exchange such that the being borrows and restores at each instant liquid or gaseous materials with the cosmic environment. What characterizes the state of chemical indifference is the suppression of this exchange, the break in the relationship between the being and the environment which then remain face to face with one another, inalterable and unaltered. Thus a piece of marble, in ordinary conditions, for example, remains in the atmosphere without appreciable change; it receives no action from it and exerts no action on it that is capable of modifying its chemical constitution.

Is it possible for living beings to fall into this degree of absolute chemical indifference? Certain physiologists have been reluctant to believe it, but there are cases where experiment obliges us to accept it. In the plant kingdom, seeds, and in the animal kingdom, certain reviviscible animals, worms, tardigrades, rotifers, show us this state of chemico-vital indifference. We already know a sufficient number of cases of latent life among the animals and plants, but beyond these characteristic examples one can say without fear of error that latent life is profusely prevalent in nature and that in the future it will explain a great number of well-known facts reputed to be mysterious at the present time.

Seeds exhibit this phenomena of latent life to us. If all do not behave in an identical manner one can understand why and by what conditions latent life is less easily maintained in some than in others. It is in consequence of a greater or lesser alterability of their constituent materials by atmospheric agents.

It can be said that the life of a seed in the latent state is purely potential: it exists ready to manifest itself if it is furnished with the appropriate external conditions, but it does not manifest itself in any way if these conditions are absent. The seed has in itself, within its organization, all that it requires for life, but it does not live because it lacks the necessary physico-chemical conditions.

It would be wrong to think that in this case the seed presents a life so attenuated that its manifestations escape observation because of the very degree to which they are weakened. This is not true, either in principle or in fact.

In principle, we know that life results from the concurrence of two factors, the one extrinsic taken from the cosmic world; the other, intrinsic, derived from the organized body. It is a collaboration that is impossible to disjoin and we ought to understand that in the absence of one of these factors, the being cannot live. It does not live when the conditions of the environment *do not exist* anymore than when they *exist alone*. Heat, humidity, and air are not life: organization alone does not constitute it any better.

In fact, we see seeds that have been stored for many years and centuries, which after this long inactivity can germinate and produce a new growth. These seeds have remained, during all this long period, as inert as if they were definitely dead. However attenuated the vital manifestations might be, the accumulation and prolongation of the exchanges would increase them somehow and make them appreciable. This reduced life ought to consume itself, yet under proper circumstances it does not do so.

Thus the seed contains within itself, within its intimate organization, all that it requires for life, but to bring this about it requires in addition the participation of external circumstances.

These circumstances are four in number.

Three *extrinsic* conditions: air (oxygen), heat, humidity.

One *intrinsic* condition: the nutritive reserve of the seed itself.

This reserve is constituted by chemical materials which enter into the constitution of the seed and which act as a reservoir of food material which the vital manifestations will later expend.

But this is not all. It is also necessary for these conditions to exist to a certain degree, in a determined quantity; then life will shine forth in all its brilliance; beyond these limits life tends to disappear and as one approaches these limits the brilliance of the vital manifestations fades and diminishes.

A. Experiments on the Latent Life of Seeds

We shall let you witness some experiments which are well known but which have a special interest here. Their object is to demonstrate that one cannot accept a free vital principle in these living beings when all the vital manifestations are closely related to physico-chemical conditions whose enumeration follows:

1. Water. In some dry earth we have placed seeds that have also been dried, and which are at a temperature and in an atmosphere appropriate for vegetation. They lack only one condition, moisture; they are therefore inert. The wheat stored in Egyptian tombs called *mummy wheat* was, one might say, in the same situation. If they were provided with the humidity that they lacked, germination soon took place. I have consulted in this regard my savant colleague Mr. Decaisne, professor of agriculture at the Museum. He told me that he considered all the examples of germination of seeds found in the hypogeous to be false, because most ordinarily (as I was able to convince myself from a sample) these seeds were impregnated with bitumen or were carbonized. Germination of specimens coming from the lake dwellings would be equally most uncertain.

Even if one must discard these poorly observed facts from science, it has been established experimentally that seeds can germinate after more than a century. Among these seeds, one should cite those of the bean, of tobacco, of the poppy, etc.

It is necessary, moreover, that humidity does not prevent the access of air. Submerged seeds do not germinate, either because the dissolved oxygen is soon used up by the seeds or because it does not act in an appropriate state, that is to say, free. Nevertheless, submersion does not destroy the germinative faculty; there are even, according to Martins, seeds that can cross the seas from one continent to another and go on to germinate.

The simple apparatus that we shall use to germinate plants consists of a test tube in which we suspend with a thread wet sponges to which adhere the seeds that one wishes to germinate. At the bottom of the test tube we place a little water so that the sponge does not become dry; then the tubes are plugged or not, according to the desired circumstances, whether one wishes to confine the atmosphere of the test tube or circulate a current of air through it.

2. Oxygen. Here are test tubes in which seeds were placed on sponges at an appropriate humidity and temperature but in an atmosphere unsuitable for development. In one there is an atmosphere of nitrogen, in the other an atmosphere of carbonic acid.

We have chosen for these experiments the seeds of the common garden cress which have the advantage of germinating very rapidly. On a humid sponge in a test tube closed and filled with nitrogen, we have seen the seed swell; they have surrounded themselves with a sort of mucillagenous coating; the ambient temperature, from 21 to 25°C, was very favorable to germination, and nevertheless germination did not take place during the two or three days since the experiment began.

In another test tube we have placed seeds of garden cress in the same way on a humid sponge, in an atmosphere of carbonic acid, and germination did not take place either.

Finally, in a third test tube we have placed the seeds of garden cress in the same way in a moist atmosphere with ordinary air, and germination is already very evident after one day.

However, the seeds that have not yet germinated in the atmosphere of nitrogen and of carbonic acid are not dead; the germination was only suspended, for if we make these gases disappear by substituting ordinary air or oxygen vegetation will soon be resumed.

These experiments show that to manifest its vitality the seed has need of all the conditions that we have enumerated above; if only one of them is lacking, water or oxygen, for example, germination does not take place.

But this air itself must contain the proper proportion of oxygen. If it has too little, germination does not manifest itself; likewise, if it has too much, either when the atmosphere possesses a percentage composition too rich in oxygen, or when with its ordinary composition this air is compressed. Then, in a given volume, the proportion of the vital gas becomes too high, as the research of Bert has demonstrated.

In addition we have observed an important fact to which we shall return later. The seeds of garden cress, for example, cannot germinate except in an air that is relatively rich in oxygen; by mixing one volume of air with two volumes of an inert gas, hydrogen, for example, germination does not take place. Strangely enough, all of the oxygen is absorbed. It seems probable that if then one were to add a new dose of oxygen to that which was insufficient at first to produce germination, it would be sufficient the second time. Respiration of the seed is thus very active and it appears up to a certain point to be relatively more intense that that of animals.

This necessity for an air rich enough in oxygen to promote germination explains to us how it happens that seeds that have been buried in the earth for a long time remain there in a state of latent life and begin to germinate when they are returned to the surface of the soil. After deep terracing one has often seen the appearance of new vegetation which could only be explained in this way. I have it from an engineer that in certain terracings executed at the time of the construction of the Chemin de Fer du Nord, there appeared on the embarkment a rich vegetation of white mustard which had not been observed before. It is probable that the changes in the terrain had exposed to the air the white mustard seeds buried in the soil and remaining in a state of latency at a depth which did not permit vegetation to take place because of the lack of oxygen.

3. Heat. Temperature must be kept within determined limits, but these limits are different for the various kinds of seeds. deCandolle has published some very interesting investigations on this subject in the *Bibliotheque Universelle* and *Revue Suisse* (Nov. 1865, Aug. and Sept. 1875). The fact that interests us here is the demonstration that for the same kind of seeds germination can be slowed or suspended, not only by too low a temperature, but also by too high a temperature. With the seeds of garden cress that have served in our experiments, the temperature that seems to be most favorable for a rapid germination ranges between 19 and 29 degrees; above that, development appears to be difficult.

1st experiment. In test tubes arranged as described we placed a few days ago some cress seeds at the ambient temperature of the month of June, varying from 18 to 25 degrees. The very next day, after 24 hours, germination was very evident, the radicles had all sprouted and the folioles were beginning to disengage.

2nd experiment. In four test tubes arranged as above we introduced the seeds of garden cress on damp sponges. We modified the experiment so that in the four test tubes we had a confined atmosphere. Instead of leaving the tubes open, we closed them with a clamped rubber tube.

Two of these test tubes were left in the ambient air of the laboratory (17–21° C). The other two test tubes were immersed in a water bath heated to between 38 and

39° C. By the next day, the seeds had germinated in the two test tubes left in the laboratory, while no development had taken place in the test tubes sunk in the water bath. By the third day, germination was complete in the test tubes in the laboratory and those immersed in the water bath were as on the first day without any indication of germination. Then I removed one of the two test tubes from the water bath and placed it on the table beside those whose seeds were in full vegetation. The next day one could see no clear indication of germination but on the second and third days germination occurred and progressed actively thereafter. As for the other test tube remaining in the water bath at 38 to 39° C, on the seventh day it still showed no trace of germination; the seeds were spoiled, and surrounded by mold. This test tube was taken out of the bath and was placed on the table beside the others. Germination started but very slowly; it did not begin to be evident until the third or fourth day. In the other experiments in which I left the test tubes more than eight days at a temperature of 38 to 39° C, germination never took place. Thus I have reason to believe that under the conditions as described, this point marks the upper limit for germination.

3rd experiment. I placed other test tubes containing seeds of garden cress in a drying oven at 32° C. They germinated very well although perhaps a little slowly. Then I raised the temperature of the oven to 34.5° C; then an arrest of germination took place. Sometimes two or three seeds grew well nevertheless, but most often none germinated. I left seeds in the oven in this way for six to seven days without result. When they were taken out, the very next day germination was proceeding actively.

In résumé, it is seen that at 35 to 40°C, germination of garden cress is slowed or suspended, but not destroyed beyond recall. There is thus a sort of anesthesia, or rather of dormancy, produced by an excessive temperature as well as by too low a temperature. Thus the manifestation of vital phenomena requires not only the participation of heat, but a fixed level of heat for each being.

I would compare these experiments with another singular fact that I have noticed for a long time, namely that one can anesthetize frogs at this same temperature of 38° C, which is nevertheless the temperature of normal life in mammals.

We ought to make a remark here: the seed cannot be compared physiologically with the egg, as has been done too often. We shall see later that the egg never falls into the state of latent life. The seed is not the ovule, the germ of the plant; it is its embryo. The essential part of the seed is in fact the miniature of the complete plant; in it are found the rudiments of the root or *radicle*, the rudiment of the stem or *tigelle*, the terminal bud or *gemmule*, and the first leaves or *cotyledons*.

It is therefore the *embryo* that remains in the state of latent life as long as the external conditions are not conducive to its development.

It results from this that what we have said previously about latent life does not apply to the plant egg but rather to the plant itself.

Water and heat are for the plant embryo the indispensable conditions for the return from latent life to manifest life. Suppression of these conditions always causes life to disappear; their return makes it reappear.

A curious experiment by de Saussure shows that even when the embryo has begun its germinative development, it can still stop and fall back into chemical indifference. Germinating wheat is taken and dried; in this state it can be stored for a long time, absolutely inert, just as the seed from which this embryo arose is stored. The air confined in the vessel that contains the dried embryo undergoes no modifications, and thus suggests that the exchange between the rudimentary being and the environment is nil. By restoring humidity and heat, that is to say, propitious conditions to it, life reappears. This can be repeated alternately a sufficiently great number of times and the result produced is always the same. The faculty of latent life will disappear only when the development is advanced enough for green matter to appear in the first leaves.

These phenomena of latent life explain certain very remarkable natural circumstances which forcefully struck the imagination of those who observed them for the first time.

A large number of true seeds or spores (simple seeds of the acotyledons) are buried in the soil or scattered on the surface in a state of inertia. Suddenly, following an abundant rain or the reworking of the soil, they enter into germination and the ground is covered with an unexpected and seemingly spontaneous vegetation.

Similarly, on garden paths after a thunderstorm, one sees green plaques formed by the development of a species of algae.

None of these growths appeared suddenly or spontaneously; the seeds existed in the depths of the soil or in a state of desiccation in the dust that covered it and they did not show any development until they found the conditions of aeration, of humidity, and of heat that are the three essential factors for vital manifestations.

B. Latent Life Among Animals

Animal organisms also afford many examples of latent life. A large number of beings are susceptible of lapsing into a state of *chemical indifference* through desiccation. Such are many of the infusoria, the colpodas among others, which have been well studied by Coste, Balbiani, and Gerbe (*Compt. Rend. de l'Acad. des Sci.* vol. LIX, p. 14). The best known of these animals are the *rotifers*, the *tardigrades*, and the *worms of rusty wheat*.

The colpoda are ciliated infusoria of quite large size, having the form of a bean, armed with vibratile cilia over all their surface. Under the microscope they are seen to take monads, bacteria, vibria, etc. into their stomach via a mouth placed in an indentation in their body, and expel the residue of their digestion by an anal aperture placed at the broad end of the body. Near this anal opening there is a contractile vesicle taken for the heart by certain micrographers, which seems to be the propulsive organ of an aqueous apparatus. At the center of the body of the colpoda there is a quite voluminous organ of reproduction.

When on the surface of infusions there forms a pellicle in which monads, vibria, and bacteria develop, the colpodas dispersed throughout the vessel can be seen to direct themselves toward this pellicle to satisfy their hunger on the animalcules which compose it, or instead to place themselves in contact with the air. Then, among these colpodas, some are seen to stop suddenly, begin to rotate in place, roll up in a ball, and continue this gyration until a secretion from their body coagulates around them into an enveloping membrane; in short, they become encysted, and then they become completely immobile in their envelope like an insect in its cocoon. At this period of their existence the smallest ones have a great similarity to an ovule; it is this which might have led to the belief in a *spontaneous* egg.

Soon the encysted and immobile colpodas separate into two, four, and sometimes twelve smaller colpodas which, once separate and distinct, enter into a gyration on their own account within their common envelope. The movements to which they devote themselves finally wear through the cyst at some point, and once a fissure is made they can be seen leaving their prison and mingling with the population whose number they increase. These are *multiplication* cysts in contrast to another encystment which is related to the conservation of the individual. This is the explanation for multiplication in infusions.

When the colpodas in infusion have exhausted their reproductive power and evaporation threatens, they encyst in order to protect themselves. They can then be dried on glass slides and preserved in this state indefinitely; they return to life as soon as moisture is restored to them. In this way Balbiani has preserved individuals for seven years, restoring them to active life and drying them up each year.

These colpoda cysts, impalpable animal seeds, attach themselves like dust to the surface of objects, on leaves, branches, the bark of trees, on the grass at the bottom of dried up ponds, in the sand or in the dried up mud. Their small size permits them to pass through filters, and one cannot get rid of them. They rupture their envelope each time rain or dew restores humidity to them, take the food within their reach, and form a new cocoon as soon as they begin to lack water. They therefore pass by turns through states of apparent death and of resurrection, under the influence of a physical condition which is present or absent.

The *rotifers* or rotators are animals that are already at a high level of organization, classified either among the worms (Gegenbaur) or as a separate group between the crustaceans and worms (Van Beneden).

These animals are from 0.05 to 1.0 mm in length; they are therefore far from being microscopic. They are found in mosses and especially in those (*Bryon*) which form in green tufts on roofs. Their organization exhibits quite a variety of apparatus to us; they possess quite complicated visceral and locomotor organs. They can creep or swim and according to whether they have recourse to one mode of locomotion or the other the aspect under which they present themselves changes. In the most ordinary state their body is fusiform, thinner at the anterior end and terminating in a sort of ciliated suction cup by means of which they attach themselves to solid bodies to progress by reptation like leeches. At other times, this prolongation is retracted to the inside and then two round lobes are seen to project in the form of discs edged by cilia. In the state of latent life they are immobile and gathered into balls.

The *tardigrades*, well studied from the point of view of their latent life by Doyère are still more highly organized animals, than the preceding. They belong to the class *Arachnida*: it is a family of the *Acarina*. They have four pairs of short legs, articulated and provided with nails. Their body, pointed in front, permits one to distinguish three or four articulations.

Strictly walkers, these animals live in the dust on roofs, or on the mosses that grow there. Exposed to excessive hygrometric variations, they live sometimes in the water which bathes the sand of the gutters like true aquatic beings, sometimes like earthworms.

When they are short of water, they retract, shrivel up, and mix with the surrounding dust; they can remain several months, and it is conceivable that they can remain indefinitely in this state of desiccation without appreciable signs of life.

But when this dust is moistened, as Leeuwenhoek did for the first time on the 27th of September 1701, within an hour the animals can be seen swarming there, active and mobile; their organs, muscles, nerves, digestive viscera, reassume their forms, in a word they resume the fullness of their vitality until they once again become desiccated.

These facts have had wide reverberations and at other times have given rise to discussions about the question of knowing whether in truth life had been suspended completely during desiccation or only attenuated as is produced by cold in hibernating animals. After a debate brought before the Société de Biologie by Doyère, Davaine, and Pouchet, it was firmly established that 1) there is no appreciable life in the inert bodies of reviviscible animals, and 2) these bodies retain their property of revival in conditions (vacuum, dry, at 100°C) *incompatible with any kind of manifest life.*

According to these facts, it seems quite certain that life is completely arrested despite the complexity of the organization of these animals. In them in fact are to be found muscles, nerves, nervous ganglia, glands, eggs, and in a word, all the tissues that constitute higher organisms. To my knowledge, however, no one has ever carried out the experiment of storing them for a very long period of time in the state of latent life. The true *criterion* which permits a decision on whether life is really arrested in an absolute manner is the indefinite duration of this arrest.

Anguillules of rusty wheat. The facts observed regarding the anguilles of rusty wheat are no less interesting than those we have examined above. They lead moreover to the same conclusions (Davaine, *Mémoires de la Société de Biologie*, 1856).

Rust manifests itself in wheat by a deformation of the grain after maturity and by a change in color. The grains are small, rounded, and blackish, and consist of a thick hard shell whose cavity is filled with a white powder. This disease is produced by the presence of very small nematode helminths, present in each seed in many thousands. These worms (*Anguillula tritici*) have no sexual organs at all and cannot reproduce, they come from eggs laid by other worms provided with genital organs which had penetrated the seed before its maturity. They introduced themselves into the developing young plant during germination, between the sheaths of the leaves which enclose the head in the course of formation.

But this introduction is only possible if the plant is moist, because only then is the worm active and able to climb the length of the stalk. If not, the worm will remain in the soil at the foot of the new head, and the wheat will be protected from its attack. Thus it is in the wet years, when the rains are abundant during the time of formation of the head that the grain is subject to rust. The farmers knew that, but they could not understand the relationship between the wetness of the season and the rust of the wheat. It can be seen that this relationship has nothing mysterious about it; it is a simple physical condition which determines whether the pathway is accessible to the parasite or not. It is generally so, and all the natural harmonies are referable to physico-chemical conditions when we know their mechanism.

All this time the grain of wheat is composed of a young and soft parenchyma in which the various parts, the paleolas, stamens, and ovaries are not at all distinct and into which the worm can easily penetrate. It is there that the animal passes from the larval to the adult state; its sexual organs, which had not yet developed, appear and attain their organic perfection; the female lays eggs which hatch and live in a larval state in the cavity which shelters the parents, doomed to perish. The larval worms soon dry up with the wheat itself and await, in a state of apparent death, the conditions necessary for their vital manifestations, humidity and air.

The larval worms present themselves in the form of a white powder, grossly resembling starch, having an average length of eight tenths of a millimeter.

The respiration of these animals when they are in the grain of wheat is nil. Davaine has maintained worms enclosed in green heads in a vacuum for 27 hours, without the activity of these animals being modified noticeably by this treatment. It can thus be conceived that it would be possible to preserve desiccated worms indefinitely in a vacuum. But living larvae in water cannot be treated in the same way. Exposed to a vacuum, they soon fall into a state of apparent death; they return to activity when the air is allowed to return again. I have shown you that it is enough to prevent the contact of air with the water in which they live, by placing oil, for example, around the cover slip of the microscope slide, to see the worms soon fall into a state of asphyxia.

Davaine, having found in the intestines of these animals neither a cellular lining to which digestive functions could be attributed, nor solid particles, concluded that apparently the nutrition of these animals, like their respiration, is accomplished in part through the skin. I think that nutrition must above all take place by means of alimentary reserves which the body of the animal contains and not by absorption of substances coming from outside.

These animals move in a fixed position without really making any progress as long as their life lasts. Their movements are not subject to interruption unless some external condition intervenes. Desiccation or withdrawal of air are the ordinary conditions that stop these movements as well as all the apparent manifestations of life.

In 1771 Baker observed that worms kept inert for twenty-seven years resumed their activity as soon as they were moistened. For my part I have seen worms come back to life after having been stored for four years in a very dry and well-stoppered flask.

Spallanzani was able to induce their revival and their dormancy as many as sixteen times in succession. These animals cannot return to life indefinitely because with each revival they consume a part of their nutritive materials without being able to restore this loss, since they do not eat. Thus in the end, the intrinsic condition constituted by the reserve of nutrient materials eventually disappears, and prevents life from manifesting itself even though the three other extrinsic conditions, heat, water, and air remain.

If the temperature of the water containing the worms is lowered progressively, they retain their movements down to zero. Then the movements die out. When the temperature is again increased, it is only at about 20° C that they can be seen to leave their state of apparent death. They revive even when they have been subjected to a considerable drop in temperature, to about −15 or −20°C. They are much less resistent than rotifers to high temperature and at 70° C they inevitably perish.

It has been observed that it is necessary to continue the action of moisture for quite unequal periods of time to produce revival in the worms. But it can be arranged so that only one of the other necessary conditions is lacking, aeration for example; when this is made to intervene after prolonged humidification, revival will take place in nearly the same length of time. To carry out the experiment I moistened rusty grains of wheat for 24 hours; then on opening them it was seen that just about the same time is necessary to restore the animals to the possession of their vital functions. At any rate, if whole rusty grains of wheat are left immersed in water for entirely too long a time, the worms eventually lose the faculty of reviving.

Other examples of latent life: eggs, ferments, brewers' yeast, etc. We have seen that the seed provides one of the clearest examples of latent life. The substratum of life is indeed present in the seed, but when the external physico-chemical conditions are lacking, all conflict, all vital motion is suspended.

An attempt has been made to look for analogous phenomena in the eggs of certain animals by comparing them to seeds. Such an assimilation is not correct. The seed is not an egg as we have already said, it does not have the properties of one; it is an embryo.

It is not surprising, by the way, that the egg connot like the seed enter a state of chemical indifference, into a state of latent life. The egg is a body in the course of development, whose development cannot be arrested completely. It is only in the state of dormant or oscillating life; as we shall see, it always remains in a relationship of material exchange with the environment. In a word, the egg breathes; it takes in oxygen and gives off carbonic acid; it does not remain inert in the unaltered, ambient medium.

The indifference or apparent inertia of the egg is only an illusion produced by the slowness, the attenuation, or the obscurity of the phenomena that take place within it. The eggs of the silkworm, for example, await the return of spring to hatch; but it must be admitted that life therein has not been completely suspended. Changes take place in them under the influence of cold, and on the return of spring warmth does not find an egg in the same state and with the same constitution that it had at the end of autumn. From this it can be understood that the heat which previously could not cause the development of the egg, can now do so.

These phenomena resulting from the influence of the physical conditions of the environment on the latent or the dormant life of beings explains to us certain harmonic adaptations of nature. What would be the purpose, for example, for the silkworm egg to hatch in the middle of winter, since the animal would not find any of the leaves upon which it must nourish itself? It is thus natural for this egg to acquire this faculty only in springtime and for it to lie dormant during the cold of the winter while slowly completing its development. Phenomena similar to hibernation no doubt take place in plants. We should not however attribute these phenomena to supernatural or magical causes. The influence of the course of the seasons and the influence of their duration are explained by the repetitions and successions of definite physico-chemical conditions. Winter has not acted upon the eggs of the silkworm as a special or extraphysical condition; winter has simply acted as a physical condition, *as cold*. This is what the experiments of Duclaux have demonstrated. A silkworm egg laid at the end of the summer will not hatch naturally until the following spring because the winter and cold bring about a physical condition favorable to certain imperceptible developments which must precede its hatching. But natural winter can be replaced by an artificial winter. If these eggs are subjected for 24 hours to the action of a temperature of zero degrees, then upon the intervention of heat, development takes place immediately and without delay.

The *ferments*, those agents of life so important and still so little known have the faculty of entering a state of latent life. However, we must make a distinction here in regard to *soluble* ferments and *formed* ferments. The first are not living beings, and the property that they exhibit of drying up, then redissolving and recovering their chemical activity does not resemble, except distantly, the phenomena of latent life. Formed ferments, on the contrary, are living beings which reproduce; after having been desiccated they revive under the influence of humidity, and manifest not only their chemical properties but also their property of proliferation or of reproduction; these are indeed true phenomena of *latent life*.

Brewer's yeast provides us a good example of this double faculty. When yeast is taken in full activity and submitted to gradual desiccation it will become reduced to the state of latent life; it can be exposed to a much higher temperature or to the prolonged action of alcohol and it will resist these conditions, then, when placed in proper conditions it will revive and be able to develop again.

Here is a tube in which we set into fermentation some brewer's yeast that had been dried at 40 degrees and stored for two years; little by little it absorbed water and produced alcoholic fermentation when sugar was added.

In another tube, we placed brewer's yeast equally desiccated and stored in pure alcohol for a year and a half. It also absorbed water little by little and successfully produced alcoholic fermentation.

In another experiment, I suspended fresh brewer's yeast in absolute alcohol, where it remained immersed for three or four days. Following this, I collected the yeast on a filter to dry; when again immersed in sugar solution, it gave rise to a very active alcoholic fermentation. I must add that in all cases where yeast has been dried beforehand, whether or not it has been subjected to the influence of alcohol, it is necessary for it to be soaked again in a preliminary maceration for 24 to 36 hours before alcoholic fermentation appears with all its characteristics; inversion of the sucrose into glucose, splitting of the glucose into carbonic acid and alcohol, etc. Thus, it is seen that the two ferments of which brewer's yeast is constituted, the soluble ferment and the *torula cerevisiae* or formed ferment, both possess the faculty of regaining their properties after desiccation.

Explanation of latent life. Desiccation is a protective state for organisms that must be exposed to atmospheric vicissitudes. We have seen colpodas, rotifers, tardigrades, and anguillules encyst, divide, roll up, etc., as soon as the water necessary for their vital state is lacking.

If now we seek to give an account of the mechanisms by which the state of latent life is produced and the return to manifest life is made, we shall see with the greatest clarity the influence of external conditions manifested upon the two orders of phenomena with which we have associated life in all the beings, *organic creation* and *destruction*.

Let us first take up the passage from manifest life to the state of latent life. The principle condition that an organism must fulfill to enter this state is *desiccation*. The other circumstances, temperature, the composition of the atmospheric gases, cannot act as effectively as desiccation to suspend life. A moist seed submitted to cold or exposed to an inert gas would probably finally end in decay. However, it cannot be concluded absolutely that the unlimited maintenance of latent life demands desiccation, because seeds buried in the earth or under water are preserved in the state of latent life for indeterminate but certainly very considerable lengths of time (at least a century).

The immediate consequence of desiccation is to cause the phenomena of *organic destruction* to disappear and to be rendered impossible, that is to say, the functional manifestations of living beings; it is equally true for other conditions that produce latent life. The physical properties of tissues, their elasticity, their density, their tensile strength, are first modified by an excessive degree of desiccation of the organized substance. Then follow the chemical phenomena of vital destruction, whose action is arrested by the very fact of desiccation; for the agents of these phenomena, the *ferments*, become inert when desiccated. Desiccation thus brings about the suppression of *vital destruction* by causing the physical and chemical properties of the tissues to disappear. Vital *creation* itself then also stops in the desiccated cells. In short, life considered in its two aspects is suspended; the organism is in a state of chemical indifference; it is inert. There is an arrest of life, or *latent life*.

The influence of desiccation on the *physical properties* of the tissues and the substances of the organism has been emphasized in a fundamental work published in 1819 by Chevreul (*Mémoires du Museum*, Vol. XIII).

It can be understood from these examples that desiccation abolishes the two orders of physical and chemical phenomena of the organism. Those phenomena characterizing vital destruction being prevented, organic creation stops in turn; the organism loses the characteristics of life.

The awakening of a being deep in the state of latent life, its return to manifest life, is explained just as simply.

First it is vital destruction that again becomes possible through the return of the physical and chemical phenomena, then creative life reappears in its turn when the animal again takes in nourishment.

As soon as humidity and heat are restored to the organism, the tissues, as the investigations of Chevreul have shown, regain the quantity of water that they had before their desiccation, and their mechanical and physical properties of resistance, elasticity, transparency, and fluidity appear. The return of the chemical phenomena takes place just as quickly; the desiccated ferments, moistened again, regain their activity, the interrupted fermentations resume their course within the living organism as well as outside of it, as direct experiment has shown us.

It is therefore primarily by the reestablishment of the acts of vital destruction that the return to life takes place. Creative life evinces itself only as a second event. This is a law that should be emphasized.

In its renascence the animal or the plant always begins by destroying its organism, by consuming the materials previously placed in reserve. This observation makes us understand the necessity for a new condition for reviviscence or return to manifest life. It is necessary for the being to possess reserves, accumulated in its tissues, to be able to nourish itself and provide for its initial expenditures, until the time when it is completely restored to life, and is able to draw from the outside, by alimentation, the materials that are necessary to make new reserves. Here again, incidentally, we encounter an application of that great law upon which we shall never cease to insist, namely, that nutrition is always indirect instead of being direct and immediate. The accumulation of reserves is therefore a necessity for beings in latent life; the resumption of vital manifestations is possible only at this price.

Once the phenomena of vital destruction have begun again within the being, which was inert only a short time ago, vital creation also resumes its course, and life is reestablished in its integrity with its two orders of characteristic phenomena.

II. Oscillating Life

The living being, considered as a complex individual, can be linked to the external environment in such a close dependency that its vital manifestations, without ever being completely extinguished as in the state of latent life, are nevertheless attenuated or enhanced in very large measure as the external conditions vary.

The beings whose vital manifestations can vary within wide limits under the influence of cosmic conditions are beings with *life* that is *oscillating* or *dependent* on the external environment.

These creatures are very numerous in nature.

All plants are of this nature; they are dormant during the winter. Life is not completely extinguished within them; the material exchanges of assimilation and dissimilation are not absolutely abolished, but they are reduced to a minimum. Vegetation is hardly visible; the vital processes are nearly imperceptible. In the spring, when warmth returns, the vital movement is enhanced, the dormant vegetation assumes its utmost activity, the sap sets in motion, leaves appear, buds open and develop, and new parts, roots and branches spread out into the soil or the air.

Similar phenomena take place in the animal kingdom. All invertebrates, and among the vertebrates, all cold-blooded animals, possess an *oscillating life*, *dependent* on the cosmic environment. Cold makes them dormant, and if they cannot be removed from its influence during the winter, life becomes attenuated, respiration slows down, digestion is suspended, movements become feeble or disappear. In mammals, this state is called the *state of hibernation*; the marmot and the dormouse serve as examples.

It is usually a lowering of the temperature that produces this decrease in vital activity. Sometimes, however, its elevation can have the same consequences. We have already seen that germinating seeds, and among animals, frogs become dormant at a high temperature; there is moreover an American mammal, the Tenrec, which, it is said, goes into a true state of lethargy under the influence of high heat.

The highest vertebrates (warm-blooded animals), which have a more perfect *internal environment*, that is to say, circulating fluids in which the temperature is constant, are not subject to this influence by the external environment. Nonetheless, at a certain period of their existence, at the beginning, they start by being creatures with oscillating lives. This happens when they are in the state of the *egg*. The developmental work that must take place in the bird's egg requires a certain level of temperature, close enough to that of the adult animal; if this appropriate temperature is not provided for the egg, it remains dormant. It is not in a state of chemical indifference, because it can be established that it respires; it absorbs oxygen and gives off carbonic acid. Nevertheless this material exchange is not very active. A new-laid hen's egg is taken and placed in a footed cylinder above a layer of baryta water; this latter slowly becomes cloudy from the deposition of barium carbonate resulting from the exhalation of respiratory carbonic acid. The egg is able to remain for a certain time in this state of dormant life, ready to develop into a new animal if the conditions for incubation are present. But it is not able to retain this aptitude indefinitely; after a few weeks it will become what we call *passé*, that is to say, dead and unfit for incubation. It was therefore not completely inert; it lived obscurely.

When on the contrary the egg is subjected to a temperature of 38 to 40° C, vital activity will be increased; respiration, evidence of this energetic activity, will become very marked; the cicatricula will divide and proliferate; the rudiments of the embryo will appear at first, and by means of a successive epigenesis will complete the pattern of a fully formed bird; then life is no longer dormant; it is on the contrary extremely active.

We ought to ask ourselves how dormancy is produced under the influence of cold, and by what mechanism the return of heat imparts a new impulse to the vital activity. Experimentation establishes that the animal enters a state of dormancy or hibernation because all its organic elements are surrounded by a cooled environment in which chemical actions and, proportionally, the functional manifestations of life are diminished. In the cold-blooded or hibernating animal mechanisms are lacking to maintain a constant environment around the elements, despite atmospheric variations. It is the cooling of the internal environment that benumbs the animal; it is the rewarming of this same environment that arouses it.

When a cold-blooded animal, a frog, for example, has become dormant, it might be believed that the action of the cold is exerted primarily on its sensibility, on its nervous system, which is the general regulator of the functions of organic life and of animal life.

This is not so at all. When the *internal environment*, that is to say, the whole of the circulating fluids, is cooled, every element in contact with the blood becomes dormant on its own account, revealing thus its autonomy and the conditions of its own activity.

In a word, every organic system, every element, is in itself influenced by cold like the whole organism. It has the same conditions of activity or inactivity as the whole and it forms a new microcosm within the living being, itself a microcosm within the universe.

In the same way, when the dormant organism comes back to life it is not the nervous system that revives the other systems; how could this be since it is in the same state of dormancy as they are? It is again the *internal environment* that receives the influence of the *external environment*, and reawakens each element in turn, according to its sensibility or excitability. An experiment that I have carried out before places these ideas in clear relief. A frog is taken, dormant from the cold. Sensibility and motion are absent, the apparatus of organic life functions obscurely, blood returns red from the tissues where vital combustion is extremely attenuated, the heart gives but four beats a minute instead of fifteen or twenty as it does during the summer.

This frog can be brought out of its lethargic state. For this it is enough to rewarm it. How then does elevation of the temperature operate? It is not at all, we have said, by a nervous action. To assure myself of this I carried out the following experiment: The paw of a dormant frog is placed in warm water, after its heart has been exposed. Whether the nerve to this limb has been sectioned, or whether it remains intact, the frog revives within the same time. The heart resumes its more rapid beats, and all the systems reawaken in turn. It is the rewarmed blood which has created around all the elements the physical condition of temperature necessary for vital activity. The blood, returning warmer from the paw, has revived the heart beats, and it is the excited heart which has aroused the animal.

The influence of temperature is thus clearly brought to light. In the frog one sees an animal with an oscillating life, that is to say, dependent upon the cosmic environment. A fall in temperature diminishes its vital activity; an elevation of temperature increases it.

The proposition, expressed in these terms is too absolute, however. In this regard we ought to recall the facts that I have already mentioned to demonstrate that there are measures, gradations, and infinite nuances in the actions of physico-chemical agents upon the organism. In a general way, it is true that by increasing the temperature vital activity is enhanced, but if the temperature exceeds certain limits, if, for example, in the frog it reaches 37 to 40°C, the animal is on the contrary anesthetized and stupefied. It is the same for seeds, which, excited to germinate at 20°C, are inactivated at 35°C. We place before you two frogs, the one which we have immersed in water at 37 degrees; you see that it is stupefied and no longer moves; it is in the same state as the second that had been placed in ice water. Let us change their jars; they will both revive except that it is the cold that revives the first, and the heat that revives the second.

Dormant or anesthetized animals and plants resist agents which would kill them if they were in a more active state of life. This resistance varies, moreover, with the nature of the toxic agents employed.

Because of the depression of their vitality, dormant animals resist conditions in which others would perish. Thus, dormancy is also a condition of vital resistance as is latent life. A frog goes the entire winter without taking nourishment; the attenuation of the vital process permits this long suspension of the supply of materials; the animal could not support abstinence for such a long time if it were at a higher temperature. A very small bird, whose vital activity is always considerable, dies of hunger if it is left without food for twenty-four hours.

In their excellent studies on respiration, Regnault and Reiset have called attention to the remarkable resistance of hibernating marmots to conditions that would cause them to perish if they were in their ordinary state of life. A marmot, which breathes weakly during hibernation, can be placed without any inconvenience in an atmosphere deficient in oxygen; awakened, it would soon die from asphyxia. Similarly, this animal, which had remained for several months without food, and withstood the deprivation without harm, would no longer be able to endure it once it is awakened. It must be provided with abundant food, which it will devour with voracity, and without which it would soon die. I have often repeated this experiment in dormice or marmots which I reawakened; if I did not give them food they soon succumbed, having rapidly exhausted the reserves derived from previous feeding.

To complete the account of the facts relating to *oscillating* life we can say that the mechanism of dormancy and the mechanism of the return to active life are as clearly explained as in the case of latent life.

The influence of cosmic conditions at first produces the incomplete suppression of the physical and chemical phenomena of vital destruction. Dormant animals no longer make any movements, their muscles undergo only a slight combustion, and their venous blood is nearly as bright as arterial blood; moreover, combustions are considerably reduced in the other tissues, heat production is weak, and carbonic acid is given off in small quantities. Thus it is the functional vital manifestation of life, which corresponds to the destruction of the organs, that is primarily attenuated. Creative life undergoes a parallel reduction. It can even be said that it is entirely suspended as regards the formation of the immediate principles that constitute the reserves. Nonetheless certain morphological phenomena, cicatrization and repair, still proceed very actively. We shall explain these facts at a later time.

The return to vital activity is further explained in the same way as reviviscence.

It is necessary for the hibernating animal to have reserves, not only to provide for the early expenditures of awakening, but also to suffice for the consumption it makes during the state of dormancy. Vital destruction, in fact, is not suspended, it is only diminished; as for vital creation, and the formation of reserves, it no longer has materials upon which to operate during hibernation, since the animal no longer takes in food from the outside.

This is why, before falling into the winter sleep, or as soon as they have an indication of its approach, animals prepare reserves in diverse forms. In the marmot, the tissues load themselves with fat and glycogen; in the frog, and in all animals, organic provisions of various substances are accumulated. It is upon these prudent reserves, prepared by nature, that the animal lives during the period of dormancy; it no longer does anything but consume, it no longer creates, it no longer accumulates. These reserves suffice for a certain time for the attenuated manifestations observed in these dormant animals, but they would be exhausted rapidly if the vital activity were renewed. Thus it is necessary that as soon as they awaken the animals find at their disposal the foodstuffs on which the creative elaboration will be exerted. Dormice place within the nest in which they hibernate provisions that they consume as soon as they awaken. I have had the opportunity of making some interesting experiments on these animals. If dormant dormice are taken and sacrificed while they are fully asleep, and their liver is analyzed, a certain store of glycogen is still found therein, but if they are not sacrificed until four or five hours after they are awakened, hardly any traces of this material are found. These four hours of active life have used up the savings that would still have sufficed for several weeks of dormant life.

Besides the prolonged dormancy about which we have just spoken, and which the animal supports only on the condition of having considerable reserves accumulated beforehand, there are more or less passing dormancies that really do not require such reserves. Insects that are dormant in the morning after a cool night can be in full activity in the daytime sun. The immobile bee that can be picked up with impunity in the morning is in a state in which it will sting vigorously toward noon. It is clear that these periods of activity and dormancy are too short, and succeed each other too rapidly, to require any considerable reserves; nevertheless one can be satisfied that the great law of nutrition by means of reserves is immutable and that things take place more or less in the same way in all the states of life.

III. Constant or Free Life

Constant or free life is the third form of life; it belongs to the most highly organized animals. In it, life is not suspended in any circumstance, it unrolls along a constant course, apparently indifferent to the variations in the cosmic environment, or to the changes in the material conditions that surround the animal. Organs, apparatus, and tissues function in an apparently uniform manner, without their activity undergoing those considerable variations that were exhibited by animals with an oscillating life. This is because in reality the *internal environment* that envelops the organs, the tissues and the elements of the tissues does not change; the variations in the atmosphere stop there, so that it is true to say that the *physical conditions of the environment* are constant in the highest animals; it is enveloped in an invariable medium, which acts as an atmosphere of its own in the constantly changing cosmic environment. It is an organism that has placed itself in a hothouse. Thus the perpetual changes in the cosmic environment do not touch it; it is not chained by them, it is free and independent.

I believe I was the first to insist upon this idea that there are really two environments for the animal, an *external environment* in which the organism is placed, and an *internal environment* in which the elements of the tissues live. Life does not run its course within the external environment, atmospheric air for the aerial being, fresh or salt water for the aquatic animals, but within the *fluid internal environment* formed by the circulating organic liquid that surrounds and bathes all of the anatomical elements of the tissues; this is the lymph or plasma, the liquid portion of the blood which in the higher animals perfuses the tissues and constitutes the ensemble of all the interstitial fluids, is an expression of all the local nutritions, and is the source and confluence of all the elementary exchanges. A complex organism must be considered as an association of *simple beings*, which are the anatomical elements, and which live in the fluid internal environment.

The constancy of the internal environment is the condition for free and independent life, the mechanism that makes it possible is that which assures the maintenance within the *internal environment* of all the conditions necessary for the life of the elements. This enables us to understand that there could be no free and independent life for the simple beings whose constituent elements are in direct contact with the cosmic environment, but that this form of life is on the contrary the exclusive attribute of beings that have arrived at the summit of complication or organic differentiation.

The constancy of the environment presupposes a perfection of the organism such that external variations are at every instant compensated and brought into balance. In consequence, far from being indifferent to the external world, the higher animal is on the contrary in a close and wise relation with it, so that its equilibrium results from a continuous and delicate compensation established as if by the most sensitive of balances.

The conditions necessary for the life of the elements which must be brought together and maintained constant in the internal environment, for the exercise of free life, are those that we know already: water, oxygen, heat, and chemical substances or reserves.

These are the same conditions as those which are necessary for life in the simple beings, except that in the more perfect animals with independent life the nervous system is called upon to regulate the harmony among all these conditions.

1 *Water*. This is an indispensable element, qualitatively and quantitatively, in the constitution of the environment within which the living elements function and evolve. In free-living animals there must exist an ensemble of dispositions regulating output and intake so as to maintain the necessary quantity of water within the internal environment. In the lower beings, the quantitative variations in water compatible with life are more extensive, but the creature is on the other hand without means of regulating them. This is why it is chained to the vicissitudes of the climate, dormant in latent life during dry weather, revived in wet weather.

The higher organisms are inaccessible to hygrometric variations thanks to means of construction and to physiological functions which tend to maintain the relative constancy of the quantity of water.

In man, especially, and in general in the higher animals, loss of water occurs in all the secretions, in the urine and the sweat especially, and secondarily in respiration, which carries off a notable quantity of water vapor, and finally by cutaneous perspiration.

As to the intake, this is accomplished by the ingestion of fluids or of foods that include water, or even in some animals by absorption through the skin. In all events, it is most likely that the whole quantity of the water in the organism comes from the outside by the one or the other of these two routes. It has not been possible to demonstrate that the animal organism really produces water; the contrary opinion appears to be nearly certain.

It is the nervous system, we have said, that provides the mechanism for compensation between intake and output. The sensation of thirst, which is under the control of this system, makes itself felt whenever the proportion of fluid diminishes within the body as the result of some condition such as hemorrhage or abundant sweating; the animal thus finds itself induced to restore by drinking the losses it has undergone. But even this ingestion is regulated in the sense that it cannot increase the quantity of water present in the blood beyond a certain level; urinary and other excretions eliminate the surplus as a sort of overflow. The mechanisms that vary the quantity of water and reestablish it are thus most numerous; they set in motion a host of mechanisms of secretion, exhalation, ingestion, and circulation which transport the ingested and absorbed fluid. These mechanisms are varied, but they cooperate to the same end: the presence of water in effectively fixed proportions within the internal environment, the condition for free life.

These compensatory mechanisms exist not only for water; they are observed also for most of the mineral and organic substances contained in solution in the blood. It is known that the blood cannot take on a considerable load of sodium chloride, for example; above a certain limit the excess is eliminated in the urine. As I have established, it is the same for sugar, which, normally present in the blood, is, above a certain quantity, eliminated in the urine.

2 *Heat.* We know that for each organism, elementary or complex, limits of external temperature exist between which its activity is possible, with a mid-point which corresponds to the maximum of vital energy. This is true not only for beings that have arrived at the adult stage but also for the egg or embryo. All these creatures are subject to oscillating life, but for the higher animals, the so-called warm-blooded animals, the temperature compatible with the manifestations of life is closely fixed. This fixed temperature is maintained within the internal environment despite extreme climatic variations, and assures the continuity and the independence of life. In a word, there exists in animals with constant and free life a function of calorification which does not exist at all in animals with an oscillating life.

For this function there exists an ensemble of mechanisms governed by the nervous system. There are *thermic* nerves and *vasomotor* nerves to which I have called attention, whose activity produces sometimes an elevation and sometimes a fall in temperature, according to the circumstances.

The production of heat is due, in the living world as in the inorganic world, to chemical phenomena; such is the great law whose understanding we owe to Lavoisier and Laplace. It is in the chemical activity of the tissues that the higher organism finds the source of the heat it conserves within its internal environment, at a nearly constant level, from 38 to 40° C for mammals, and 45 to 47° C for birds. The regulation of heat takes place, as I have said, by means of two kinds of nerves; the nerves that I call *thermic*, which belong to the sympathetic system and serve as brakes, so to speak, on the chemico-thermic activities taking place within the living tissues. When these nerves act, they diminish the interstitial combustions and lower the temperature; when their influence is weakened by suppression of their action or by the antagonism of other nervous influences, then combustions are increased, and the temperature of the internal environment rises considerably. The *vasomotor* nerves, by accelerating circulation in the periphery of the body, or in the central organs, intervene also in the mechanisms for the equilibration of animal heat.

I will add only this last fact. When the action of the cerebrospinal system is considerably attenuated, while that of the sympathetic (*thermic nerve*) is permitted to remain intact, temperature is seen to fall considerably, and the warm-blooded animal is so to speak converted into a cold-blooded animal. I have carried out this experiment on rabbits, cutting the spinal cord between the seventh cervical vertebra and the first dorsal. When, on the contrary, the sympathetic is destroyed, leaving the cerebrospinal system intact, the temperature is noted to rise, at first locally and then generally; this is the experiment I carried out in horses by cutting the sympathetic trunk, especially when they were weakened beforehand. A true fever then follows. I have elsewhere developed the history of all these mechanisms at length (see Leçons sur la chaleur animale, 1873). I recall them here only to establish that the calorific function characteristic of warm-blooded animals results from the perfecting of the nervous mechanism which, by an incessant compensation, maintains an apparently fixed temperature within the *internal environment*, within which there live the organic elements to which ultimately we must always attribute all the vital manifestations.

3 *Oxygen.* The manifestations of life require for their production the intervention of air, or better, its active portion, oxygen, in a dissolved form and in an appropriate state for it to reach the elementary organism. It is moreover necessary for this oxygen to be in proportions that are to a certain degree constant within the internal environment; too small a quantity or too great a quantity are equally incompatible with the vital functions.

Thus in animals with a constant life appropriate mechanisms are required to regulate the quantity of this gas which is assigned to the internal environment, and to keep it more or less constant. In the highly organized animals the penetration of oxygen into the blood is dependent upon the respiratory movements and the quantity of this gas present in the ambient environment. Moreover, the quantity of oxygen that is found in the air depends, as physics teaches it, on the percentage composition of the atmosphere and its pressure. Thus it can be understood that an animal could live in an atmosphere less rich in oxygen if an increase in pressure compensated for this decrease, and inversely, that the same animal could live in an environment richer in oxygen than ordinary air if a diminution in pressure compensated for the increase. This is an important general proposition, resulting from the work of Paul Bert. It can be seen in this case that the variations in the environment compensate and balance each other without the intervention of the animal. If the percentage composition diminishes or increases in the opposite direction, when the pressure rises or falls, the animal ultimately finds the same quantity of oxygen in the environment and its life goes on under the same conditions.

But there can be mechanisms within the animal itself that establish this compensation when it is not accomplished on the outside, and which insure the penetration into the internal environment of the quantity of oxygen required by the vital functions; we mention the different variations that can take place in the quantity of hemoglobin, the *active absorbing* material for oxygen, variations that are still little known but which certainly also intervene for their own part.

All these mechanisms, like the preceding, are without effect except within rather restricted limits; they are perverted and become powerless in extreme conditions. They are regulated by the nervous system. When the air becomes rarefied for some reason, such as during ascension in a balloon or on mountains, the respiratory movements become deeper and more frequent, and compensation is established. Nevertheless, mammals and man cannot sustain this struggle for compensation very long when rarefication is extreme, as when for instance they are transported to altitudes above 5000 meters.

We cannot enter here into the particular details that the question deserves. It suffices for us to propose it. We call attention only to an example related by Campana. It is relative to the high-flying birds, such as the birds of prey and particularly the condor, which rises to heights of 7000 to 8000 meters. They remain there, moving around for long periods of time, although in an atmosphere that would be fatal to a mammal. The principles set forth above permit the prediction that the internal respiratory environment of these animals ought to escape, by some appropriate mechanism, from the depression of the external environment; in other terms, that the oxygen contained in their arterial blood ought not to vary at these great heights. In fact there are in the birds of prey enormous pneumatic sacs, connected to the wings, which do not operate except when these move. When the wings lift, they are filled with external air, when they fall, they pump the air into the pulmonary parenchyma. So that, as the air is rarefied, the work of the bird's wings which supports it there is necessarily increased, and the supplementary volume of air that traverses the lungs is also necessarily increased. The compensation for the rarefication of the external air by an increase in the quantity inspired is thereby assured, and with it the constancy of the respiratory environment characteristic of the bird.

These examples, which we could multiply, demonstrate to us that all the vital mechanisms, however varied they might be, always have one purpose, that of maintaining the integrity of the conditions for life within the internal environment.

4 Reserves. Finally, it is necessary for the maintenance of life that the animal have reserves that assure the constancy of the constitution of its internal environment. Highly organized beings draw the materials for their internal environment from their food, but as they cannot be subjected to an identical and exclusive kind of diet, they must have within themselves mechanisms that derive similar materials from these varied diets and regulate the proportion of them that must enter the blood.

I have demonstrated, and we shall see later, that nutrition is not *direct* according to the teaching of accepted chemical theories, but that on the contrary, it is *indirect* and carried out by means of reserves. This fundamental law is a consequence of the variety of the diet as compared with the constancy of the environment. In a word, *one does not live by his present food, but by that which he has eaten previously*, modified, and in some way created by assimilation. It is the same with respiratory combustion; nowhere is it *direct*, as we shall demonstrate later.

Thus there are reserves, prepared from the food, and consumed at each moment in greater or lesser proportions. The vital manifestations thus destroy the provisions which no doubt had their primary origin from the outside, but which have been elaborated within the tissues of the organism, and which, added to the blood, insure the constancy of its chemico-physical constitution.

When the mechanisms of nutrition are disturbed, and when the animal finds it impossible to prepare these reserves, when it only consumes those that it had accumulated beforehand, it is on its way to ruin, that can end only in the impossibility of life, in death. It would then be of no use for it to eat; it would not be nourished, it would not assimilate, it would waste away.

Something of the kind takes place when the animal is in a state of fever; it uses without restoring, and this state becomes fatal if it persists to the complete exhaustion of the materials accumulated through previous nutrition.

Thus, the nutritive substances that enter an organism, whether animal or plant, do not participate in nutrition directly or immediately. The nutritive phenomenon takes place in two stages and these two stages are always separated from one another by a longer or shorter period, whose duration is a function of a host of circumstances. Nutrition is preceded by a particular elaboration that is terminated by a *storage of reserves* in the animal as well as in the plant. This fact permits one to understand how a being can continue to live, sometimes for a long time, without taking food; it lives on its reserves, accumulated within its own substance; it consumes itself.

These reserves are of variable importance depending upon the creatures concerned and the various substances, in different animals and plants, and in annual or biennial plants, etc. This is not the place to analyze such a vast subject; we have wanted to show that the formation of reserves is not only the general law of all forms of life, but that it constitutes also an active and indispensable mechanism for the maintenance of a constant and free life, independent of variations in the ambient cosmic environment.

Conclusion

We have examined in succession the three general forms in which life appears: *latent* life, *oscillating* life, and *constant* life, in order to see whether in any of them we might find an internal vital principle capable of producing its manifestations of external physico-chemical conditions. The conclusion to which we find ourselves led is easy to draw. We see that in latent life the being is dominated by external physico-chemical conditions, to the point that all vital manifestations can be arrested. In oscillating life, if the living being is not as absolutely subject to these conditions, it nevertheless remains so chained to them that it is subject to all their variations. In constant life, the living being seems to be free, and vital manifestations appear to be produced and directed by an inner vital principle free from external physico-chemical conditions; this appearance is an illusion. On the contrary, it is particularly in the mechanism of constant or free life that these close relations exhibit themselves in their full clarity.

We cannot therefore admit the presence of a free vital principle within living beings, in conflict with physical conditions. It is the opposite fact that is demonstrated, and thus all the contrary concepts of the vitalists are overthrown.

10

Excerpts from

The Teleologic Mechanism of Living Nature

E. PFLÜGER

1. Introduction

A superficial consideration of the totality of the processes in the living being immediately shows that in the strictest sense a dynamic equilibrium is never present. Changes occur as in a rippling mountain stream which carries away rocks and digs into the bed of the river every moment to the beating of the waves and the current. During the constant change of the work of the life-giving force, only a general focal point can be found which controls the law of effectiveness. Thereafter, only such combinations of causes enter reality which favor the welfare of the animal. This becomes true even when totally new conditions have been artificially introduced into the living organism.

What is more noteworthy than the fact that in the highly organized mammals the excised gallbladder duct reproduces itself, that a considerable part of the nerve roots in a higher animal which has been removed through a bloody operation is reestablished and the many thousand nerve fibers which belong together unite in spite of the fact that neither microscopy nor chemistry nor physiologic experimentations were successful in finding a trace of a material difference in the various nerve fibers? What is more wonderful than the fact that the organism becomes accustomed, to a degree, to the various organic and inorganic poisons which can cause a multitude of changes, i.e., at times as through inoculation or vaccination? After the initial effect has taken pace, the body experiences various combinations of "living factors" which are especially suited to withstand harmful influences. Endless would be the number of facts which could be listed to support the theory that the variation of the numerous factors differing, depending on the circumstances, according to the rule are controlled by no other principle than the appropriate security of existence.

2. Psyche and Instinct

When we see that an animal, similar to the human, constantly adapts his actions to prevailing circumstances of the environment as is advantageous to his welfare, we conclude that these actions are motivated and determined by thought, that is, the effluence of a psychic power endowed with consciousness. I use the word "consciousness" in the most general sense, however obscure, which is unthinkable without involvement of the soul and spiritual powers of the so-called "self."

I consider the assumption that there is a consciousness combined with spiritual endowment present in animals as relatively certain. However, the "self" possesses no means to prove with absolute necessity that another consciousness exists outside its own. Nobody doubts this as far as the human is concerned. The animal is not of a different nature.

We further note that the organs in animals which are not influenced by the conscious soul are regulated like the entire animal in an analogous manner according to their changing circumstances, and we are not surprised that the great hero of ancient times, Aristotle, searches for a psychic power in all organs, and Alexander von Humboldt states that he does not wish to determine if the functions of the living being are interrelated with imaginary spiritual activity.

According to Aristotle, a psyche exists which governs the development and nourishment of all organs as an effective power; concomitantly it does not exist outside the living body and has no individuality. Aristotle states: "A soul cannot be present in a being other than in the one it is; a part can exist which is without a soul." Furthermore, it is enlightening that "something" exists that builds the organs, but not in the same manner as an individual being.

It is of interest that Aristotle concentrates on a higher level—in spite of all soul activities including his nourishing soul—rather than the general material. However, only conscious thought—soul is purely divine and unlike the closer soul which has developed simultaneously with and in the body—enters the body from the outside.

Aristotle concluded that the endowment of intellectual capacity alone penetrates the body (the developing embryo) from the outside and is divine, because these activities have nothing in common with the other functions and activities of the body. Each soul appears to be bound to another body of higher form than the so-called elements. The substrates of the soul segregate exactly like the souls segregate among each other in accordance with their higher or lower rank.

Whoever wishes to justify criticizing me—and this, of course, I expect only of a certain kind among the investigators of nature—I nevertheless refer to the literature of millenniums ago. I ask one to consider that this issue deals with general questions which were examined by "Him," one of the greatest geniuses of all times. "He" who was so powerful that his name lives on in the memory of the most renowned with great brilliance after more than two thousand years.

The most imposing part of the thoughts stated above rests in the recognition of the analogy of the processes which exist between the appropriate and rational living organs of the body and the activity of the conscious reasoning soul. I do not wish to pursue "Him" further. There is no justification for the assumption that the active powers of living organs are bound to anything other than the general matter which constitutes these organs. The scientific analysis of the life processes supports the theory that these powers do not consist of the kind inherent in general matter. Since in my opinion, the countless life phenomena—in spite of every indication of profound variation—are only variations of one and the same basic phenomena. The conclusion appears to be imminent that the various souls of Aristotle, inclusive of the conscious thinking soul, are sisters of the same kind. In accord with this conclusion we are left with the question, which has no solution today, whether the wonderfully practical or rational processes performed by all cells are only illuminated by the bright light of consciousness in the ganglion cells of the control nervous system; whereas the specific analogous process of the sister cells in the organism are deprived of the weak dawning of the consciousness which remains concealed from the brain while there is no direct communication between the two. As soon as it is possible to lead the appropriate activities of the organs back to an absolute mechanical process, which is my objective, the cause for assumption of psyche is lacking. This leaves us with the most difficult of all problems, whether the conscious psyche is a natural phenomena of analogous kind like the rational functions of all organs. I further question whether, on the contrary, all psychic and spiritually stimulated activities of the brain are illuminated by consciousness.

Instructive and convincing examples supply us above all with the manifestations of the so-called "instinct" of the animal. Rational, instinctive behavior is determined by the conscious "self," but in accord with its meaning is not caused or motivated by advanced conscious thought. With discriminate precision the singing bird will build its artistic nest and the bee its honeycomb. One could assume a secret genius stirs in these beings because they lack conscious logic power of the proper intellect, desire, and aspiration at the appropriate time. Hardly has the butterfly left its larvae when it ascends in the air—a skilled master among the flyers, never having learned this art, it swarms around flowers it has never seen which offer food and descends upon them; it finds and sucks the honey, the existence of which had been previously denied. If we had more accurate knowledge of the kingdom of atoms and molecules in the living cell, we would find in details everywhere what presents us here with so much amazement on a larger scale. As delicate as this interplay may be, it nevertheless obeys the law of causality. On several occasions I performed research on the manifestations of the instinct and achieved several characteristic results.

I raised a hen in isolation for a year surrounded by high walls. In spring the animal which had never been mated began to lay 16 eggs. Of course, the eggs were not capable of development. This leads one to the conclusion that the oviducts of the female bird assume responsibility of the totally purposeless function to equip an unfertilized egg as though it were determined for development. The egg receives the albumin, the skin of the shell, and the calcium shell. My virgin hen searched for a certain high location which was covered by bushes, scratched and dug a flat ditch where she layed her eggs as though they were in need of future care. After the hen had laid her 16th egg it began to hatch. I only left it a few eggs and waited to see how long it would take. The bird sat night and day and only left the nest for very short periods of time in order to get food and immediately returned to the business of hatching. After this lasted for a week I removed all the eggs from the hen; however, this did not disturb the hatching job of the hen, but it warmed the naked, blank earth. The animal sat for several weeks becoming increasingly industrious in its occupation and considerably thinner because it left less frequently for food. When I removed it several times from the empty nest and placed it in remote areas of the yard, it returned with great speed to its nest to satisfy its instinct. I repeated this extraordinary experiment on a turkey the following year with essentially the same results. The oviduct provides the egg, capable of development, as though it contained the seed of a new bird. The virgin hen constructs a nest and instinctively assumes the responsibility of breeding even though this is totally unfamiliar to her though in this case the eggs did not need any attention. Here one clearly recognizes a mechanism in a sequence of different vegetative and psychic activities of the bird the cause of which is the basis found in the production of the egg; regardless of whether the egg needs it or if it exists after it has been produced. The fact that in the more intelligent bird the instinctive drive can and does regulate the intellect does not contradict the principal importance of my observation with the breeding virgin turkey. This attempt demonstrated that the instinctive activities of breeding are not only deliberate and spontaneous but also aspired to with passion. At the same time one recognizes that the actual purpose of the instinctive act does not affect consciousness since it continues when the purpose no longer exists. Apparently the stimulation to breeding was activated by the ovaries and oviducts which engaged in regeneration because they underwent important loss of substance. However, the question is not so simple because many male birds also practice hatching when one necessarily has to assume other motives. I have the impression that the human being also experiences thoughts and wishes which result in highly rational actions that are motivated by very changeable reasons without the purposes being the conscious motive of the soul. With the changing of seasons and the physiologic circumstances of our body our appetite adjusts to various means of nourishment. Each monotonous diet causes a change in the direction of the appetite which evidently is designed to restore the normal nutritional balance of the body. We spontaneously ingest different dishes in varying amounts, our purpose is to remove the feeling of hunger, and the achievement of the greatest pleasure. The purpose of nature is preservation of life and promotion of health. What applies to hunger in principle also applies to thirst.

Soon the human escapes from noise and work, and soon he resumes work because he delights in it. On one hand he requires a substitute for the loss, on the other hand, he needs the variations filled with tension and energy of work. For both are necessary for good health and are the purpose of nature. When we approach the edge of a cliff and look down we become dizzy which prevents us with magic power from approaching the extreme edge. One also has this feeling when in such a position where falling down is absolutely impossible, and this feeling of fear can vary in degree, which because it becomes repulsive can stimulate the soul without experiencing a trace of dizziness. Therefore, this protects life and causes the avoidance of dangerous cliffs and heights. Certainly, some, like many rope dancers, lose their life through dizziness (vertigo) because they disregarded danger contrary to instinct, which only confirms the importance of the latter. For an explanation of dizziness consider looking up to the top of a very high tower which can also cause dizziness, even though the feet are on even ground at the bottom of the tower; exactly similar to glancing at a balloon high in the sky, or even as one imagines an enormous height or depth when the body is in a certain position which causes dizziness. It follows that it is the imagination of horrendous height which necessarily becomes interrelated with the feelings of dizziness. The fact that the view of the clouds or the moon usually does not produce this effect can be explained by the fact that these experiences, which are familiar to us, attract the viewer's attention to their form, color, and degree of light; i.e., the differences which are demonstrated from one day to the next, but never calls attention to the distance from earth for we lack all means to judge this. However, when we deliberately look at a star and consider imagining the tremendous distance, we can also experience the feeling of dizziness.

Nearly everyone experiences repugnance at the sight of reptiles, amphibious animals, and spiders; this feeling is, however, not identified with fear but rather with disgust and aversion. This does not disappear even when the certainty is present that the repulsive animal is totally harmless. That the human views these kinds of animals with detestation and avoids all contact with them, among which the most dangerous and poisonous creatures are to be found, should not be considered as coincidental.

The temperature of the air provides the feeling of cold and because of the resulting dissatisfaction of cold, shivering creates the urge to locate places which provide maximum protection from coldness. For this reason humans as well as animals seek refuge in their dwellings, caves, and nests during the winter, or even migrate to the south. The purpose is avoidance of an unpleasant peculiar feeling. The purpose of nature is security of life, because warmth is life. Precisely because the preservation of high temperatures in the body of higher animals and the human constitutes the fundamental condition of all life, nature engaged a multitude of means to accomplish this objective safely.

The blocking out of light and closing of the eyes produce the feeling of darkness. The human, and most creatures, do not like darkness, but rather light, and retreat to secure places at the beginning of darkness. Where sight is lacking, danger can approach us more readily before we realize it and before we are in a position to avoid it. The feeling of darkness or "blackout" therefore represents a demand for light because it is absent, and urges one to seek protection from danger of one's own personality when the hunger for light cannot be satisfied. The spooky feeling in dark unknown places still has another origin in the impossibility of satisfying another intense desire. Every human is compelled to interpret and comprehend his impressions of the senses. This is readily recognized when one becomes excited upon hearing or seeing something which is not comprehensible. Probably all of us can remember the impression when in a railway compartment of a train, which is located between two other trains engaged in different motions, how one experiences the sensation of being in motion, and suddenly is convinced, by a glance at an adjacent house, that the train is at a standstill. The reverse is also true. The commanding drive to understand the perception of senses in general extends to everything we could possibly come in contact with and which could be either harmful or useful to us.

A male while in serious conversation with a female is instinctively a different person in thought and deed than he is when in contact with his own sex. As long as the power of procreation exists, both sexes shall adorn themselves with all kinds of ornaments and decorate themselves with finery and trinkets. For the service of appeal they endure the greatest discomfort or even serious pain, which is demonstrated by the piercing of ears, nose and mouth and the practice of extracting teeth for the sake of pure beautification. In the animal, nature provides for the preparation of the wedding process and the male is cared for in a particularly generous manner. Even nature frequently loves an elaborate wedding dress, burdening the male which is proud nonetheless. This is exhibited by the peacocks, the turkey, and others which complement the female bird. The female birds must find extraordinary pleasure in such an inflated character. In the human, the individual's conscious purpose of dressing is to present a pleasing appearance which is characteristically called alluring, which attracts attention and stimulates sensuality. This, however, is nature's chosen and unfailing means to preservation of its kind. The desire to dress up is therefore a normal physiologic desire.

It remains to be noted that especially the human being, more so than other creatures, finds himself in a position with his equals in the struggle for existence, which often makes satisfaction of his instincts impossible and the gratification of passionate instincts very difficult. This frequently leads to a variety of confusions, and to misrepresentation of perhaps originally very normal directions and instincts. With the normal instinct to regulate procreation, animals mistreat the cripples of their kind, or even kill them, because nature does not wish to promote further production of cripples, and even places a curse on them. Nature loves and favors the strong and normal in every way. Who could doubt that this instinctive trait is also present in us. No one shies away from the magic of beauty, youth, and power. Nothing raises the hearts of people higher than the thought of a hero. Everyone is appalled by the sick, the weak, and the crippled, and the instinct is against the welfare of such poor creatures. The newborn infant spontaneously delights in drinking the mother's milk, not as a mechanical reflex, as is the opinion of most current physiologists, because when it is satisfied it does not suck. When the infant is nursed by a mother rich with milk, he soon stops sucking; but when he obtains little with each pull, he sucks for a long time. The newborn starts sucking immediately by nature, when he has the nipple between his lips, because otherwise the sucking would be meaningless. Frequently this urge is so powerful that the infant starts sucking even when he has nothing between his lips. The first sucking action is as little a reflex action as the first flight of a butterfly, the first search, finding and drinking honey from a blossom, as little as the first mating of a couple raised in captivity or prison. In a word, the first sucking is an expression of instinct.

Nature provided for the tender care of the newborn by another instinct: mother love which controls the soul of the mother with such power that she is capable of every heroic deed and sacrifice. The hatching pigeon will attack with great courage the human who touches her nest and even bite and beat him with her wings.

It is clear, that from birth until death, even the healing process in the human is much more dependent on the instinct than one is usually inclined to admit.

In the actions performed for the first time by an individual which were guided by instinct, like the first outing of a butterfly, memory, of course, does not play an important role.

3. The Law of Teleologic Mechanism

Since the investigator of nature tries to comprehend the rationality in the efficacy of all these obscure powers, he must, in order to communicate with Helmholtz, start with the assumption of comprehending the processes of nature and has to recognize as valid the law of contributing cause.

The following elaborations should smooth the path to explore the mechanism of appropriate life processes.

The "law of teleologic causality" which I wish to discuss at this time, and which in spite of its simplicity, so far has not been recognized, is as follows:

The cause of every need of a living being is concomitantly the cause of the satisfaction of that need.

The word "cause" which is used in a diverted sense, was deliberately selected by me in order to emphasize the necessary interrelationship in accord with the law where "cause of need" and "satisfaction of need" are present. More correct, but less descriptive would be the phrase "inducement" instead of "cause."

I designate every altered condition of the living organism, which has to be transformed into a different condition in the interest and welfare of the individual or its kind, as cause of need.

To elaborate my definition I refer to several examples.

Food and drink restore the depleted condition of the organism to normal.

The sex drive, the need to procreate, is the normal consequence of the ardent condition of the female who changes to a pregnant condition.

The development of a permanent heart defect resulting in insufficiency of the mitral valve requires for preservation of life that a second defect associates with the first, namely, that the abnormal enlargement of the right and left chamber gradually develops according to the rule. In this case the need is the creation of an intrinsically defective condition in one of the most vital organs.

Experience permits the establishment of two rules for the practical application of our principles:

I. When the need applies to only one certain organ, this organ alone induces satisfaction.

II. When the same need involves many organs simultaneously, frequently only one organ induces satisfaction of all.

4. The First Rule of the Teleologic Mechanism

Some examples may clarify the first rule.

It is known that objects to be seen clearly must project images to the inner structure of the eye; in order to regulate the images which are produced by visible objects under given light conditions, nature permits entrance of light into the eye only through the pupil, which becomes narrow by contraction of a circular muscle, and the inverse, in darkness it enlarges. That is how blindness is avoided under intense light, and clear vision is permitted under weak light. The need therefore is to provide the pupil with an opening corresponding with the strength of light.

It is now understood that light has a stimulating effect on some cells. Therefore a mechanism is conceivable as a result of which the cells surrounding the pupil are stimulated to stronger contractions by the encountered light depending on its intensity. This kind of influence does appear to exist among some animals. Indeed, I have observed that of two eyes, which were removed from the head of a frog that had just been killed, the eye which had been exposed to more light showed a narrower pupil.

The principle of the teleologic mechanism demonstrates that nature must have solved the problem in a manner essentially different and more complete in action. The need here actually is the proper degree of stimulation of the optic nerve as is suitable for sharp perception. Therefore, the stimulation of the optic nerve regulates the width of the pupil. For this reason we observe that the strongest light of the sun leaves the pupil motionless when the optic nerve is blinded, whereas under normal conditions each stimulation would immediately result in the corresponding narrowing of the pupil. It is known that the mechanism is as follows: the nerve endings of the constricting muscle in the pupil and the endings of the optic nerve in the brain are in such close relationship that stimulation of the nerve fibers is immediately transferred to the motor fibers of the pupillary muscle in the brain, which necessarily causes narrowing of the pupil. The approach nature pursues for solution of the problem is essentially so much better than the one conceivable by us because the ability to stimulate sensitive tissue is subject to extraordinary fluctuations during a life time. Therefore it appears impossible that two qualitatively different kinds of living matter, like the retina and the contractile substance of the surroundings of the retina which is sensitive to light, could under all conditions present the same ratio of the extent of stimulation to the same strength of light. In other words it would be necessary that in both different tissues the sensitivity always changes and varies in equal, corresponding proportions, which would require a stable, pre-established harmony which cannot exist in the independent processes of nature. Variations in sensitivity are more extensive than one usually believes, which can be deduced from the following experience.

One day, when I was involved in the examination of the phosphorescence of a dead seafish, a colleague of mine told me that he made the discovery that the development of the light is a periodic appearance, because during the day luminosity is not present, but at night the fish appears to be brightly illuminated. I asked my colleague to go to the physiology laboratory with me during the day. We spent 15 minutes in a room which was darkened and then went to the basement where the fish were kept, which now—during the day—were already shining brightly.

If one has spent some time outdoors, especially during the summer, my experience shows that it is necessary to spend half an hour in a completely dark room until the retina again contains maximum sensitivity. I presume that this time is shortened in younger individuals.

These observations have taught us that strong stimulation of the optic nerve is the cause of diminution, and weak stimulation is the cause of enlargement.

The cause of the need is also the cause of satisfaction of the need. The mechanism dealt with above is analogous to many familiar situations.

A foreign body which penetrates into the eye (the conjunctiva) is removed by the flow of tears and blinking of the eyelids. What previously affected the optic nerve is now provoked by the stimulated sensory nerve of the membrane transmitting stimulation through involvement of the brain to the sensory nerves of the tear ducts and the motor nerves of the muscle in the eyelids. It is recognized that damage is the cause for removal of damage, because it aroused the sensory nerve.

Entirely the same thing occurs with the expulsion of foreign bodies which penetrate the nose, throat, and stomach via sneezing, coughing, and vomiting.

When we consider the digestive process, we know that the food components are chemically altered by the glands in the stomach and intestines through peculiarly assembled juices and therefore dissolve in the watery liquid. One could assume that it would be appropriate for the food to encounter the carefully prepared gastric juices in large amounts when passing through the stomach and intestine, in order to avoid prolonging this process which is so vital to life. In line with this thought the digestive glands would have to work periodically or continually, to deliver a certain amount of gastric juice on a daily basis, sufficient to digest the food which is necessary to preserve the body. This is conforming to the principles of teleologic mechanism.

The stomach, intestine, and pancreatic glands, even if they last until death from hunger occurs, do not release a single drop of digestive juice. It would be a needless production during the time of denial of food. Here the profound words of Aristotle find justification. "God and nature do not do anything in vain."

As soon as the first bite reaches the stomach the excretions generously flow into the stomach and intestine. They also flow forth when one touches the inner surface of the stomach with a mechanically activated object as is possible through a stomach fistula. The same response is evoked by food present in the mouth cavity and also by every stimulation of the membranes in the mouth. The flow of saliva is released reflexly. Therefore we conclude that the food itself attracts the necessary juices through activation of the nerves and the membranes of the stomach and intestines. If the food intake is particularly large, the mass stretches the membranes of the digestive system, overloads it, and the digestive process takes longer because it takes less time to dissolve smaller amounts and longer for larger. Accordingly, richer nourishment requires stronger and longer lasting secretions because stimulation is longer and stronger. This mechanism is precisely calculated to the varying needs of man and animals.

We do not know whether a mechanism exists which would adjust the relative compositions of the digestive secretions to the various compositions of food.

Only one case can be demonstrated: the dryness of food requires secretions rich in water. The dry materials, as has been experimentally proven, stimulate the nerves of the membranes in the mouth particularly strongly, which at this time cause the salivary gland to secrete vigorously in a reflex way, i.e., mechanically transmitted through the brain; this secretion, however, is almost pure water. The cause of the need for water—namely, the dryness of the food—brought water on the way.

Let us consider the expulsion of excrement—urine and discharge of the bowel: one could assume that after 3–12 hours, repetitive, periodic contractions of the bladder and the bowel, which probably paused during sleep, would be sufficient to cause normal evacuation. We know many organs which in fact work regularly in periods of short or long intervals. Nature does not perform useless work, but it loves to work harder than is the rule under normal circumstances, as soon as the welfare of the body demands it.

As soon as the bladder and rectum are filled, and this is the only reason, they activate expulsion of their content by reflex nerve transmission. The cause of the need, i.e., the fullness of bladder and rectum with urine and waste alone are the cause of satisfaction of need. It is very interesting that there are organs which contract through muscle power to expell their contents and also do the same even when no contents are present. This is the case in the excised heart. Here it has to be realized that the heart, within the human body never is confronted with this situation which therefore does not have to be regarded from the standpoint of mechanism.

The analogy with which we have so far considered regulating mechanisms for expulsion of excrements, which represents a consequence of the teleologic law of causality, permits us to judge the motives for the release of mature male sperm and copulation. The existence of mature sperm in the testicles, which nature produces to fertilize mature eggs, must be the primary motivating factor for the origin of sexual desire, which excites the nerves of the testicle which are stimulated by the sperm. It cannot be determined with certainty whether stimulation of these nerves is produced by the tension of the sperm, or whether by the motion of the sperm threads, i.e., the rhythm of tails beating against the wall, which is rich with nerves.
against the wall, which is rich with nerves.

Especially in this case one could be of the opinion that periodically the entire "sex life" awakens and sperm formation and the sexual urge are two coordinated appearances. It is, however, difficult to deny, and I have conducted extensive evaluations, that castrated males in the prolific years consummate cohabitation. In Italy they even had public houses where castrated males were held to satisfy aberrant females without any consequences. To explain this remarkable fact, one has to remind oneself that the sensation of sensual pleasure is called forth by stimulation of certain nerves.

Just as after removal of both eyes, stimulation of the optic nerves still produces the sensation of light, because the central visual faculty of the blind still produces visual images, the stimulation of the sexual nerves after removal of the testicles can still cause sensual pleasure and the memory of aroused sensual pleasure, with the allied results. That this is so noticeably prominent in the human rests in the vitality of his fantasy, the strong intellect and memory, and the strong development of his intellectual powers, because in the animal this is seldom found and usually only observed for a short time after castration. We know that in general all traces of the sexual drive disappear in them, when they do not produce any seminal fluid in their testicles and when in early youth their sexual glands were removed, i.e., with only a small part of the normal substance of the testicle. I wish to emphasize that in the entire literature only one case is described where the presence of the sex drive of long duration and consummation of cohabitation of a castrated human could be proven with scientific certainty. No doubt, he was equipped with a small remainder of normal testicular substance.

For this reason I do not hesitate to search for the cause of normal cohabitation in the presence of sperm.

The rule established by me—that the living cell regulates the supply of oxygen—is a simple deduction from the principle of teleologic mechanism. The different organs have different oxygen requirements, and the same organ does not have the same requirement at all times. Therefore the organ has to instigate satisfaction of this requirement by itself. The more oxygen used up by the organ, the more it is deprived of; as a result of the difference in the oxygen content of the organ the blood increases and the flow of oxygen diffusion from the blood to the organ is increased. However, the more oxygen lost by the blood in this way to the capillaries of the large circulatory system, the more it has to absorb in the lungs because here it will always be satisfied if the reduction has not progressed abnormally. Therefore the consumption is the cause for regaining that lost—or the cause of the need (lack) induced satisfaction.

This easily leads us to the recognition that when a cell has used up all substance and power through hard work, once more the loss must be the cause of regaining them back. Those areas where building blocks are removed from the structure of living organisms are provided with strong molecular attraction which allows for capacity to introduce new material. It is a fact that when greater losses are encountered due to harder work, such conditions exist where a little more is always gained than lost. The continuing stronger use of the organ allows it to take on additional strength and size. Therefore muscles become larger from heavier work and are capable of considerable effort.

When the left heart chamber undergoes heavy strain, as in pathologic disturbance of the blood circulation, as a result of which the blood pressure of the aortic system is abnormally high, it immediately performs heavier work with each contraction.

When a kidney is excised, the materials in the blood, which stem from other organs and previously had been transferred to both kidneys, are no longer removed from the kidney as quickly, and the one remaining kidney is compelled to work harder and longer. In view of this, the kidney, like the muscle, increases in size, becomes hypertrophic and compensates for the loss. Lack of kidney substance produced additional growth of kidney substance.

If it is true that the pancreatic gland, a very important digestive organ, can be wasted without disturbance to the digestive process, it follows that the other digestive glands which function in analogous fashion, can take over the work of the missing pancreatic gland in a compensatory way. The mechanism is unknown.

5. The Second Law of the Teleologic Mechanism

I now turn to the elaboration of the second law of teleologic mechanism.

When nutrition is withheld over a longer period of time from man or animal, the cells suffer under these circumstances and have to live at the expense of their own existence. A nerve, such as the vagus, brings its damaged nutritional status to the surface, in the form of hunger pains, which necessarily results in the intake of food. Therefore deficiency is the cause for removal of the same.

The same applies to water deficiency in the tissues of the organism. A nerve acts as an aid to all components of the body in that scarcity of water is manifested in the consciousness as thirst.

We now proceed with the discussion of the motions of respiration the purpose of which lies in the intake of oxygen and delivery of carbon dioxide of the lungs. Therefore, it appears as though these needs are satisfied when the respiratory center in the spinal cord is compelled to work continuously at periodic intervals compatible with its organization like the motor center of the heart in order to make possible the exchange of gas in the lung. It is recognized that this mechanism is inconsistent with our principle. In reality, quite frequently sufficient cases occur where the breathing motions are exercised to the maximum in order to permit the normal current of events of the life processes. In other cases, like in the embryo in the womb or in so-called apnea where the need to breath is completely satisfied, the motion of respiration can be dispensed with and therefore actually stop. A mechanism, therefore, which is not designed for fluctuating needs, would be useless in one situation and in another insufficient; and as a consequence could not prevent choking. If this impractical mechanism would nevertheless exist, man and animal would instinctively avoid all actions which would require an increased need for breathing. The ability to perform would be lowered considerably.

The respiratory motions are primarily regulated by need and demand. This need is designated intake of oxygen, output of carbon dioxide. Because consumption of oxygen and formation of carbon dioxide are not in a constant relationship, it is immediately apparent that each motivation causes regulation of the breathing motions.

When we hold our breath or inhale nitrogen, or when the newborn has just left the mother's womb, all tissues proceed to use oxygen until the supply present in the body has been exhausted, and threatens the life of all cells. Once again, certain groups of cells act as an aid to their sister cells. The chemical balance which has been altered as a result of oxygen deficiencies, as a necessary consequence, produces strong stimulation of the nerve cells, which in their turn again cause respiration. Therefore, lack of oxygen produces inflow of oxygen.

All experience leads to the conclusion that a condition which quickly and significantly increases the susceptibility for stimulation of the nerve substance at the same time "stimulates." This is not subject to question because all living nerve substance is in a constant state of activity.

The analogy further speaks for my concept.

Is the feeling of cold not a stimulus because of lack of warmth?

Is the sensation of black not a stimulus because of lack of light?

Is hunger not a stimulus which is caused by lack of certain material, namely, food?

Is thirst not a stimulus produced by lack of certain material, namely, water?

Is it therefore not striking that the urge to breathe air, essentially hunger for oxygen, is a stimulus caused by the lack of oxygen?

The fact that absence of a substance or ability can have such drastic results cannot be a paradox when one considers that removal of a beam can cause an enormous building to collapse; that removal of a catch can stop a clock; that evaporation of oil from an axle of a railway car can cause the axle to catch fire; or that famine drives animals and humans to eat their kind. The last example should be the most appropriate.

Among the findings, which represent examples of the second law of the teleological mechanism, are none which can be compared, with significant meaning, to the work which the nervous system performs for other organs and the rest of the body. Infinitely varied is the efficacy tailored to the needs of the individual. The conscious psyche continuously tries (frequently in the most complicated ways) to secure the welfare of the organism, and to lead to the most advantageous conditions for the satisfaction of all needs. In many cases the instinct serves as a wise counselor, as we earlier pointed out. The mechanism of this rule is relatively simple and based on the principle of pleasure and displeasure. Because all these conditions, which are advantageous to the satisfaction of the individual needs, in general evoke pleasure and the reverse, those conditions which harm the welfare are avoided because they produce displeasure.

An interesting example which is still suitable to application of my concept of the "second law" is the regulation of body temperature in the warm-blooded animal, when, under unusual artificial circumstances, i.e., overheating, the blood temperature is forced up above normal. In this case not only one organ alone, but many serve as aids to the need. Elevation of the temperature in the chest cavity results, as Goldstein demonstrated, in exceptionally strong respiratory action, increased evaporation of water in the lungs, and cooling of the circulating blood.

As known from ancient times, abnormal elevation of the heart temperature causes increased activity of this organ, or faster circulation of the blood, which now flows in larger amounts through the cooled-off skin.

According to the latest findings of Luchsinger, the elevated temperature in the spinal marrow stimulates the central organs which secrete perspiration as well as the inhibiting nerves of the blood vessels in the skin. The elevated marrow of the spine therefore produces skin rich in blood and dripping with sweat, and presents a powerful aid to overcome life-threatening danger.

6. Conclusion

While physiologic research has progressed considerably, it leads to the realization that the rational accommodation of the living being are always subordinate to the teleologic laws of causality.

Perhaps a simile can clarify the mechanism of appropriate reactions in the living being. One should imagine a large musical box with the interior structured such a way that, after the clock mechanism has been wound up, it can play a thousand different songs. There are a thousand buttons on the box which are connected with the mechanism. By movement of a certain button with the finger of a hand, a certain song is played. We can further imagine the arrangement that the volume of the melody is dependent on the degree of movement of the button. The various melodies represent the different processes in the body of an animal which are necessary to satisfy the needs that may occur during the normal course of life, or to balance the more frequently occurring disturbances.

The mechanism in the animal is arranged in such a way that each cause of a need consisting of a material or functional change in living matter, activates a certain button which plays the right melody; i.e., need satisfying. It follows that disturbances could occur, for the removal of which no suitable mechanism exists, causing the organism to work uselessly or to perish; or that a button producing a melody has been put in motion through a cause, other than the normal one, at a time when the resulting reaction was unnecessary or harmful, etc. It is certain that the appropriateness of work is not absolute, but only exists under definite requirements. Especially in this situation, the purely mechanical (removed from the will) character is manifested.

One should consider that man constructs his musical boxes, which play many songs, from relatively rough parts, like metal or wood. Nature, however, works with atoms and can produce a musical box to play a million different melodies in a very small space; and which is designed and calculated to meet the needs during the course of life in a million different ways.

How the teleologic mechanism evolved remains one of the greatest and most obscure problems. It appears to me that, in the strictest sense, no living being can be considered fit to live, because after a long, and often very short, period of time all of them perish with internal necessity. The death of the individual is a law of nature.

I have assembled a hypothesis about the essential intrinsic characteristics of the processes involved in the creation of living nature which at least will provide the possibility of opening up an understanding of *how*, in agreement with the laws of causality and all known experiences, the greatest event of the world may have occurred.

11

The Influence of the Environment on the Composition of Blood of Aquatic Animals

LEON FREDERICQ

In all living beings, the internal environment, which is a product of the organism, maintains necessary relations of exchange and equilibrium with the external, cosmic environment, but as the organism becomes more perfect, the organic environment delimits and isolates itself more and more from the environment (Cl. Bernard, *Introduction à l'étude de la médecine expérimentale*, p. 110, 1865).

This internal environment of which Claude Bernard speaks, composed chiefly of blood and lymph, presents in the higher animals, in vertebrates, a remarkable constancy in its properties. In fact, the physical and chemical conditions of the internal environment are regulated by complicated nervous mechanisms, which function by automatic or reflex means.

The respiratory centers, for example, maintain the normal content of oxygen and CO_2 in the blood; they reach this goal by accelerating or slowing down the breathing movements of the animal. The kidney is responsible for the maintenance of the proportion of water and salts within the correct limits and for the elimination from the blood of harmful substances which might accumulate there. Other organs replace in the internal environment the nutrients used up by tissues, etc.

The living being is organized in such a way that each disturbing influence releases the activity of the compensatory apparatus which should neutralize and repair the damage.

As one goes up in the living scale, these regulatory mechanisms become more numerous, more perfect, and more complicated, they tend to free the organism completely from noxious influences and from changes occurring in the external environment. In the invertebrates, on the contrary, this independence from the external environment is only relative.

In this connection it is interesting to determine the effect which a stay in more or less salty water has on the salt content of blood and tissues of invertebrates. Here are the results of experiments made on the blood of crustaceans living in fresh water or in more or less salty water.

Fresh Water Crustaceans

Seven big crawfish (*Astacus fluviatitis*) were drained of blood by severing the legs; they produced a rather large quantity of blood with a slightly salty taste. 23.453 g of that blood weighed in a platinum crucible were dried on a water bath, then incinerated at a moderate heat until the carbon was free of odors. The carbon was then extracted with hot water. The washings, filtered through a very small filter, as well as the wash water of the filter were collected in a small platinum crucible and evaporated on a water bath. The residue, strongly heated and then cooled in the desiccator weighed 0.221 g.

Therefore the blood contained 0.94 percent or slightly less than 1% of soluble ashes.

Brackish Water Crustaceans

Crabs (*Carcinus moenas*) bought alive in Liège (coming from the brackish waters of l'Escaut). 6.48 g of blood yielded 0.096 g of soluble ashes, e.g., 1.48%.

Sea Water Crustaceans

One big female lobster (*Homarus vulgaris*) coming from the lobster beds of Ostende, bought alive at Liège. 26.49 g of blood yielded 0.8055 g of soluble salts, e.g., 3.040%. The taste of that blood appeared identical to that of the water of the North Sea. The latter, according to my analysis, contains 3.41% of soluble salts (25.028 g of water taken at la Panne, at high tide, evaporated in a platinum crucible on the water bath yielded a residue weighing 0.355 g, e.g., 3.41%.)

Crabs (Carcinus moenas) from Roscoff (Britanny). 23.01 g of blood, preserved dry, yielded 0.708 g of soluble salts, e.g., 3.07%.

Crabs (Carcinus moenas) from Roscoff, living in sea water of 1025 density, 14.78 g of blood yielded 0.445 g of soluble salts, e.g., 3.001%.

Crabs "tourteau" (*Platycarcinus pagurus*) from Roscoff. 13.54 g of blood of a density of 1037 yielded 0.419 g of soluble salts, e.g., 3.101%.

Crab "tourteau" from Roscoff. 31.08 g of blood of a density of 1036 yielded 0.965 g of soluble salts, e.g., 3.004%.

Crayfish (Palinurus vulgaris) from Roscoff. Blood serum 22.94 g yielded 0.666 g of soluble salts, e.g., 2.9%.

Maja squinado from Roscoff. 15.60 g of blood yielded 0.476 g of salts, e.g., 3.045%.

The sea water of Roscoff in which these crustaceans live was also analyzed. 27.312 g of water yielded after evaporation 0.929 g of salt residue, e.g., 3.401%. 26.266 g of the same water yielded 0.894 g of salt residue, e.g., 3.407%.

Crustaceans of Highly Salty Sea Water

Maja squinado of Naples. Blood collected in a sealed glass tube. 14.807 g of blood yielded 0.498 g of soluble salts, e.g., 3.37%.

A sample of the sea water in which the *Maja squinado* lived was likewise analyzed: 20.669 g of sea water yielded 0.821 g of salt residue, e.g., 3.9%.

The content of salts contained in the blood of crustaceans varies therefore within wide margins (from 0.94 to 3.37% or more than one to three). The proof that it is truly a question of the influence exerted by the salt content of the external environment is provided by the fact that one and the same animal species, the *Carcinus moenas*, presents similar differences in the chemical composition of its blood, according to whether the animal lives in brackish water or sea water (1.48% for the crabs from the Escaut and 3.07% for those in sea water).

Also the blood of the Maja from Naples, living in a very salty water yields 3.37% of soluble salts while, at Roscoff, the blood of the same animal yielded only 3.045% of salts.

Moreover, one can, within short intervals, cause variations within very large limits in the composition of blood of *Carcinus moenas* by transporting them successively in more or less diluted sea water. The *Carcinus moenas* from Roscoff have more than 3% of salts in their blood as we have seen (3.001% and 3.07%).

Placed in sea water (with a density of 1026) diluted with fresh water to 1015, they freed themselves of salt to such an extent that their blood contained only 1.99% of soluble salts (11.83 g of blood yielded 0.236 g of soluble salts).

After a stay in even more diluted sea water, with a density of only 1010, their blood contained only 1.56% of soluble salts (15.17 g of blood yielded 0.239 g of soluble salts).

Kept in water with 1007 density, the crabs provided blood containing 1.65% of soluble salts (13.13 g of blood yielded 0.217 g of soluble salts).

The following table assembles the figures cited:

Table I
Salt Content of Blood of Crustaceans

Animal species	Blood		Water in which the animal lived	
	Density	Salt Content	Density	Salt Content
Astacus fluviatilis	–	0.940%	–	fresh water
Carcinus moens		1.480	?	brackish
	–	1.650	1007	approx. 0.90
	–	1.560	1010	approx. 1.30
	–	1.990	1015	approx. 1.90
	–	3.001	1026	3.40
	–	3.007	–	–
Homarus vulgaris	–	3.040	1026	3.41
Patycarcinus pagurus	1037	3.101	–	3.40
	1036	3.104	–	–
Palinurus vulgaris	–	2.900	–	–
Maja squinado	–	3.045	–	–
	–	3.370	–	3.90

The influence of the salt content of the external environment on the content of salts contained in the blood of crustaceans therefore cannot be denied. It is probably across the gills that the exchange of salts takes place between the blood and the external water. The thin branchial membrane might play a role similar to the role of the membrane in a dialyzer. However, salt equilibrium is never completely reached between the two liquids present. In the crawfish and the crabs living in brackish water the blood contains notably more salts than the external water. On the contrary, the blood of sea water crustaceans is always poorer in salts than the water which bathes the gills.

Other aquatic invertebrates seem to feel the influence of the salt content of the external environment in the same way as the crustaceans.

The blood of fresh water mollusks is poor in salt while the blood of sea mollusks has exactly the same taste as the sea water in which they live. The blood of the octopus contains nearly 3% salt.

The aquatic vertebrates, fish, behave differently. In their case, the gills, so permeable to respiratory gas exchanges, seem on the contrary to constitute an almost impassable barrier for the salts dissolved in sea water. The blood of sea fish is no more salty to the taste than the blood of fresh water fish. The blood of a big shark yielded only 1.3% soluble salts.

It has been known for a long time that the flesh (muscles, glands, etc.) of sea fishes is no more salty than the flesh of fresh water fishes. This follows clearly from numerous figures of analyses of ashes of fish muscles published by Almea. I have likewise verified in a good many sea invertebrates that the muscles, glands etc. have only a faintly salty taste.

Table II
Content of Salts in Tissues of Sea Invertebrates

Lobster muscles	1.127 per 100 of soluble salts
Octopus muscles	1.750 per 100 of soluble salts
Octopus muscles	1.910 per 100 of soluble salts
Mollusk muscles	1.950 per 100 of soluble salts

This work was done at the Laboratory of Physiology of the University of Liège.
Part of the material was collected by the author at the Laboratory of Experimental Zoology at Roscoff in 1882.
Several samples of blood and tissues of sea animals were later sent from Roscoff by M. Charles Marty, keeper of the station at Roscoff.
The author is happy to be able to thank him. He wishes also to express his gratitude to Prof. de Lacaze-Duthiers, creator and Director of the Laboratories at Roscoff.

12

Functions of Defense

C. RICHET

Conclusion

To summarize these facts (so numerous that we have really only been able to provide a list) we see that all life functions contribute to defend the organism against the enemies who assault it.

Preventive, immediate or consecutive defenses, active or passive defenses, general or special defenses, living beings have everything to allow them to resist the environment and to maintain themselves, in spite of all hindrances, in an almost perfect state of equilibrium and homogeneity.

On the whole, the living being is stable and it has to be, so as not to be destroyed, dissolved, disintegrated by the colossal, often hostile forces which surround it. But, by a kind of contradiction which is only apparent, it maintains its stability because it is excitable, able to modify itself according to outside irritations, and to adapt its responses to the irritation; thus it is stable only because it is modifiable. Defense is only compatible with a certain instability. The latter must play ceaselessly but within narrow confines and this moderate instability is the necessary condition of the true stability of the living being.

Life is a perpetual self-regulation, an adaptation to changing external conditions. The level has to shift constantly but it has to oscillate around a mean nearly invariable.

Therefore, if I could try to give a formula to this defense of the organism which, looked at in this way constitutes the whole physiology, I would say of the living being:

It is subject to all impressions and it resists them all; it renews itself always and it is always the same.

Editor's Comments on Papers 13–20

Homeostasis

III

The concept of the constancy of the internal environment did not spread rapidly. Leon Fredericq in Belgium certainly understood it and contributed to its development. Sir Michael Foster, in *A Textbook of Physiology* published in 1877, made mention of it but one really has to wait until the twentieth century to pick up the trail once again.

A considerable number of physiologists, particularly in England, Germany, France, and the United States, were making major contributions in various facets of physiology in the early years of the new century. The experimental method, urged so consistently by Bernard, was now the rule. Paper after paper was published which added to the sum total of our knowledge of regulatory mechanisms and could be included in this collection. Certainly those of Bayliss, Starling, Haldane, Henderson, and Cannon, to name a few, are significant to the development of this concept. But, as explained earlier, there are space restrictions; thus the decision was made to include only those publications that address themselves directly to the concept.

Lawrence J. Henderson (1878–1924) was a graduate of Harvard and also received degrees from Cambridge University and the University of Grenoble in France. His major work was accomplished while professor of biological chemistry at Harvard. In 1921, he was an exchange professor at the University of Paris. He also was guest lecturer at Yale, the University of California, and the University of Berlin. He was decorated with the French Legion of Honor in 1948 and, most pertinent to the development of the concept of homeostasis, he took every opportunity to publicize in the United States the work and conclusions of Claude Bernard. He was instrumental in having Bernard's *Introduction à l'étude de la médecine expérimental* translated into English in 1927.

In 1913, Henderson published *The Fitness of the Environment.* A short section from this book is reproduced to show how he considered the interrelationship between the external and internal environments.

Amusingly, Henderson is criticized by J. S. Haldane (1892–1964) for statements made in his book, *Blood, A Study in General Physiology*. Haldane was a noted English biochemist and, as his article in *Science* in 1929 shows, he did not think that Henderson really understood Bernard's "le milieu interiéur."

Walter Bradford Cannon (1871–1945) was born in Wisconsin but received his A.B., A.M., M.D., and Sc.D. degrees all from Harvard. Not surprisingly, he stayed at Harvard where in 1906, only six years after receiving his M.D. degree, he succeeded his mentor, the noted H. P. Bowditch.

Cannon first directed his attention to movements of the gastrointestinal tract. This led him to an investigation of the autonomic nervous system. During his long career at Harvard, he published not only voluminously, but each publication was an entity rather than a fragment, and in each he furthered our knowledge of basic physiologic mechanisms.

As Cannon's work on the autonomic nervous system progressed, he came to recognize it as an important factor contributing to the ability of the organism to maintain the internal environment relatively constant. Finally, in 1926, he first used the term "homeostasis," a term he coined. In this respect, mention should be made of the fact that although, as John Fulton states, Cannon outlined for the first time his classic concept of homeostasis in 1926 in an article included in a very obscure volume, a year earlier he had published almost the identical material in the now defunct and little known *Transactions of the Congress of American Physicians and Surgeons*. Both publications are included here. A close comparison proves revealing.

Finally, in 1929, in an article in *Physiological Reviews*, entitled "Organization for Physiological Homeostasis," Cannon develops completely the concept of homeostasis. Here we see that he first mentions Pflüger, Fredericq, and Richet, but he does give credit to Claude Bernard using the date 1878 which, as previously documented, is much too late. Nonetheless, Cannon's 1929 publication is a true landmark and if only one paper had to be selected, this would be it.

Alfred James Lotka (1880–1949) was a mathematician. He was born in Austria of American parents. He was educated in England, the United States, and Germany. He gradually switched from mathematics to physical biology, and then devoted his later years to working with Louis Dublin on the mathematical analysis of population, mathematical theory of evolution, and actuarial problems. It is his book, *Elements of Physical Biology*, published in 1925, that interests us here. Therein he discusses the principle of Le Chatelier because certain biologists had attempted to apply the principle to biological systems. This principle states that a system tends to change so as to minimize an external disturbance. If it could be applied to biological systems then indeed it would have direct relevance to homeostasis. Lotka analyzes that proposition, concludes in the negative, and then adds, "Facts are stubborn things; it seems a pity to demolish the idol of a pretty generalization, but in such things we cannot permit the wish to be the father to the thought."

13

The Fitness of the Environment: An Inquiry into the Biological Significance of the Properties of Matter

LAWRENCE J. HENDERSON

C

THE CHARACTERISTICS OF LIFE

Under the circumstances it is certainly no rash enterprise to seek a definition of some of the essential characteristics of life. Although it is probably far beyond our present power to make a complete study of the problem, I feel sure that a brief analysis will justify certain very definite conclusions. Life as we know it is a physico-chemical mechanism, and it is probably inconceivable that it should be otherwise.[1] As such, it possesses, and, we may well conclude, must ever possess, a high degree of complexity, — physically, chemically, and physiologically; that is to say, structurally and functionally. We cannot imagine life which is no more complex than a sphere, or salt, or the fall of rain, and, as we know it, it is in fact a very great deal more complex than such simple things. Next,

[1] I mean, of course, for the purposes of physical and chemical study. With such qualifications the statement is probably no longer open to objection from any quarter.

living things, still more the community of living things, are durable. But complexity and durability of mechanism are only possible if internal and external conditions are stable. Hence, automatic regulations of the environment and the possibility of regulation of conditions within the organism are essential to life. It is not possible to specify a large number of conditions which must be regulated, but certain it is from our present experience that at least rough regulation of temperature, pressure, and chemical constitution of environment and organism are really essential to life, and that there is great advantage in many other regulations and in finer regulations. Finally, a living being must be active, hence its metabolism must be fed with matter and energy, and accordingly there must always be exchange of matter and energy with the environment.

Returning to the concept of the organism as a durable form through which flow matter and energy, it is now possible to make these ideas more vivid. The complex structure of the living being is relatively stable, alike in the chemical composition of its individual constituent molecules, in their proportions and amounts, in their aggregation into the invisible structural elements of protoplasm, in the visible parts of the cell, in the organs and tissues, and finally *in toto*, as a man or a tree. Similarly stable are the physical conditions within this structure: temperature, pressure, alkalinity, and osmotic pressure. Finally, that which surrounds it, the immediate environment, possesses also a high degree of stability, or if the organism be very complex, it may be that it has an efficient protection against change of environment; a skin which insulates, for instance. But in this case it has also acquired an environment, a *milieu intérieur* for its cells, — like the blood and lymph, — which serves the same purpose as stability of the external environment, and exercises the further function of supplying food.

It is through this structure, in the process of metabolism, that matter and energy flow. Entering in various forms and quantities, they are temporarily shaped exactly to the form and condition of the organism; they conform to the characteristics of the kingdom, class, order, family, genus, species, and variety to which it belongs, and they assume even the characteristics of the individual itself.[1] Then they depart through the various channels of excretion.

When these ideas are reduced to their very simplest forms, it appears that life must be highly complex in structure and function; that the conditions of the environment must be regulated, and that there must be very exact regulation of conditions, both structural and functional within the organism, and finally,

[1] Science is, of course, still at a loss for an adequate general explanation of such processes.

that, while life is active, there must be exchange of both matter and energy with the environment. Complexity, regulation, and food are essential to life as we know it, and in truth we cannot otherwise conceive of life, or indeed of any other durable mechanism. For my part I do not doubt that these postulates are quite as true of the world of our senses as are the fundamental laws of matter and energy, space and time.

Obviously these few conclusions can make no claim to completeness. Fully to describe life, the discovery of many other fundamental characteristics is necessary, including such as are related to inheritance, variation, evolution, consciousness, and a host of other things. But in the formation and logical development of such ideas there is danger of fallacy at every step, and since the present list will suffice for the present purpose, further considerations of this sort are best dispensed with. This subject should not be put aside, however, without clear emphasis that the postulates which have been adopted above are extremely meager. The only motives for abandoning further search are the economy and the security which are thus insured, and the very great difficulty of extending the list. Any one who is familiar with similar efforts to elucidate the essential characteristics of life, such as that of Wallace,[1] cannot,

[1] A. R. Wallace, "Man's Place in the Universe." New

I fear, fail to perceive the extreme limitations which are imposed upon inquiry by assuming complexity, regulation, and metabolism exclusively. Perhaps in reality these postulates are only two. Metabolism might without difficulty be included under regulation, but the consideration of such purely logical questions is beside the present purpose. However, these are probably the characteristics of the organism which are best fitted for discussion in relation to the physico-chemical phenomena of matter and energy, and it is barely possible that no others bear the same simple relations to the outside world.

York, 1903, Chaps. X and XI, especially the following statement: —

"The physical conditions on the surface of our earth which appear to be necessary for the development and maintenance of living organisms may be dealt with under the following headings: —

"1. Regularity of heat supply, resulting in a limited range of temperature.

"2. A sufficient amount of solar light and heat.

"3. Water in great abundance, and universally distributed.

"4. An atmosphere of sufficient density, and consisting of the gases which are essential for vegetable and animal life. These are Oxygen, Carbonic-acid gas, Aqueous vapor, Nitrogen, and Ammonia. These must all be present in suitable proportions.

"5. Alternations of day and night."

It must be remembered, however, that such conclusions depend upon reasoning from analogy, a dangerous proceeding.

V

THE PROBLEM

We may now return to the problem of the fitness of the environment. So long as ideas of the nature of living things remain vague and ill-defined, it is clearly impossible, as a rule, to distinguish between an adaptation of the organism to the environment and a case of fitness of the environment for life, in the very most general sense. No doubt there are clear instances of both phenomena which require no close analysis for their interpretation. Thus the hand is surely an instance of adaptation, and the anomalous expansion of water on cooling near its freezing point an instance of environmental fitness. But how much weight is to be assigned to adaptation and how much to fitness in discussing the relations between marine organisms and the ocean? Evidently to answer such questions we must possess clear and precise ideas and definitions of living things. Life must by arbitrary process of logic be changed from the varying thing which it is into an independent variable or an invariant, shorn of many of its most interesting qualities to be sure, but no longer inviting fallacy through our inability to perceive clearly the questions involved.

Such is the purpose, and the justification, for setting up the postulates of complexity, regulation, and metabolism as inherent in that mechanism which is called the living organism. With them, at length, we face the problem which awaits us. To what extent do the characteristics of matter and energy and the cosmic processes favor the existence of mechanisms which must be complex, highly regulated, and provided with suitable matter and energy as food? If it shall appear that the fitness of the environment to fulfill these demands of life is great, we may then ask whether it is so great that we cannot reasonably assume it to be accidental, and finally we may inquire what manner of law is capable of explaining such fitness of the very nature of things.

14

Claude Bernard's Conception of the Internal Environment

J. S. HALDANE

DISCUSSION

CLAUDE BERNARD'S CONCEPTION OF THE INTERNAL ENVIRONMENT

PROFESSOR L. J. HENDERSON entitles his valuable recently published book on "Blood" as "A Study in General Physiology," and at the same time treats blood as a physico-chemical system. It may escape notice that he thus makes a very far-reaching fundamental assumption; and the matter is so important that I ventured to bring it before the British Physiological Society on March 16. He refers to the authority of Claude Bernard in justification of his procedure; but in so doing he seems to me to have altogether misunderstood Bernard's conclusion. Bernard was the first to formulate the extremely fruitful idea that the blood of a living animal is an internal medium kept remarkably constant as regards its physico-chemical conditions by the coordinated influence upon it of the various organs of the body. He accepts as fundamental the coordination thus displayed. L. J. Henderson, on the other hand, treats the blood as simply something which, as the result of various "buffer" reactions occurring within itself, is not as readily disturbed in its physico-chemical conditions as other liquids would be. We can, for instance, add a good deal of acid or alkali to blood without much disturbing its reaction. Or if we simultaneously add carbon dioxide and abstract oxygen from it there is a similar diminution of the disturbance which would be produced by either addition of carbon dioxide alone or abstraction of oxygen alone.

These buffer reactions are of great importance and interest, but they were unknown to Bernard, and do not in any way modify his conception of the coordinated activity of organs by which the conditions in the blood are kept constant. This coordinated activity is an essential part of his conception of blood in the living body, whereas L. J. Henderson leaves it out of account, thus turning blood in the living body into what for a physiologist is a mere artifact, and completely disregarding Bernard's principle. It seems to me that if we disregard the coordination we have disregarded all that is characteristic of life, and that therefore the book in question can not be regarded as a study in general physiology, but only as a study in physical chemistry.

To come to details, L. J. Henderson treats the constancy of reaction in the living body as if it depended on the physico-chemical properties of blood. In actual fact this constancy depends during health on the coordinated activity of the kidneys and respiratory organs, in accordance with Bernard's principle; and in various individual parts of the body the constancy depends on the coordinated or regulated influence of the circulation. Not all the buffering in the world would keep the reaction constant otherwise, though the buffering greatly smooths the regulation. In the human body acid in excess is being continuously produced, partly as ionized sulphuric and other non-volatile acids, and partly as ionized carbonic acid. The formation of acid is constantly being exactly compensated by the excretion of acid urine and formation of ammonia on the one hand, and on the other by the washing out of carbon dioxide through the lungs. The exact coordination or regulation of these activities is the essential matter, and the quantitative investigation in various directions of physiological coordination in recent times has separated the old mechanistic physiology of last century from recent physiology. The normal responses of the kidneys and respiratory organs depend on the simultaneous maintenance of many conditions included under the comprehensive word "health"; but we assume this maintenance in quantitative investigations of physiological function.

If, following L. J. Henderson, we neglect active organic coordination, we are, it seems to me, taking a step backwards. As one who has been closely connected during the last thirty years with the development of Bernard's conception, as well as with the development of knowledge as to the physical chemistry of blood, I wish, therefore, to express my dissent from what appears to me to be L. J. Henderson's misinterpretation of Bernard. In my book, about to be published, on "The Sciences and Philosophy," I have discussed the subject from a wider standpoint, but before I had seen L. J. Henderson's book. It seems to me that apart from the central biological conception of specific coordination we can not make even a beginning in the scientific treatment of general physiology, whether we start from the unicellular organisms which Henderson unjustifiably assumes to consist of a physico-chemical system called "protoplasm," or from compound organisms with a well-defined internal environment between individual cells.

J. S. HALDANE

UNIVERSITY OF OXFORD

15

The Physiological Basis of Thirst

WALTER B. CANNON, M.R.C.

A custom which has usually been respected by investigators who in years past have had the high honour of delivering the Croonian Lecture is that of reporting and interpreting a group of related researches upon which they have been engaged and which they have already made public. That is a custom which I should have been happy to follow on the present occasion if military service had not sharply broken in on my studies months ago and made them seem now very remote and the summarising of them a difficult occupation. And, after all, is it not natural for us as investigators to hold the forward look, to consider the problems before us rather than those that have been solved? May I, therefore, be permitted to bring to your attention some ideas and observations which have not yet been published and which, though incomplete, may prove interesting and suggestive.

In regarding the human body as a self-regulating organisation we observe that, so far as mere existence is concerned, it depends on three necessary supplies from the outer world,—on food, to provide for growth and repair and to yield energy for internal activities and the maintenance of body heat; on oxygen, to serve the oxidative processes essential to life; and on water, as the medium in which occur all the chemical changes of the body. These three supplies are of different orders of urgency. Thus a man may live for 30 or 40 days without taking food, as professional fasters have demonstrated,* and suffer no apparent permanent injury to his bodily structure or functions. On the other hand, lack of oxygen for only a brief period may result in unconsciousness and death. Indeed, certain nerve cells in the cerebral cortex cannot withstand total deprivation of oxygen for more than 8 or 9 minutes without undergoing such fundamental changes that they do not again become normal when they receive their proper supply.† Intermediate between the long survival without food and the very brief survival without oxygen is the period of existence which is possible without water. Records of men who have missed their way in desert regions and who, with no water to drink, have wandered in the scorching heat have proved that they rarely live under these circumstances of struggle and torrid atmosphere for more than three

* Luciani, 'Das Hungern,' Leipzig, 1890.
† See Gomez and Pike, 'Jour. of Exp. Med.,' vol. 11, p. 262 (1909).

days, and many die within 36 hours. An exceptional instance has been reported, of a Mexican, who, lost in the dry plains of the south-western part of the United States, walked, or crept on his hands and knees, between 100 and 150 miles, repeatedly drinking his own excretions, and succeeded, after nearly 7 days wholly without water, in reaching a habitation.* This is a record which, for its conditions, has no parallel. If the thirsting man is not subjected to heat or exertion his life may continue much longer than 7 days. Viterbi, an Italian political prisoner, who committed suicide by refusing food and drink, died on the eighteenth day of his voluntary privation. After the third day the pangs of hunger ceased, but, until almost the last, thirst was always more insistent and tormenting. He records again and again his parched mouth and throat, his burning thirst, his ardent and continual thirst, his thirst constant and ever more intolerable.† Thus though the period of survival varies, death is sure to come whether food, or oxygen, or water is withheld.

Normally these three supplies—food, oxygen, and water—are maintained in more or less constant adjustment to the bodily needs. Food material is being continually utilised in building body structure, and in providing energy for bodily activities, but it is periodically restored. Oxygen is continually combining with carbon and hydrogen and leaving the body in CO_2 and H_2O, but the loss is compensated for with every breath. And water, likewise, is always being discharged in expired air, in secretion from the kidneys, and in the sweat. So great is the escape by way of the lungs and skin alone that it is estimated that approximately 25 per cent. of the heat loss from the body is due to evaporation from these surfaces.‡ This continuous lessening of the water content must be checked by a new supply, or important functions will begin to show signs of need.

The evidence for the absolute necessity of water in our physiological processes requires no elaboration. Water is a universal and essential ingredient of all forms of organisms. Without it life disappears or is latent—the dry seed awakens only on becoming moist. Because we may have it at almost any moment we are likely to overlook its absolute necessity in our lives. Among inhabitants of desert regions, however, water is the central nucleus of thought about which all other ideas revolve; it is an ultimate standard of things, incomparably more stable and more exalted than the gold of civilised commerce, the constantly remembered basis of existence.§ In

* McGee, 'Interstate Med. Jour.,' vol. 13, p. 279 (1906).

† Viterbi, quoted by Bardier, Richet's 'Dictionnaire de Physiologie,' article "Faim," vol. 6, p. 7 (1904).

‡ Gephart and Du Bois, 'Arch. Int. Med.,' vol. 17, p. 902 (1916).

§ McGee, "The Seri Indians," '17th Annual Report of the Bureau of American Ethnology,' p. 181.

our bodies the presence of water as the main constituent of the digestive secretions, its *rôle* in the chemical changes of digestion, its service as a vehicle of absorption, its importance in the composition of blood and lymph, its use, together with other substances, in body fluids as a lubricant, its action in regulating body temperature—these functions need merely to be mentioned to illustrate how water influences every activity which living beings display.

Because water is a fundamental essential to life, and is continually escaping from the body, and because there is consequent need for repeated replenishment of the store, an inquiry into the mechanism of the replenishment is a matter of interest.

That such a mechanism exists is indicated by the fact that all our essential functions, leading to preservation of the individual and of the race, are controlled not through memory and volition, but by insistent sensations and desires. The unpleasant sensation of thirst causes us to drink. Not towards the subjective aspect of these automatic arrangements, however, is the special attention of the physiologist directed. He is primarily concerned with the bodily states which give rise to the sensation. Only when these states and their relations to the needs of the organism are known is the automatic control explained.

About six years ago I called attention to some graphic records of motions of the stomach in man which showed that the sensation of hunger is associated with powerful contractions of the empty or nearly empty organ. And because the hunger pang began to be experienced after the contraction had started, the conclusion was drawn that hunger is not a "general sensation," as was formerly held by physiologists and psychologists, but has its immediate origin in the stomach, and is the direct consequence of the strong contraction.* This conclusion has since been abundantly confirmed by Carlson and his collaborators in observations on themselves and on a man with a gastric fistula.†

Even more imperious than hunger as an insistent and tormenting sensation, accompanied by a dominant impulse which determines our behaviour, is thirst. Indeed, these two experiences—hunger and thirst—are such impelling motives in directing our conduct that from early times they have been used as supreme examples of a strong desire. The ancient prophet spoke of a "hunger and thirst after righteousness" to express the eagerness

* The Harvey Lectures, New York, 1911–1912, p. 130—Cannon: Harvey Lecture, December 16, 1911, "A Consideration of the Nature of Hunger"; also Cannon and Washburn, 'Am. Journ. Physiol.,' vol. 29, p. 441 (1912).

† Carlson, 'Control of Hunger in Health and Disease,' Chicago, 1916.

of his yearning. And the common acquaintance of mankind with the potent demands of hunger and thirst for satisfaction renders these similes easily understood.

In undertaking a discussion of thirst it is necessary at the start to distinguish clearly between the primitive sensation itself and appetite. The same distinction had to be drawn in considering the nature of hunger. The hunger pang is a disagreeable ache or gnawing pain referred to the lower mid-chest region or the epigastrium. Appetite for food, on the other hand, is related to previous experiences which have yielded pleasurable sensations of taste or smell. Thus associations become established between particular edible substances and the delights they convey, with the result that a wish develops that the delights may be renewed. In either circumstance, whether for the satisfaction of appetite or for the satisfaction of hunger, the body is supplied with food.

Similarly in the case of drink, the appetite for this or that peculiar potable substance develops from former experience and from established associations of an agreeable character. We drink not only because we are thirsty, but also because we relish a certain aroma or bouquet, or a peculiar taste, and wish to enjoy it again. In respect to appetite the taking of fluid differs from the taking of food, in that fluid, which leaves the stomach rapidly, may not occasion a sense of satiety as does food, which accumulates in the stomach. In this possibility of continuing pleasurable sensations associated with drinking lie the dangers that arise from the excessive use of beverages. Under normal conditions, however, it is through the satisfaction of appetite for a particular drink, *e.g.*, for tea, or coffee, or light alcoholic beverages, that the body may be supplied with sufficient water for its needs before thirst has had occasion to manifest itself. But just as there is provided, back of the appetite for food, in readiness to become imperious if necessary, the sensation of hunger; so likewise, as a final defence against a too great depletion of the water content of the body, there may appear the urgent and distressing sensation of thirst.

There is a general agreement that thirst is a sensation referred to the mucous lining of the mouth and pharynx, and especially to the root of the tongue and to the palate. McGee, an American geologist of large experience in desert regions, who made numerous observations on sufferers from extreme thirst, has distinguished five stages through which men pass on their way to death from lack of water.* In the first stage there is a feeling of dryness in the mouth and throat, accompanied by a craving for liquid. This is the common experience of normal thirst. The condition

* McGee, 'Interstate Med. Jour.,' vol. 13, p. 279 (1906).

may be alleviated, as everyday practice demonstrates, by a moderate quantity of water, or through exciting a flow of saliva by taking into the mouth fruit acids such as lemon or tomato juice, or by chewing insoluble substances. In the second stage the saliva and mucus in the mouth and throat become scant and sticky. There is a feeling of dry deadness of the mucous membranes. The inbreathed air feels hot. The tongue clings to the teeth or cleaves to the roof of the mouth. A lump seems to rise in the throat, and starts endless swallowing motions to dislodge it. Water and wetness are then exalted as the end of all excellence. Even in this stage the distress can be alleviated by repeatedly sipping and sniffing a few drops of water at a time. "Many prospectors," McGee states, "become artists in mouth moistening, and carry canteens only for this purpose, depending on draughts in camp to supply the general needs of the system." The last three stages described by McGee, in which the eyelids stiffen over eyeballs set in a sightless stare, the distal tongue hardens to a dull weight, and the wretched victim has illusions of lakes and running streams, are too pathological for our present interest.

The fact I wish to emphasise is the persistent dryness of the mouth and throat in thirst. Direct testimony is given by King, a medical officer in a United States Cavalry troop, which for 3½ days was lost without water in the torrid "Llano Estacado" of Texas. He records that, on the third day, salivary and mucous secretions had long been absent, and that mouths and throats were so parched that food, on being chewed, gathered about the teeth and in the palate, and could not be swallowed. "Sugar would not dissolve in the mouth."*

Further evidence of the relation between local dryness of the mouth and throat and the sensation of thirst is found in some of the conditions which bring on the sensation. Breathing hot air free from moisture, prolonged speaking or singing, the repeated chewing of desiccated food, the inhibitory influence of fear and anxiety on salivary secretion, have all been observed to result in dryness of the buccal and pharyngeal mucous membrane and in attendant thirst. On the other hand, conditions arising in regions remote from the mouth and involving a reduction of the general fluid content of the body, such as profuse sweating, the excessive diarrhœa of cholera, the diuresis of diabetes, as well as such losses as occur in hæmorrhage and lactation, are well recognised causes of the same sensation. There appear to be, therefore, both local and general origins of thirst. In correspondence with these observations, two groups of theories have arisen, just as in the

* King, 'Amer. Jour. Med. Sci.,' vol. 75, p. 404 (1878).

2 A 2

case of hunger—one explaining thirst as a local sensation, the other explaining it as a general and diffuse sensation. These theories require examination.

The view that thirst is a sensation of local origin has had few advocates, and the evidence in its favour is meagre. In 1885 Lepidi-Chioti and Fubini* reported observations on a boy of 17, who, suffering from polyuria, passed from 13 to 15 litres of urine daily. When prevented from drinking for several hours, this youth was tormented by a most distressing thirst, which he referred to the back of the mouth, and at times to the epigastrium. The observers tried the effect of brushing the back of the mouth with a weak solution of cocaine. Scarcely was the application completed before the troublesome sensation wholly ceased, and the patient remained comfortable from 15 to 35 minutes. If, instead of cocaine, water was used to brush over the mucous membranes, thirst was relieved for only two minutes. The temporary abolition of a persistent thirst by use of a local anæsthetic, in a human being who could testify regarding his experience, is suggestive support for the local origin of the sensation. The evidence adduced by Valenti is also suggestive. He cocainised the back of the mouth and the upper œsophagus of dogs which had been deprived of water for several days, and noted that they then refused to drink.† One might suppose that the refusal to take water was due to inhibition of the swallowing reflex by anæsthetisation of the pharyngeal mucosa, as reported by Wassilief.‡ But Valenti states that his animals are quite capable of swallowing.§

Though these observations are indicative of a local source of the thirst sensation, they leave unexplained the manner in which the sensation arises. Valenti has put forward the idea that all the afferent nerves of the upper part of the digestive tube are excitable to stimuli of thirst, but that suggestion does not advance our knowledge so long as we are left unenlightened as to what these stimuli are. A similar criticism may be offered to Luciani's theory that the sensory nerves of the buccal and pharyngeal mucosa are especially sensitive to a diminution of the water-content of the circulating fluid of the body; indeed, that these nerves are advance sentinels, like the skin nerves for pain, warning the body of danger.|| No special features of the nerves of this region, however, are known. No special end-organs are known. The intimation that these nerves are peculiarly related to a general bodily need is pure hypothesis. That they

* Lepidi-Chioti and Fubini, 'Giorn. d. R. Accad. d. Med.,' Turin, vol. 48, p. 905 (1885).

† Valenti, 'Arch. Ital. de Biol.,' vol. 53, p. 94 (1910).

‡ Wassilief, 'Ztschr. f. Biol.,' vol. 24, p. 40 (1888).

§ Valenti, 'Cbl. f. Physiol.,' vol. 20, p. 450 (1906).

|| Luciani, 'Arch. di Fisiol.,' vol. 3, p. 541 (1906).

mediate the sensation of thirst is unquestioned. But the problem again is presented, How are they stimulated ?

The view that thirst is a general sensation was well stated by Schiff. It arises, he declared, from a lessened water-content of the body, a condition from which the whole body suffers. The local reference to the pharynx, like the local reference of hunger to the stomach, is due to association of experiences. Thus the feeling of dryness in the throat, though it accompanies thirst, has only the value of a secondary phenomenon, and bears no deeper relation to the general sensation than heaviness of the eyelids bears to the general sensation of sleepiness.* The conception of thirst, as a general sensation, is commonly accepted, and is supported by considerable experimental evidence. The interpretation of this evidence, however, is open to question, and should be examined critically.

First among the experiments cited are those of Dupuytren and the later similar experiments of Orfila.† These observers abolished thirst in dogs by injecting water and other liquids into the dogs' veins. And Schiff quotes Magendie as having treated successfully by the same procedure the thirst of a patient suffering from hydrophobia. In these instances the treatment was no doubt general, in that it affected the body as a whole. But the assumption that thirst is thus proved to be a general sensation is unwarranted, for the injection of fluid into the circulation may have changed local conditions in the mouth and pharynx, so that the local sensation no longer arose.

A classic experiment repeatedly cited in the literature of thirst was one performed by Claude Bernard. He opened a gastric fistula which he had made in a dog, and allowed the water which the animal drank to pass out. As the animal became thirsty, it would drink until "fatigued," as the report states, and when "rested" it would begin again. But after the fistula was closed, drinking quickly assuaged the desire for water. The inference was drawn that thirst must be a general sensation, for the passage of water through the mouth and pharynx wet those surfaces, and yet the animal was not satisfied until the water was permitted to enter the intestine and be absorbed by the body.‡ This evidence appears conclusive. The expressions "fatigued" and "rested," however, are interpretations of the observer, and not the testimony of the dog. Indeed, we may with equal reasonableness assume that the animal stopped drinking because he was not thirsty, and started again when he became thirsty. The only assumptions necessary for such an interpretation of the animal's behaviour are that appreciable time is

* Schiff, 'Physiologie de la Digestion,' Florence and Turin, vol. 1, p. 41 (1867).

† See 'Dictionnaire des Sci. Méd.,' Paris, vol. 61, p. 469 (1821).

‡ Bernard, 'Physiologie Expérimentale,' Paris, vol. 2, p. 49 (1856).

required to moisten the buccal and pharyngeal mucosa sufficiently to extinguish thirst—a point made by Voit*—and that these regions become dried rapidly when there is absence of an adequate water-content in the body. This interpretation is consistent with the view that thirst is a sensation having a local source. Furthermore, this interpretation is not contradicted by the satisfaction manifested by the dog after the fistula was closed, for the water which is absorbed, like that injected into veins, may quench thirst by altering local conditions. We cannot admit, therefore, that Bernard's experiment is proof that thirst is a general sensation.

Another set of observations cited as favourable to the theory of the diffused character of the origin of thirst are those of Longet. After severing the glosso-pharyngeal, the lingual and the vagus nerves on both sides in dogs, he observed that they drank as usual after eating.† If thirst has a local origin in the mouth and pharynx, why should the animals in which the nerves to these regions were cut still take water? Two answers to this question may be given. First, as Voit has pointed out,‡ Longet did not cut all branches of the vagi and trigemini to the mouth and pharynx, and, consequently, some sensation persisted. And second, even if all nerves were cut, the fact that the animals drank would not prove that thirst exists as a general feeling, for one may drink from the sight of fluid, or from custom, without the stimulation of a dry mouth, just as one may eat from the sight of food without the stimulus of hunger. In other words, the element of appetite, previously considered, may enter, and as a matter of habit and associated experience determine present reactions.

The remaining evidence in favour of the diffused origin of thirst is found in studies of blood changes. These changes, by altering the "milieu intérieur" of the body cells, must affect them all. In 1900, Mayer published reports on the increase of osmotic pressure of the blood, as determined by depression of the freezing point of the serum, which he noted in conditions naturally accompanied by thirst. Dogs deprived of water for several days had a blood serum in which the osmotic pressure was increased, and rabbits kept in a specially warmed chamber showed the same change. Thus, conditions in which the water supply to the body was stopped, or the loss of water from the body by sweating or pulmonary evaporation was increased, either of which is known to cause thirst, were associated with a rise of osmotic pressure. And Mayer argued that all other circumstances in which thirst appears—in diabetes with increased blood sugar, in renal disease with

* Voit, 'Hermann's Handbuch der Physiologie,' Leipzig, Abth. 6, p. 566 (1881).

† Longet, 'Traité de Physiologie,' Paris, vol. 1, p. 35 *et seq.* (1868).

‡ Voit, *loc. cit.*.

accumulation of waste material in the body fluids, in acute rabies with total deprivation of water, in cholera with excessive outpouring of water into the intestine—the osmotic pressure of the blood would be augmented. Moreover, when a thirsty dog drinks, the hypertonicity of his serum disappears, his normal condition is restored, and he stops drinking.

By these observations Mayer was led to the conclusion that whenever the osmotic pressure of the blood rises above normal, thirst appears; whenever it returns to normal, thirst vanishes; and as the pressure varies, thirst also varies. Since intravenous injections of hypertonic salt solution cause, by stimulation of the bulbar centres, according to Mayer, a rise of arterial pressure and renal and intestinal vasodilation—both operating to lower the abnormally high osmotic pressure of the blood—he infers that other agencies are present in the organism besides the desire for water, which tend to keep the blood normal. Thirst, he declares, is the last of a series of mechanisms acting to protect the organism against hypertonicity of its fluids.

In summary, then, the thirsty individual has a blood with high osmotic pressure. This condition affects all the cells of the body. It disturbs the cells of the central nervous system, and thus leads both to protective circulatory reactions and, in case these fail, to malaise and irritability, and a reference of unpleasantness to the region of the pharynx. Accompanying this, there is the impulse to drink, and when that is satisfied, the water taken in restores the normal state.*

Mayer's observations were soon confirmed, but his inferences were challenged. In 1901, Wettendorff, working in Brussels, reported that if dogs are deprived of water their blood does, indeed, develop a hypertonicity, as Mayer had found, but that this is a phenomenon which does not occur to any marked degree in the first days of the deprivation. In one instance there was no change in the freezing point of the serum during three days of thirst. Serious alteration of the osmotic pressure of the blood, therefore, is comparatively tardy in its appearance. Since the organism is continually losing water, and, nevertheless, the blood remains for a day or two unchanged, Wettendorff concluded that the consistency of the blood is preserved as long as possible by withdrawal of water from the extravascular fluids and the tissues. Further, thirst is clearly demonstrable long before any considerable change in the blood is evident. One animal in which the freezing point of the serum had been lowered only 0·01° C. by four days' deprivation of water, drank 200 c.c. of physiological salt solution, a liquid which to the dog in normal condition is quite repugnant. Again, when the blood has become

* Mayer, 'C. R. Soc. de Biol.,' vol. 52, pp. 154, 389, 522 (1900); also 'Essai sur la Soif,' from the Laboratory of Experimental Pathology, Faculty of Medicine, Paris, 1900.

slightly hypertonic, a dog may drink normal salt solution without lowering his osmotic pressure and afterwards, by refusing further drink, act quite as if he had slaked his thirst. But if an animal with a very hypertonic blood is placed before hypertonic salt solution he takes it again and again—an action which may be explained by a draining of water from the tissues with increasing intensity, and a consequent increasing thirst.

From all these observations Wettendorff concluded that the origin of thirst does not reside in alterations of the blood itself, but in the act of withdrawing water from the tissues. The liquids bathing the cells, therefore, would be first to concentrate as water is lost from the organism. And since the conditions of cellular life would thus be modified in all the tissues, the peculiar state would develop which occasions the sensation of thirst. This effect is generally diffused, and is independent of any peculiar influence of the process of dehydration on the nervous system itself.

In accounting for the localising of the sensation in the mouth and throat Wettendorff distinguished between a "true thirst" and a "false thirst." "True thirst," he declared, is dependent on an actual bodily need, and is persistent until the need is satisfied. "False thirst" is only a dryness of the mouth and pharynx. Dryness in this region occurs, to be sure, in true thirst, but it is then an expression of the general dehydration of the tissues, exaggerated perhaps by contact with the outer air. Through experience the two conditions—buccal dryness and general dehydration—have become associated. Even in true thirst we may temporarily abolish the sensation by moistening the pharyngeal mucous membrane, but the result is only a "false satisfaction," a self-deception, made possible because long and pleasant experience has proved that moistening this region by drink leads to the satisfaction of an instinctive need.*

The foregoing review of observations and theories has revealed that the attitude of physiologists with reference to thirst has been much as it was with reference to hunger. In each condition a general bodily need has arisen from a lack of essential bodily material and is signalled by a well-defined sensation. In each the testimony of ingenuous persons regarding their feelings has been carefully set down, and then explained away. Thus in the case of thirst the primary sensation is described universally as an experience of dryness and stickiness in the mouth and throat.† Instead of attempting to account for the experience as such,

* Wettendorf, 'Travaux du Laboratoire de l'Institut Solvay,' Brussels, vol. 4, pp. 353–484 (1901).

† Foster, 'Textbook of Physiology,' London, p. 1423 (1891); Ludwig, 'Lehrbuch der Physiclogie,' vol. 2, p. 586; Voit, 'Hermann's Handbuch der Physiologie,' Abth. 6, p. 566.

however, attention has been paid to the bodily need which accompanies it; apparently, since the need is a general one, the sensation has been supposed to be general, and the thirst which everybody experiences and knows about has been classed as an associated secondary phenomenon or the peripheral reference of a central change. The really doubtful feature in this view of thirst, just as in the older conception of hunger, is the "general sensation." That even the early stages of a need of water may be accompanied by increased irritability, and a vague sense of weakness and limpness, is not denied. But the thirsty man does not complain of these general conditions. He is tormented by a parched and burning throat, and any explanation of the physiological mechanism for maintaining the water content of the body must take into account this prominent fact.

In looking for a mechanism which would automatically keep up the water supply of our bodily economy, we may follow two clues; first, that there may be a peripheral arrangement which in the presence of a general bodily need for water would lead to dryness of the mouth and throat; and second, that a peripheral arrangement of this nature should be especially characteristic of animals which are constantly and rapidly losing water and require repeated renewal of the supply. These two clues offer a biological approach to the explanation of thirst which I wish to utilise.

In one sense all animals are constantly losing water, for even in the simplest forms waste material is excreted in solution. With respect to water loss, however, we should expect to find a marked difference between animals living in water itself and those living in air. Indeed, it is difficult to conceive of an animal living in water as experiencing thirst. The entire body surface and the mouth and throat are throughout active life continuously bathed in a moving flow. The food is taken wet from a wet medium. Probably renal activity and the secretion of the digestive glands are the only important ways for water to leave the economy; and the digestive secretions are soon largely re-absorbed. In contrast, the land animals, mammals, for example, lose moisture not only in these ways but also by the moistening of dry food, by evaporation from the extensive surface of the lungs, and by the action of innumerable sweat glands. It is because of the possibility of great and rapid loss of water from its body that the land animal has special need for an assurance of adequate supply.

In the water inhabitant the skin, and the mouth and gullet, are all kept wet by the medium in which he lives and moves. In the process of evolution, however, as organisms changed their habitat from water to air, the skin became dry and scaly. Of the parts which in marine animals were constantly bathed by water, only the mouth and throat continue to be moist. These regions are now exposed to air, however, instead of being flushed by a flowing stream, and consequently they tend to dry. The structural lining of these parts probably renders them especially liable to desiccation in the presence of dry air, for the mucosa of the mouth and also of the pharynx, below the level of the floor of the nasal chambers, is composed of squamous epithelium. Some scattered mucous glands are present, but they are not capable of keeping the surfaces satisfactorily wet, as any one can readily prove by breathing through the mouth for only a few minutes. When air passes to and fro by way of this watercourse, as in prolonged speaking or singing, and in smoking, it is to be expected, therefore, that feelings of dryness and stickiness, which we call thirst, should arise.

Contrast this condition of the mouth with the condition of the respiratory tract, in which the lining membrane consists of columnar epithelium and is richly provided, particularly in the nose, with multitudes of mucous glands. Through this tract air moves to and fro constantly with no sign of inducing desiccation except in extreme and prolonged deprivation of water. But there is one portion of this normal pathway for the air which, in the absence of sufficient moisture, is peculiarly liable to become dried. It is the pharynx, where the respiratory tract crosses the digestive tract—*i.e.*, where the inbreathed air, which may be insufficiently moistened in the nose, passes over surfaces of the ancient watercourse. Here, even with nasal respiration, unpleasant feelings may be excited, if the water-content of the body is reduced, and, in cases of marked thirst, the dryness of this region may stimulate tireless swallowing motions.

The central questions now appear: Why do not the mouth and pharynx feel dry and uncomfortable under normal conditions? and why do they feel so when the body stands in need of water? Again, a comparison of conditions in the water inhabitants, in which the buccal and pharyngeal regions are kept moistened by the surrounding medium, with conditions in the air inhabitants, in which these regions tend to be dried by the surrounding medium, will offer pertinent suggestions. A characteristic difference between these two animal groups is the possession, by the air inhabitants, of special buccal glands. They are not present in fishes, but are found in the rest of the vertebrate series from the amphibia onwards. At first little differentiated, they develop in mammals into the three pairs of salivary glands—the parotid, sub-maxillary, and sub-lingual. For the purpose of considering thirst in man, we may deal solely with this salivary group. The action of these organs is to secrete a fluid which is normally more than 97 per cent., and may be more than 99 per cent., water.* The

* Becker and Ludwig, 'Ztschr. f. Rat. Med.,' vol. 1, p. 278 (1851).

theory of thirst, on which I wish to offer evidence, may now be stated. In brief, it is that the salivary glands have, among their functions, that of keeping moist the ancient watercourse; that they, like other tissues, suffer when water is lacking in the body—a lack especially important for them, however, because their secretion is almost wholly water, and that, when these glands fail to provide sufficient fluid to moisten the mouth and throat, the local discomfort and unpleasantness which result constitute the feeling of thirst.

That one of the uses of buccal glands is to keep wet the surfaces over which their secretion is distributed is indicated by the fact that these structures first appear in air-inhabiting vertebrates. This indication receives support from the conditions seen in the cetacea, the mammalian forms which have returned to an aquatic existence, and in which both the water-loss from the body and the need for wetting the mouth and throat are greatly reduced. It is a remarkable fact that in these animals the salivary glands are either lacking or are very rudimentary. The appearance and disappearance of the buccal glands in large animal groups, in correspondence with the exposure or non-exposure of the mouth and throat to desiccating air, point to these glands as protectors of the buccal mucosa against drying.

Experimental evidence as to the protective function of the salivary secretions was provided incidentally many years ago by Bidder and Schmidt. They were interested in studying any fluid secretion which might appear in the mouth apart from saliva. To this end they tied in dogs all the salivary ducts. The first effect was such a striking diminution of the fluid layer over the buccal mucosa that only when the mouth was held closed was the surface kept moist, and, when the animal breathed through the mouth, a real drying of the surface was hardly prevented. The eagerness for water, they state, was enormously increased, so that the animal was always ready to drink.*

Related to this service of saliva in moistening and lubricating the mouth parts is the presence of a special reflex for salivary secretion when the buccal mucosa is exposed to conditions which tend to dry it. Thus, as Pavlov's† researches demonstrated, with dry food in the mouth, much more saliva is secreted than with moist food. And Zebrowski‡ found, in the course of observations on patients with a parotid fistula, that, whereas no saliva flowed with the mouth closed, as much as 0·25 c.c. in five minutes came from the duct when the mouth was opened. This reflex is readily

* Bidder and Schmidt, 'Verdauungssäfte und Stoffwechsel,' Leipzig, p. 3 (1852).

† Pavlov, 'The Work of the Digestive Glands,' London, 2nd ed., pp. 70, 82 (1910).

‡ Zebrowski, 'Arch. f. d. ges. Physiol.,' vol. 110, p. 105 (1905).

demonstrated. If one closes the nostrils and breathes through the mouth for five minutes, usually nothing happens during the first minute. The mucosa then begins to feel dry, and at once the saliva starts flowing, and continues for the rest of the period. I have thus collected as much as 4·7 c.c. in four minutes. Chewing motions, with the mouth empty, yielded in five minutes only about 1 c.c. In these observations precautions were taken against any psychic effect due to interest, by adding long columns of figures during the test. It seems clear, therefore, that if the mouth tends to become dry, the salivary glands are normally stimulated to action, and, if there is sufficient outflow from them, the affected surfaces are moistened. The act of swallowing favours the process, for the fluid is thereby spread backwards on the tongue and wiped down the back wall of the pharynx.

The question whether there is a relation between the existence of water-need in the body and diminished flow of saliva I have examined in two ways—by going without fluid for a considerable period and by profuse sweating, combined with measurements of salivary secretion under uniform stimulation. The method of determining salivary output was that of chewing for five minutes and at a uniform rate a tasteless gum, collecting the saliva which flowed during this period, and measuring its volume. All these observations are best made when one is inactive, and in my experience more nearly uniform results are obtained if one lies quiet during the tests.

The influence on salivary flow of going without fluid for some time may be illustrated by an example. The chewing to evoke salivary action was started at 7 o'clock in the morning, and repeated each hour until 8 o'clock in the evening. A breakfast consisting of a dry cereal preparation was taken between 8 and 9 o'clock, and a luncheon of dry bread between 12 and 1 o'clock. Nothing had been drunk since the previous evening. From the first test at 7 o'clock until 11 there was little change in the output of saliva; the average amount secreted in 5 minutes was 14·1 c.c., with variations between 13 and 16·4 c.c. Then the output began to fall, and at 2 o'clock only 6·4 c.c. was secreted. The average amount for the two observations at 2 and 3 o'clock was 7·7 c.c.—only little more than half that poured out in the morning. Between 3 and 4 o'clock a litre of water was drunk. The effect was soon apparent. At 4 o'clock the output was 15·6 c.c., and during the next 4 hours, in which more water was taken, and a supper with thin soup and other fluid was consumed, the average amount secreted was 14·6 c.c., a figure closely corresponding to the 14·1 c.c. of the morning hours. These results are illustrated graphically in fig. 1. Other tests of this character gave similar results, though there was variation in the rate of decrease in the amounts of saliva secreted.

A similar diminution of the salivary secretion occurs after the loss of water from the body by sweating. In one instance, the loss in about one hour

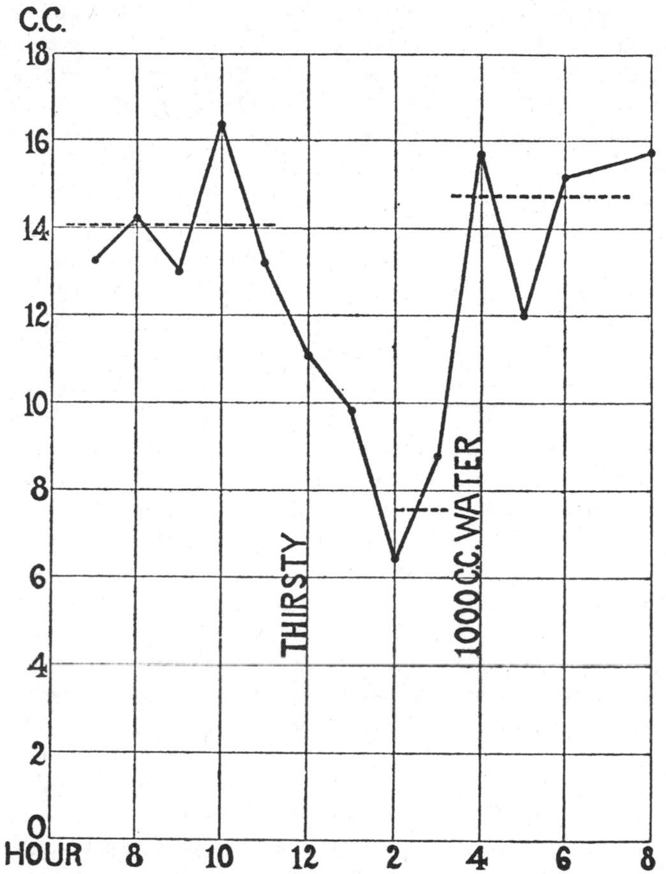

Fig. 1.—Chart showing saliva secreted each hour from 7 A.M. until 8 P.M. in consequence of chewing a tasteless gum five minutes. No fluid was drunk between 7 o'clock the previous evening and 3 o'clock P.M. For further description, see text.

of approximately 500 c.c. of body fluid as sweat was accompanied by a reduction in the salivary output of almost 50 per cent.

Corresponding to the diminution of the salivary output as the result of chewing was a diminution in the reflex flow as a consequence of letting the mouth become dry. The reflex flow has fallen, in my experience, from 3 or 4 c.c. in five minutes under normal conditions to a little more than 1 c.c. during thirst.

The relation between the decrease of salivary flow in these experiments and the sensation of thirst was quite definite. In the experiment illustrated in fig. 1, for example, the feeling of being "thirsty" was absent until the secretion of saliva began to decline, after 11 o'clock. From that time onward the back of the throat began to feel dry; there was frequent swallowing, and both the movements of the tongue and the act of deglutition were associated with a sense of "stickiness," a lack of adequate lubrication of the parts. All of this unpleasantness and discomfort disappeared after the restoration of the saliva flow by drinking water.

The increased spontaneous activity of the tongue and the repeated swallowing motions as "thirst" became more marked are noteworthy. These movements are a slight stimulus to salivary secretion, and they have, furthermore, the obvious effect of spreading about any fluid that might be present. In the absence of sufficient fluid, however, they augment the disagreeableness of the condition by making prominent the friction due to lack of lubricant. The "lump in the throat," which is complained of by persons who suffer from extreme thirst, can be explained as due to the difficulty encountered when the epiglottis and root of the tongue are rubbed over the dry back wall of the pharynx in attempts to swallow.

The only statement that I am aware of, which is contradictory to the evidence just presented, is that made by one of a group of psychologists, reported by Boring.* This one observer testified that when he was beginning to be thirsty the saliva flow was still copious. The eight other observers of the group speak of thirst as being characterised by dryness of the roof of the mouth, dryness of the lips, the sensation of having a "dry sore throat," feelings of stickiness, and uncomfortable "puckery" pressure localised in the middle and back of the tongue and in the palate—in other words, as one of them summed up his experience, "dryness expresses the complex as a whole." This body of testimony agrees closely with that presented earlier and suggests that there may have been error in the one observation that thirst was associated with free secretion of saliva.

Other evidence on the relation between absence of saliva and the presence of thirst as a sensation was obtained through checking salivary secretion by atropine. Before the injection the amount secreted during 5 minutes by chewing averaged 13·5 c.c. After the full effect of the drug was manifest, the amount fell to 1 c.c. All the feelings that were noted in ordinary thirst—the sense of dry surfaces, the stickiness of the moving parts, the difficulties of speaking and swallowing—all were present. These disagreeable experiences, constituting the thirst sensation, disappeared as soon as the mouth and throat

* Boring, 'The Psychological Review,' vol. 22, p. 307 (1915).

were washed out with a weak novocaine solution. The immediate effect in these circumstances was doubtless due to the water in the solution, but since the relief lasted much longer than when water was used, the anæsthetic was also a factor. This experience agrees with that of Lipidi-Chioti and Fubini, mentioned earlier. No water was drunk by me during the period of atropine effect, and yet when that effect disappeared, and the saliva flow was re-established, thirst also was abolished. The relation between thirst and such drug action has been noted before, but so strong has been the theory that thirst is a "general" sensation, that the drug has been supposed to produce its effect not by local action but by central changes and by alteration of the blood.*

Similar in character to the thirst which results from the action of atropine is that which accompanies anxiety and fright. The effect of such emotional states in causing inhibition of salivary secretion is well known. It was the basis of the ancient "ordeal of rice" employed in India as a means of detecting the guilty one in a group of suspected persons. It is illustrated in these days by Hoche's report of the effects of air raids on the people of Freiburg-in-Baden, in whom the signs of great fear—chattering of the teeth, pallor, and diarrhœa—were attended by intense thirst.† The unquenchable nature of the thirst which results from terror is a large part of the torment suffered by the novice in public speaking.

On the basis of the foregoing evidence I would explain thirst as due directly to what it seems to be due to—a relative drying of the mucosa of the mouth and pharynx. This may result either from excessive use of this passage for breathing, as in prolonged speaking or singing, or it may result from deficient salivary secretion. In the latter case "true thirst" exists, but it is not to be distinguished, so far as sensation is concerned, from "false thirst." True thirst is dependent on the fact that the salivary glands, which keep the buccal and pharyngeal mucosa moist, require water for their action. According to the observations and inferences of Wettendorff, the osmotic pressure of the blood is maintained, in spite of deprivation of water, by the withdrawal of water from the tissues. The salivary glands are included under "tissues," and they appear to suffer in a way which would support Wettendorff's view, for in the presence of a general need for water in the body, they fail to maintain the normal amount and quality‡ of secretion. The same is doubtless true of other glands. The importance of this failure of

* See Sherrington, 'Schäfer's Textbook of Physiology,' London, vol. 2 p. 991 (1900).

† Hoche, 'Med. Klinik,' vol. 13, p. 906 (1917).

‡ There is evidence that, as the quantity of saliva diminishes, its water content is less; *i.e.*, it is more viscous. (See Tezner, 'Arch. Intern. de Physiol.,' vol. 2, p. 153.)

action of the salivary glands, however, to the mechanism of the water supply of the body, lies in the strategic position of these glands in relation to a surface which tends to become dry by the passage of air over it. If this surface is not kept moist, discomfort arises and with it an impulse to seek well tried means of relief. Thus the diminishing activity of the salivary glands becomes a delicate indicator of the bodily demand for fluid.

The foregoing explanation is in agreement with the suggestions which have been offered to account for thirst as having a local origin. But it does not require specialised nerves, or peculiar sensitiveness of the first portion of the digestive tract, which have been assumed to be present by the upholders of this theory. And by calling attention to the arrangement by which the salivary glands are made to serve as indicators of the general bodily need for water, it presents a reasonable account of the manner in which a widespread condition of the organism may exhibit itself locally.

The experiments which have long been the chief support of the theory that thirst is a general sensation can also be explained by the evidence above adduced. The abolition of thirst by injecting fluid into the veins of thirsty animals would be expected, for, as shown in the experiment illustrated in fig. 1, by providing an adequate water supply the saliva flow is promptly re-established, and the parched mouth and throat are again continuously moistened. In the classic experiment of Claude Bernard the animal with an open gastric fistula continued to drink until the fistula was closed. This was not because there was a general demand for water throughout the body, so long as the fistula remained open, but because only when escape through the fistula was stopped did the body receive the water needed to provide the output of saliva which prevented local drying. And the dogs with salivary glands tied, described by Bidder and Schmidt, were always ready to drink, just as are persons who are terrified or who have been given atropine, because of thirst—because there is local drying of the mouth—from lack of saliva, though the body as a whole may not be in any need of water. The application of cocaine to the mucous surfaces of the mouth abolishes the torment of thirst, not by any central effect, and clearly not by satisfying any general bodily requirement for water, but by rendering the surfaces anæsthetic. The miraculous virtues of coca leaves, as a balm for the distress of the thirsty, a fact long ago observed, is explicable on these grounds. The thirst of those who suffer from loss of fluid from the body—the diabetic patient, the victim of cholera, the subject of hæmorrhage, the perspiring labourer, and the nursing mother—can be accounted for by the reduction of salivary flow as the water-content of the body is lowered, and by the consequent discomfort arising from the sticky buccal mucosa.

I am aware that many questions arising from the views which I have just developed remain to be solved—questions as to the effects which other glandular activity, removing fluid from the body, may exercise on the functions of the salivary glands; the alteration of properties of the blood and lymph other than osmotic pressure as affecting secretion; the relation between the so-called "free water" of the body fluids and salivary secretion when water is withheld; the influence of strong alcoholic beverages in producing thirst; and the nature of pathological states in which thirst seems to disappear. But these and other pertinent questions must await more peaceful times for their answers.

From the evidence presented, however, it seems to me that we are now in a position to understand the mechanisms by which all three of the essential supplies from the outer world are provided for in our bodily economy. The oxygen supply is arranged for by the control which changes in the blood, brought about mainly by variations in the carbon dioxide content, exert on the centre for respiration. The proper food supply ultimately is assured, because we avoid, or check, by taking food, the distressing pangs of hunger which powerful contractions of the empty stomach induce unless food is taken. And the water supply is maintained because we avoid, or abolish, by taking water or aqueous fluid, the disagreeable sensations which arise and torment us with increasing torment if the salivary glands, because of a lowering of the water-content of the body, lack the water they need to function, and fail therefore to pour out their watery secretion in sufficient amount and in proper quality to keep moist the mouth and pharynx.

16

STUDIES ON THE CONDITIONS OF ACTIVITY IN ENDOCRINE GLANDS

XIII. A Sympathetic and Adrenal Mechanism for Mobilizing Sugar in Hypoglycemia[1]

W. B. CANNON, M. A. McIVER[2] and S. W. BLISS

From the Laboratories of Physiology in the Harvard Medical School

Received for publication March 17, 1924

When insulin in an amount which reduces blood sugar below the common physiological percentage is administered to persons suffering from diabetes, characteristic symptoms occur which have been called "hypoglycemic reactions." They include pallor, rapid pulse, dilatation of the pupils and profuse sweating (1). These are indications of activity of the sympathetic division of the autonomic system and, as is often the case when that system is excited, there are tremors in skeletal muscle. Similar signs—dilatation of the pupils, erection of the hair, salivation—have been reported as occurring in cats after insulin injections (2), and these too are explicable as results of sympathetic nervous discharge. A natural inference from this evidence that the reduction of blood sugar by insulin involves sympathetic impulses is that adrenal secretion, known to be subject to such impulses, might be increased. Stewart and Rogoff have reported that in three cats, in which the influence of insulin on adrenal secretion was investigated, "no definite effect of any moment could be made out" (3). Since their method, however, gives negative results in their hands and positive results in other hands (4), their failure should not be regarded as significant.[3]

[1] A preliminary report of this work was published in the Boston Medical and Surgical Journal, 1923 (July 26), clxxxix, 141.

[2] Medical Fellow of the National Research Council.

[3] Stewart and Rogoff continue to trust their "straightforward way" of estimating adrenal secretion (collecting blood from a "cava pocket" and assaying its adrenin content outside the body), although other methods, in the hands of different observers, have yielded concordant results quite contrary to theirs. An illuminating commentary on the reliability of their procedure is afforded by Kodama (4) who has used it carefully and in many tests. Stewart and Rogoff declare that the average output from the adrenals of the cat under their experimental conditions is between 0.00021 and 0.00025 mgm. per k. per minute (5); Cannon and Rapport (6) measured it and found approximately 0.0007 mgm. (an amount not very different from the average of 0.0006 mgm. which Stewart and Rogoff (7) had estimated previously by use

If adrenal secretion is increased as blood sugar falls after injection of insulin, the increase would indicate special activity of the sympathetic division of the autonomic system, and the two together—extra adrenin and sympathetic impulses—would have as a natural consequence a mobilization of sugar from the liver. Thus an automatic recovery of a disturbed equilibrium would be provided for. The interesting possibilities in these conjectures led us to put them to experimental test.

THE METHOD. As an indicator of adrenal secretion we have used the heart denervated by removal of both stellate ganglia and section of both vagus nerves. Sympathetic supply to the thyroid gland is thereby eliminated. Previous studies have shown that in animals with the hepatic nerves severed, the acceleration of the heart thus denervated, when an afferent nerve is stimulated or asphyxia is produced, is due solely to increased discharge of adrenin into the blood stream (6) (9), and that, if animals are fasting, the hepatic factor is so slight compared with the adrenal that it may be disregarded (10). The observations on the denervated heart of cats have been confirmed by Searles in studies on the denervated heart of dogs (11).[4]

of methods which they later repudiated); now Kodama, using the "cava pocket" method, reports the average figure as 0.00065 mgm.

Stewart and Rogoff affirm that 0.00025 mgm. is the steady unvarying secretion, not changed by afferent stimulation (8); Cannon and Rapport quantitated the secretion when increased by afferent stimulation and observed that it ranged between 0.0032 and 0.0049 mgm. per k. per minute; Kodama's assays under the same conditions, but on "cava pocket" blood, though averaging considerably less, ranged as high as 0.0047 mgm.

Rogoff, in a recent comment on Kodama's paper, declares that Kodama, "although he has evidently spent much time" on the method, has not mastered it, and that one of the best proofs that he does not know how to make the assay is the "excessive outputs obtained by him" (28). There is assumption here that Stewart and Rogoff's low figures are correct, and that Kodama's higher figures are not. The same method brought forth the two kinds of results, and therefore cannot be used to decide which is correct. Stewart and Rogoff have had no support from other observers for their constant assays and for their conclusions therefrom. The first observer to employ their method outside their laboratory obtains results which agree, quantitatively, with the results found by Cannon and his collaborators.

[4] In a recent paper (12) Stewart and Rogoff have reiterated that they have caused reflex acceleration of the denervated heart after suppression of the adrenal output. We call attention again to the proof *a*, that the residual increase of rate after adrenalectomy is due to a discharge from the liver, and *b*, that if the liver nerves have been severed the effect from the adrenals persists, but disappears as soon as these glands are removed (6). Furthermore, as Cannon and Carrasco-Formiguera showed (9), after section of the hepatic nerves the denervated heart can be reflexly accelerated when the blood flow from the adrenals is free, cannot be accelerated when the flow is blocked, and can be accelerated again when the block is removed. These two lines of evidence for reflex increase of adrenal secretion, which are irreconcilable

Our experiments, which were performed on cats, were at first done under anesthesia—a condition which may profoundly affect the metabolism of sugar in the body. Griffith (15) has proved, however, that chloralose anesthesia is satisfactory for studies of physiological factors influencing the glycemic (i.e., the blood sugar) concentration. Deep chloralose anesthesia, in our experience, though not preventing a fall of the glycemic percentage after insulin, does greatly reduce or abolish the hypoglycemic reactions. Since they are signs of disturbance in nerve centers, their absence under deep anesthesia is readily comprehensible. A satisfactory dose of chloralose for our purposes was found to be 0.1 gram per kilo by mouth. We have followed Griffith's method of administering it in milk which the animal drank, or we have given it in solution through a stomach tube (10 cc. of 1 per cent per kilo). During the short time while the chest was opened for removal of the stellate ganglia, ether was commonly administered to a degree which prevented reflex movements. Only a slight amount of ether was needed, and it was discontinued as soon as the operation was ended. The animals were well nourished and had been fed meat the previous afternoon.

Observations were made also on unanesthetized animals in which the heart had been denervated aseptically, and which were living normally in the laboratory.

The insulin (Lilly's) was injected into the jugular vein, 4 units per kilo being the usual dose. The blood for assay of its sugar concentration was taken from a carotid artery in the anesthetized and from a jugular vein in the unanesthetized animals, and was tested by the Folin-Wu method (14).

The temperature of the anesthetized animals was kept uniform within a degree centigrade throughout each experiment.

The heart rate was recorded in blood-pressure records in the observations on anesthetized animals (see fig. 5); it was counted by use of the stethoscope, for 15-second intervals, in the unanesthetized animals.

with Stewart and Rogoff's denial of that secretion, they have not mentioned, though published many months before the appearance of their paper.

The method used by Cannon and Carrasco-Formiguera to prove *reflex* control of adrenal secretion was so close a repetition of Stewart's method of proving *direct* splanchnic control that Zunz and Govaerts (13) overlooked the difference and stated that Cannon confirmed Stewart. The identity of methods and results raises the question whether the proof for direct nervous control of adrenal secretion is to be rejected or the evidence for reflex control is to be accepted.

Rogoff has just published the observation that asphyxia of bulbar centers induces an increased rate of adrenal secretion, but there is "no evidence," he states, "that asphyxia of the entire animal is capable of augmenting the output of epinephrin" (26). May we call attention again to the paper by Cannon and Carrasco-Formiguera (9) which reported observed facts contradictory to this conclusion and which Rogoff does not mention.

RESULTS. We shall report first the observations on animals under chloralose anesthesia, and thereafter, at each point, the corresponding observations on unanesthetized animals.

1. As the glycemic concentration falls after injection of insulin, it reaches a critical point, at which the rate of the denervated heart begins to be accelerated, and as the sugar percentage continues to fall the heart rate continues to rise until a maximum is reached (13 cases). This observation is illustrated in figures 1, 2 and 5. As shown in figures 1 and 2, the acceleration is usually sharp when the standard dose, stated above, is given. When a smaller dose is given (see fig. 7) the rise of rate is more gradual. The maximal increase in the anesthetized cases was 48 beats per minute, the minimum, 24. The faster rate may continue for 2 hours or more.

In unanesthetized animals the acceleration was more marked for the standard dose than in the anesthetized. For example, in the case illustrated in figure 2, the increase was as much as 80 beats per minute, and the rate continued high (more than 30 beats per minute above the basal rate of 112) for more than three hours. With smaller doses the increases were smaller—as low as 20 beats per minute. The protocol of the observations illustrated in figure 2 is as follows:

Cat 232, with heart denervated. November 30, 1923, brought from animal room; 12:00 noon, heart rate (h.r.) while quiet in lap, 112 per minute. 12:15, jugular bared under ethyl chloride, cat quiet; blood sample no. 1 taken, 129 mgm. sugar. 12:24, 10 units insulin (4 units per k.) injected subcutaneously on one side. 12:27, squatting on floor, h.r. 112. 12:40, quiet in lap, h.r. 112, resp. 13 per minute. 1:00, squatting on table, h.r. 112, resp. 16 per minute; nictitating membrane (denervated) one-fourth over eye. 1:20, squatting, h.r. 172, resp. 41 per minute, 1:30, blood sample no. 2 taken. 65 mgm. sugar. 1:40, squatting, h.r. 140. 1:50, h.r. 176. 2:00, h.r. 176. 2:10, h.r. 160. 2:20, h.r. 156 to 164, varying up and down during observation. 2:30, h.r. 164. 2:40, h.r. 172. 2:50, h.r. 160–172, varying fast and slow. 3:00, h.r. 160:172. 3:10, h.r. 148. 3:20, blood sample no. 3 taken, 34 mgm. sugar. 3:25, lying on side, pupils dilated, tail hairs lifted, h.r. varying between 176 and 192. 3:30, weak, unable to stand. 3:40, lying on side, resp. 140 per minute, pupils dilated, nictitating membrane disappeared, salivation. 3:45, h.r. 164, with frequent dropped beats. 4:00, lying on side, legs stretched out, tail curved over back, h.r. 164. 4:10, same state, occasional stretching out, very rapid breathing, h.r. 188. 4:20, same state, h.r. 140 (?) paired beats. 4:23, convulsion: blood sample no. 4 taken, 32 mgm. sugar. 4:28, glucose injected under skin left side. 4:30, h.r. 158.

2. The critical point at which the denervated heart of the animal under chloralose begins to beat faster appears to lie within a range between 110 and 70 mgm. of glucose per 100 cc. of blood. We say "appears" because in judging the matter we have to assume that the initial drop in the glycemic concentration follows a fairly straight line between the first two assays. It has been reported that after insulin injection in rabbits the blood sugar falls at a practically uniform rate (18). As

shown in figure 1B and in figure 5, the first three assays lay in a line, and in those cases we may be fairly sure of the critical range; in figure 1B the faster rate started at a level of 70 mgm. and in figure 5 at a level of about 110 mgm. Other cases (e.g., fig. 1, A and C, and fig. 7) lie within these limits.

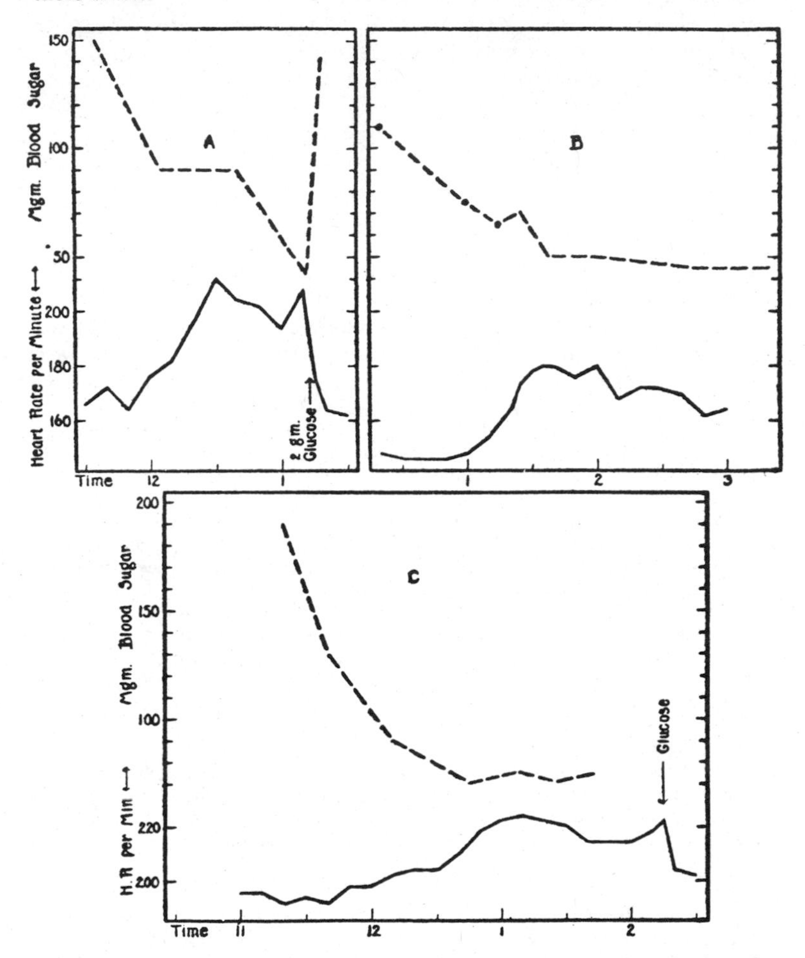

Fig. 1. Increase of rate of the denervated heart (solid line), in animals under chloralose anesthesia, when the falling blood-sugar concentration (dash line) passed a critical point. In case A, the insulin was injected into the jugular vein at 11:33; in B, at 11:08; and in C at 9:30. In each case 4 units per kilo were injected.

Figure 2 shows well the remarkable change which occurs at the critical point in the unanesthetized cat. At 1 o'clock the heart rate was 112 and the respiration 16 per minute; at 1:20 the heart rate was 172 and the respiration 41. If we may assume that the blood sugar was dropping uniformly the concentration was between 70 and 80 mgm. per 100 cc. when the heart began to speed. This coincides with clinical observation, for Fletcher and Campbell testify that "when a reaction has already been experienced the onset of a subsequent one is usually recognized by

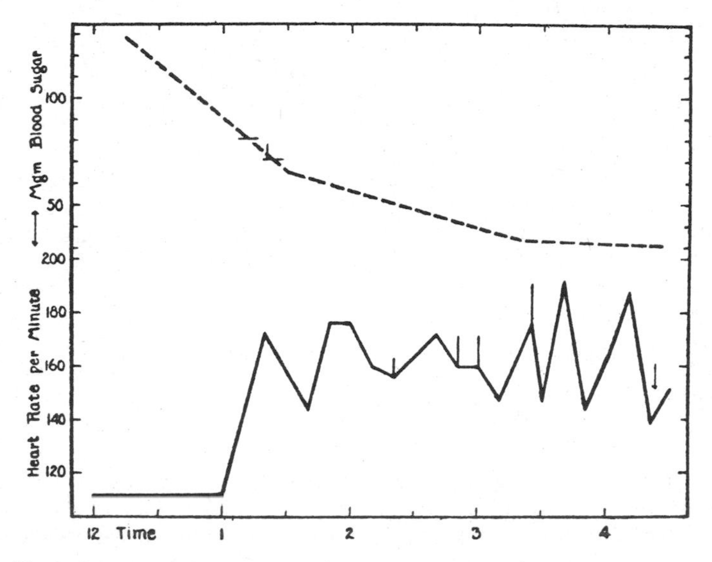

Fig. 2. Increase of the rate of the denervated heart (solid line) in an unanesthetized animal, when the falling blood-sugar percentage reached about 0.07. The rate continued high and after an hour or two was spasmodically increased still further, as represented by the height of vertical lines at several counts. A convulsive seizure (see arrow) occurred at 4:23, 4 hours after the subcutaneous insulin injection (4 units per k.), and about 3 hours after the quick rise in the heart rate.

the patient when the blood-sugar percentage falls to some point between 0.08 per cent and 0.07 per cent." And again, "when the blood sugar percentage falls to 0.07 per cent under the influence of insulin, the patient becomes aware of it" (1).

Why, in our cases, the critical level was higher as a rule in animals under chloralose than in the unanesthetized animals is not clear. One would suppose that the anesthetic would decrease rather than increase sensitiveness. If the anesthetic is given to excess, it may decrease the sensitiveness to the extent of wholly abolishing the reaction, as already noted. Animals lightly chloralosed, however, are hypersensitive to sounds and jars, as indicated by spasmodic response to such stimuli, and they frequently jerk in an incoördinate manner. These are signs of instability in motor centers. Possibly in some instances the nerve cells responsible for the hypoglycemic phenomena are rendered hypersensitive to effects of insulin by chloralose.

In our experience the acceleration of the denervated heart appears before other striking signs of sympathetic activity. In the case illustrated in figure 2, for example, the faster rate was first noted at 1:20; for the first time at 3:25 were the pupils seen to be dilated and the hairs of the tail standing erect. Unfortunately we have no convenient means of recording early effects on the hairs; and the iris is subject to control by way of the short ciliary nerves, even though the stellate ganglia have been removed—a control which may overcome the action of an opposing factor. The absence of obvious changes in the hairs and in the iris need not be regarded, therefore, as very significant. It is possible, however, to obtain evidence of sympathetic activity in the heart itself. In two cases under chloralose the adrenal glands were removed and the vagi cut, but the stellate ganglia were left intact so that the heart could be influenced directly by sympathetic impulses. In both cases the heart accelerated when the glycemic concentration fell below 80 mgm.—in one case 10 beats and the other 16 beats per minute. These results show that the cardioaccelerator nerves themselves may be stimulated at the critical range.

3. If the adrenal glands have previously been removed, or if one has been removed and the other denervated, a fall of the glycemic percentage below the critical range is not accompanied by an increased rate of the completely denervated heart (4 cases). This observation is illustrated in figure 3. It represents graphically the conditions under chloralose in an animal from which the left adrenal gland had been removed and in which the right splanchnic and the hepatic nerves had been cut 19 days previously. The concentration of blood sugar fell to 50 mgm. with no increase in the heart beats per minute as the critical range was traversed. In other instances, levels of 40 and 44 mgm. have been reached under such conditions, or with both adrenals absent, without calling forth any noteworthy acceleration of the pulse.

In figure 4 are represented the changes in an unanesthetized cat with denervated heart and with one adrenal previously removed, and the opposite splanchnic nerves severed, occurring after intravenous injection of 4 units of insulin per kilo. This figure should be compared with figure 2. As the blood sugar fell through the critical range, and further to less than 50 mgm., the heart rate increased 6 beats per minute—an insignificant change— instead of increasing 60 or more beats per minute, as in figure 2. The protocol of the observations illustrated in figure 4 is as follows:

Cat 239, with heart denervated, and with right adrenal removed and left splanchnics cut. Weight, 3.1 k. January 11, 1924. 10:40, h.r. 144. 10:50, h.r. 144. 10:52, blood sample no. 1 taken from jugular vein under ethyl chloride, 200 mgm. sugar. 10:54, 12 units insulin in jugular. 11:00, h.r., 144. 11:10, h.r. 144. 11:20, h.r. 144. 11:26, blood sample no. 2 taken, 95 mgm. sugar. 11:30, h.r. 148. 11:40, h.r. 148, intestinal gurglings heard. 11:50, h.r. 148. 11:58, blood sample no. 3 taken, 57 mgm. sugar. 12:00, h.r. 148, hairs of back and tail lifted. 12:10, h.r. 148, hairs very fuzzy, animal restless (change from previous quiet). 12:20, h.r. 150, cat weak, staggered while walking. 12:26, blood sample no. 4 taken, 51 mgm. sugar, cat weaker, unable to stand. 12:30, lying on side. Periodic rapid panting, mouth wide open, h.r. 144. 12:33, convulsion. 12:38, blood sample no. 5 taken, 46 mgm. sugar; 10 cc. 10 per cent glucose injected under skin on one side. 12:45, h.r. 144, still panting.

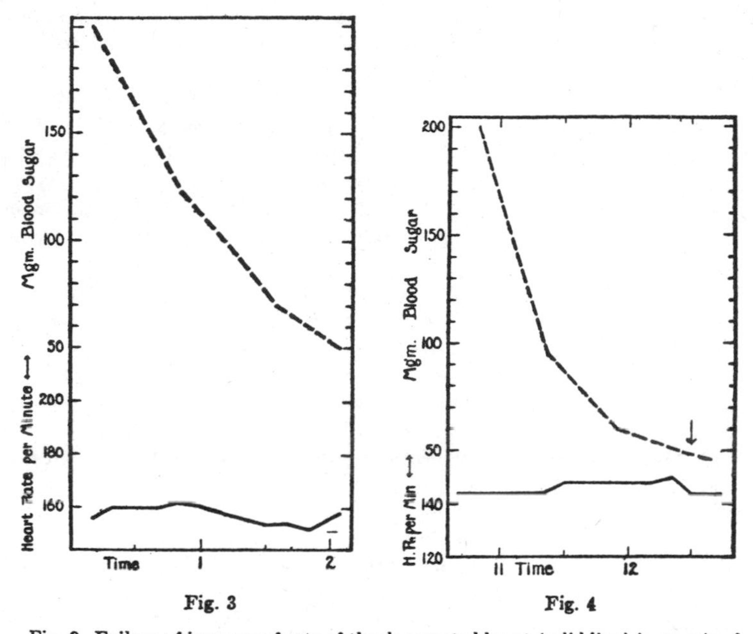

Fig. 3 Fig. 4

Fig. 3. Failure of increase of rate of the denervated heart (solid line) in an animal under chloralose when the falling blood-sugar concentration (dash line) passed the critical range. The left adrenal had been removed and the right splanchnic and the hepatic nerves severed 19 days before. Insulin (4 units per k.) was injected intravenously at 12:19.

Fig. 4. Failure of noteworthy increase of rate of the denervated heart (solid line) in an unanesthetized animal when the falling blood-sugar concentration (dash line) passed the critical range. The left adrenal had been removed and the right splanchnic nerves severed months before. Insulin (4 units per k.) was injected intravenously at 10:54. A convulsive seizure (see arrow) occurred at 12:33, an hour and 39 minutes after the insulin injection.

These observations indicate that the cardiac acceleration reported in previous sections and shown in figures 1 and 2 is not due to direct action of insulin, or to the effects of any changes due to insulin, on the heart itself. This evidence coincides with the observations of Hepburn and Latchford on the excised heart (17). Further, neither insulin, nor any disturbance produced by it, acts directly on the adrenal gland for, as shown in figures 3 and 4, one adrenal may still be present, and, if denervated, it does not exercise any notable influence on the heart rate. Moreover, in the case represented in figure 4 as well as in other similar cases, although one adrenal gland had been removed and the other denervated, *hepatic* nerves were still existent. As the falling glycemic percentage passed the critical range, however, the rate did not become markedly faster, i.e., nothing was given off from the liver that accelerated the pulse to any considerable degree.

The conclusion which we feel justified in drawing from the combined results thus far reported is that the acceleration of the denervated heart is due to a discharge of adrenin from the adrenal glands in response to nervous impulses. In other words, these glands take part in the hypoglycemic reactions; the sympathetic discharges which cause dilatation of the pupil, acceleration of the heart, etc., also evoke increased adrenal secretion.

4. If the rate of the denervated heart has been increased because of hypoglycemia, intravenous injection of glucose promptly reduces the rate. This observation is illustrated in figure 1, A and C, and in figure 5. In the case represented in figure 5, a record was taken during the intravenous injection of 15 cc. of 5 per cent glucose, and at one-minute intervals thereafter. In 65 seconds after starting the injection the rate had fallen from 174 to 160 beats per minute, after another minute it had dropped to 144, and two minutes later it was down to 141—a rate below the original. In 20 minutes it had recovered from this depression and was beating 152 per minute. In our experience, the injection of glucose (1 gram) into an animal without adrenal secretion does not thus promptly decrease the heart rate. Since the faster rate is ascribable to increased adrenal secretion called forth by disturbed cells in the central nervous system, the restorative action of glucose can be accounted for on the assumption that it provides material which these cells need, or modifies factors which have come into play because the glucose supply has been reduced. The cells, no longer excited or rendered hyperexcitable, cease to discharge impulses via the sympathetic. Extra adrenal secretion quickly stops and the heart rate therefore quickly falls.

5. As the rate of the denervated heart increases (indicating adrenal secretion), the rate of drop in the glycemic percentage decreases, i.e., the sugar curve tends to flatten. This observation is illustrated in the

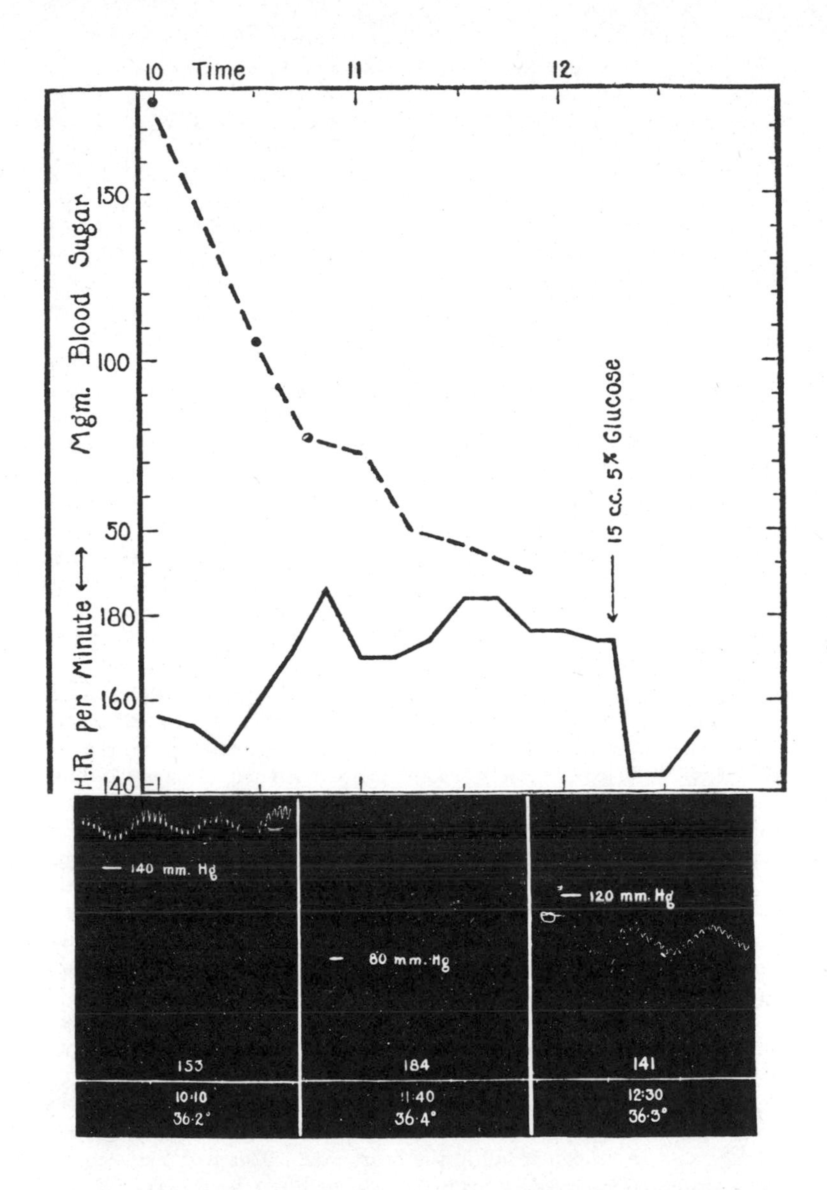

Fig. 5. Increase of rate of the denervated heart (solid line) in an animal under chloralose anesthesia, when the falling blood-sugar concentration (dash line) passed a critical point. Insulin given intravenously at 10:01. At 12:15, 15 cc. of 5 per cent glucose were injected intravenously. Portions of the original records show the method of recording the heart rate in this and in other experiments under anesthesia; the numbers above the base line (zero blood pressure; 5-second intervals) represent the heart rates per minute and the figures below, the time and rectal temperature.

55

three cases represented in figure 1 and in figures 2, 5 and 7. In clinical cases (27) and in laboratory experiments (16) tests of blood sugar after injection of insulin have shown that the flattening of the curve, following an initial drop, is a characteristic feature. As shown in figure 1C, the sugar percentage may remain fairly constant at a given level; or it may remain constant or nearly constant for a time and later fall, as in figures 1A and 1B; or it may continue falling but at a slower rate than before, as in figures 2 and 5. These variations in the blood-sugar curves are probably to be explained chiefly by differences in the glycogen stores in the different animals, for Macleod and his co-workers have published curves showing similar variations in rabbits that were glycogen-rich and glycogen-poor (16, fig. 6). In our cases, however, the check in the rate of fall in the glycemic concentration was observed when adrenal secretion occurred, as was manifested by an accelerated heart beat.

6. If the adrenal glands have previously been removed, or if one has been removed and the other denervated, the rate of drop in the glycemic percentage after the standard dose in animals under chloralose is usually not checked at the critical level; it may be retarded slightly or not at all at that point, or it may flatten at a very low concentration. These observations are shown graphically in figure 3 and in figure 6, A, B and C. An exception to the general statement is illustrated in figure 6D. In that case the curve flattened between 70 and 80 mgm., although the adrenal glands were absent. The heart rate, however, had been unaccountably high (never less than 254 beats per minute) throughout the experiment, the blood pressure when the curve flattened was only 70 mm. Hg, and the animal became so asphyxiated that artificial respiration had to be started. It is quite possible that in this case, as well as in others (cf. fig. 3, with hepatic nerves cut) aspyhxia played a rôle in setting free sugar from the liver, for we have noted not infrequently on taking the *late* samples in the course of an experiment that the arterial blood was surprisingly venous. The asphyxial condition might act either by stimulation of the hepatic cells from the sympathetic nerve centers (tail hairs were erect in 6D at 12:30) or by direct action on the hepatic cells.

The conditions in unanesthetized animals with inactivated adrenals but with hepatic innervation, when insulin has lowered the blood sugar, will be considered later.

7. If not too much insulin has been given, the increased rate of the denervated heart may be followed by an increase of the glycemic percentage, and an attendant fall in the heart rate. This reaction is shown in the record reproduced in figure 7. It has often been observed that the decrease of blood sugar due to insulin is followed fairly promptly, or after remaining at a low level for a time, by a rise to normal (16). We have frequently had occasion to note the reciprocals of such changes

in counting the rate of the denervated heart in unanesthetized animals which had been given sub-convulsive doses of insulin. The rate has continued uniformly for a varying time—from 50 to 105 minutes after

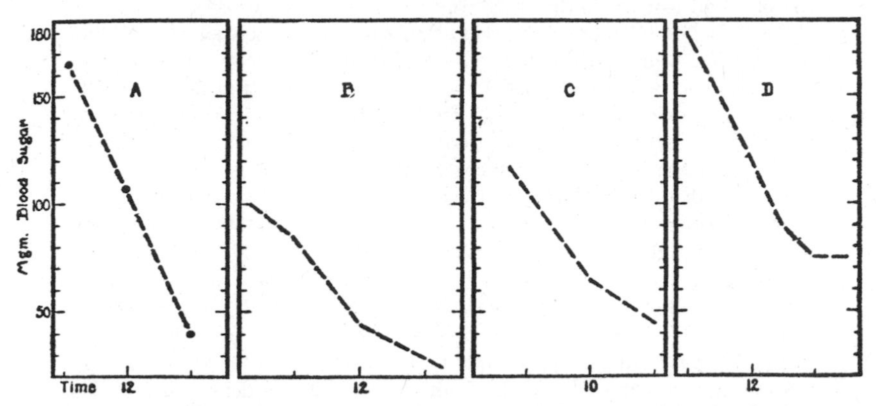

Fig. 6. The rate of decline of the blood-sugar concentration in chloralosed animals, which had been deprived of the adrenal glands. In case A the insulin (4 units per k.) was injected intravenously at 11:36, in B, at 11:06, in C, at 9:27; and in D, at 11:36. There was no check in the decline except in D (for explanation see text).

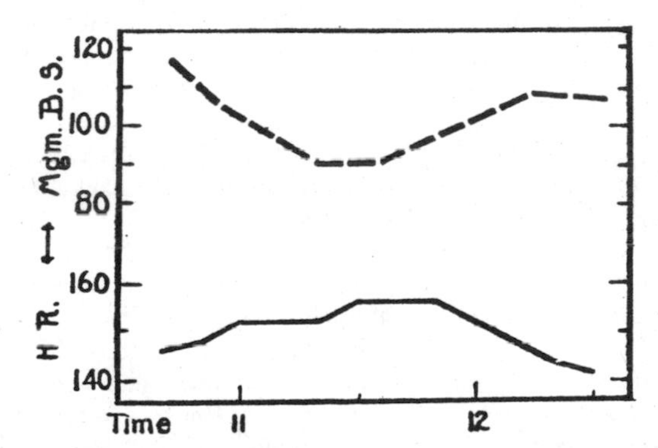

Fig. 7. Increase of rate of the denervated heart (solid line) in an animal under chloralose, as the blood-sugar concentration (dash line) fell below the critical level, and decrease of the rate as the concentration rose. One unit of insulin was given intravenously at 9:53.

the injection—whereupon it has gradually increased between 20 and 36 beats per minute, as in figure 7, and then gradually subsided again to its former level. The increased rate has not lasted longer than 120 minutes, and in one instance was over in 60 minutes. The drop in the heart rate as the blood sugar increases is just what would be expected from the observation, already noted, that introduction of glucose into the blood stream promptly brings the pulse down (see p. 54).

THE RÔLE OF ADRENAL SECRETION IN THE HYPOGLYCEMIA FROM INSULIN. The considerations presented in the foregoing pages raise the question as to the part taken by adrenal secretion in resisting the fall of blood sugar. The evidence in hand clearly indicates that this resistance and the recovery from insulin injections depend largely on the glycogen available in the liver (16). Glycogenolysis can be produced, as is well known, by direct stimulation of the hepatic nerves. Also, as Griffith has shown, it can occur reflexly in consequence of afferent stimulation, but if the adrenal glands have been excluded from action the glycemic percentage averages lower and the effect of reflex stimulation averages less than in normal animals (15). Furthermore, Trendelenburg has presented evidence that such outpouring of adrenal secretion as can readily occur in physiological states is quite capable of causing a mobilization of sugar (18). When in the early stage of an experiment the denervated heart becomes accelerated, the faster rate indicates not only that adrenal secretion is being augmented, but also by that fact that splanchnic impulses are being discharged. It is altogether probable that the liver receives these impulses just as the adrenals do. The liver, however, would be subject to both nervous and humoral stimuli—the nervous impulses and the secreted adrenin coöperating to influence it. Under chloralose anesthesia the humoral factor seems to be a necessary complement of the nervous, for, as comparison of figures 1 and 6 indicates, the course of the descent of the glycemic concentration is as a rule strikingly different, after standard doses of insulin, according to whether adrenal activity is present or absent. The character of this difference has already been pointed out (pp. 54, 56).

In the unanesthetized animal, likewise, the secreted adrenin appears to be important for the process of mobilizing sugar from the liver. In table 1 are presented the results of injecting insulin into healthy animals, with and without active adrenal glands, that had been kept under the same conditions of feeding and temperature and freedom from disturbance. (In our experience with a few cats which had just previously been subjected to prolonged excitement or which were suffering from infection, they were more sensitive to insulin than are serene and healthy animals.) In a number of cases injections were repeated in the same animal. Rabbits seem to become less sensitive to insulin as doses are repeated (25). In cats with intact adrenals this possibility was met by a larger dose at the second injection. In cats with adrenals inactivated a second dose equal to the first should have been less likely to produce convulsions, if a sort of immunity develops. This result was perhaps indicated in cats 107 and 112, and in other cases in which the convulsive seizure occurred after a longer delay in the second than in the first test. The seizures occurred, nevertheless, and strikingly differentiated this group from the other. Among animals with normally innervated adrenals doses of insulin varying from 2 to 3 units per kilo caused convulsions in only one instance. This was in cat 228 which had had a major operation 10 days previously, and had not fully recovered from it. A dose of 2.5

TABLE 1

Presence or absence of convulsions after subcutaneous injections of insulin into (A) cats with normally innervated adrenals, and (B) cats with one adrenal removed and the other denervated

A. ADRENALS BOTH INNERVATED				B. ADRENALS, 1 OUT, OTHER DENERVATED			
Date	Cat	Insulin units per kilo	Convulsion	Date	Cat	Insulin units per kilo	Convulsion
Nov. 10	228	2.5	After 3 hrs. 30 min.	Jan. 31	239	2	After 1 hr. 30 min.
Nov. 28	228	3	0	Feb. 8	239	1	After 1 hr. 55 min.
Jan. 9	228	3	0	Feb. 2	107	2	After 1 hr. 10 min.
Nov. 28	227	2	0	Feb. 13	107	2	After 1 hr. 23 min.
Feb. 8	227	3	0	Mar. 1	107*	2	0
Nov. 17	229	2.5	0	Feb. 2	235	2	0
Jan. 7	107	2	0	Feb. 8	235	3	After 1 hr. 32 min.
Feb. 8	245	3	0	Feb. 21	90	2	After 1 hr. 40 min.
Feb. 19	Bl	2.5	0	Mar. 1	90	2	After 1 hr. 55 min.
Feb. 21	W	2	0	Feb. 21	110	2	After 1 hr. 7 min.
Feb. 21	Y	2	0	Mar. 1	110	2	After 1 hr. 30 min.
Mar. 1	B-W	2	0	Feb. 21	112	2	After 1 hr. 35 min.
Mar. 1	G-W	2	0	Mar. 1	112†	2	0
Mar. 1	G-T	2	0	Feb. 21	116	2	After 1 hr. 18 min.
Mar. 1	M	2	0	Mar. 1	116	2	After 2 hrs. 8 min.

* Panting, weakness and vomiting after 1 hour and 24 minutes.

† Salivation, mewing, weakness and muscular twitching, beginning after 1 hour and 25 minutes, and continuing for about a half hour.

units per kilo brought on a convulsive seizure after three and a half hours; later, after complete recovery, this animal twice withstood doses of 3 units per kilo without any such disturbance. On the other hand, among animals with one adrenal removed and the other denervated, insulin doses lying between 1 and 2 units per kilo induced convulsions within two hours and ten minutes in all but three cases. In the first of these (cat 235) a dose of 3 units was later effective. In the other two cases the premonitory signs appeared, but the animals passed the crisis without a seizure. The liver was still innervated, it should be remembered, and therefore some protection against hypoglycemia still existed. It should be noted that a dose of 2 units per kilo had no convulsive effect in cat 107 when the adrenal nerve supply was intact—indeed the only influence was an increased rate of the denervated heart by 24 beats per minute for about sixty minutes during an observation lasting three and a half hours; later, after one adrenal gland had been removed and the other denervated and the animal had recovered and was well nourished, 2 units of insulin per kilo twice produced a convulsion, in seventy and in eighty-three minutes respectively, and on a third trial caused panting and vomiting in eighty-four minutes. Comparison of figures 2 and 4 brings out similar testimony. In each case 4 units of insulin per kilo was the dose. In each case the nerves to the liver were intact, but the case of figure 4 differed from that of figure 2 in having only one adrenal and that denervated. In the animal with active adrenal glands the glycemic percentage fell more slowly, the descent began to flatten just below 70 mgm., and the convulsion appeared much later, as compared with the animal lacking active glands.

The only difference between the animals in the two groups of table 1 was presence of active adrenals in one and absence in the other. All the numbered animals had undergone a major operation (denervation of the heart), and all were living together, eating the same food and manifesting the same good health. It could be objected, nevertheless, that inactivation of the adrenals might lower the glycogen storage and therefore, in spite of other similarities, the animals without adrenals might be more sensitive to the action of insulin. According to Stewart and Rogoff, however, the formation and storing of glycogen in the liver in cats is not affected by removal of one adrenal and section of the nerves of the other (19).

From observations on rabbits which had had adrenals removed (for intervals varying from 28 days to more than 8 months) Stewart and Rogoff concluded that the action of insulin does not differ in animals with and without these glands (20). They report tests on only three adrenalectomized animals: one was given 2.7 units, a second 8.8 units of insulin per kilo and a third was given a crude preparation which caused strange and unaccountable symptoms. Only one control was reported. Opposed to this slight and frail support for a rather large inference are the results reported by Lewis (21). He made tests on 27 normal rats and on 11 rats without adrenals, that had received the same food before and since the operation, 12 days earlier. The fatal dose of the insulin he used proved to be 10 mgm. per 100 grams for the normal, and 1 mgm. per 100 grams for adrenalectomized animals. A dose of 2 mgm. per 100 grams, twice the lethal amount for the latter group, did not even produce the typical hypoglycemic symptoms in the normal animals. Similar testimony has come from Sundberg, who studied the effects of insulin on normal rabbits and on rabbits deprived of the adrenal medulla (29). He found that a given dose caused a larger reduction of blood sugar and a more certain appearance of nervous symptoms (asthenia and convulsions) in the animals without medulla than in the normal controls. It is clear that all the evidence which we have presented above as to the importance of adrenal coöperation in protecting the body against the dangers of a too great hypoglycemia is in harmony with the results reported by Lewis and by Sundberg and offers no support for the view expressed by Stewart and Rogoff.[5]

Discussion. The typical symptoms seen after an overdose of insulin are probably due almost entirely, if not entirely, to depriving the tissues of a necessary supply of sugar. That they are not dependent on the action of insulin as such is proved by their occurrence in hypoglycemic states produced by other means. About ten years ago Fischler and his collaborators called attention to a group of symptoms—excitement, convulsions, collapse and coma—accompanying the low blood sugar produced by phlorhizin, and Fischler designated the condition as "glycoprival intoxication" (22). More recently Mann and Magath have reported that the progressive decrease of blood sugar resulting from removal of the liver is attended by characteristic symptoms, prominent among which are, first, muscular weakness and, later, exaggerated reflexes, twitchings and convulsions (23). The very rapid soothing of the disturbed nerve cells by intravenous injections of glucose, whether the disturbance has been occasioned by phlorhizin, by removal of the liver, or by insulin, points to a sugar shortage as its cause.

The deficiency of sugar is probably local, and is not necessarily a consequence of a low concentration in the blood. A number of observers have called attention to a hypoglycemia which follows an abundant ingestion of sugar (24). Folin and Berglund have noted a glycemic concentration as low as 54 mgm. per 100 cc. of blood shortly after the ingestion of glucose. This low level which is considerably under the critical range described above, they explain by assuming that because the local stores are well filled there is no need for sugar to be in transport, and consequently the blood as a carrier is lightly loaded. It seems probable that the

[5] In an article which appeared while the present paper was in galley proof, Stewart (Physiological Reviews, 1924, iv, 183) remarks on the "flimsy foundation" of Zuelzer's theory that "epinephrin is the physiological antagonist of the pancreatic hormone and that when the pancreas is removed diabetes follows owing to the unchecked action of epinephrin given off from the adrenals." He later states that "Cannon, McIver and Bliss have put forward the hypothesis, which may be considered as an elaboration of the theory of Zuelzer, that when the blood sugar tends to fall the output of epinephrin is stimulated." Readers of the present paper are in a position to judge whether our evidence is properly described as an "hypothesis," and whether in any sense it can be correctly considered as an elaboration of Zuelzer's theory.

condition of hypoglycemia after insulin is quite the opposite. The effect of insulin is to increase the utilization of sugar by active tissues; we may reasonably suppose that the local supply is the first to be reduced; the sugar in the blood probably soon passes into the tissues depleted of their sugar stores until what may be called the sugar pressure in the blood is so greatly diminished that it no longer meets the demands. In these circumstances the hypoglycemia would be caused, not by an oversupply in the local stores, but by an undersupply and an excessive demand in them. The main source of the needed material is in the liver. If the sugar is not called forth promptly from that reserve and conveyed by the blood, the organism may be in serious peril.

The observations detailed in the foregoing pages have revealed a mechanism, or set of mechanisms, having the function of maintaining the physiological percentage of blood sugar when there is danger of deficiency. If that percentage falls below a critical point in consequence of deficiency, splanchnic neurones of the sympathetic system are set in action, as indicated by increased adrenal secretion. This is a reaction which, as seen in figure 2, may occur hours before a convulsion occurs, or may occur without a subsequent convulsion (see fig. 7). It coincides with the appearance of subjective feelings of anxiety reported by patients, and shortly thereafter with objective signs of increased sympathetic innervation. Both the nerve impulses and the secreted adrenin have the effect of liberating sugar from the liver into the circulation, and thus tending to restore the disturbed equilibrium. This may be regarded as a first line of defense against a falling glycemic concentration.

Whether this first defense is effectual or not will depend in part on the reserves which can be called upon. Macleod and his collaborators have pointed out that in fasting animals with a meager glycogen store a given dose of insulin produces convulsions more frequently and earlier than in well-fed animals with an abundance of this material (16). Since adrenin is an important factor in the humoral and nervous coöperation which influences the liver, however, it is quite possible that a low adrenin content may be the occasion of weakness in the first defence. In one of our animals, already noticed on p. 59, which received a dose of 2.5 units of insulin before it had fully recovered from a major operation, the denervated heart began to beat faster within sixty minutes after the injection and eighteen minutes later it had increased 62 beats per minute; two hours later still the rate had slowly fallen to the original level, indicating that the adrenal glands had ceased to be especially active. One might suppose that this implied that the call for sugar had been answered; but a very bushy tail revealed great activity of sympathetic neurones, and the animal, weak and panting, had a convulsion a few minutes thereafter. There may have been, of course, a using up of the glycogen reserve in this case, as well as the adrenin reserve, but if so, the two must have run down simultaneously. On the other hand, in the case illustrated in figure 2, adrenal secretion was still being poured forth in extra amount, as shown by the accelerated heart rate, when the convulsion occurred. It is clear, then, that there may be a separation of the adrenal and hepatic factors. Unfortunately it is difficult in any case to discriminate between the two, for the conditions which might lead to depletion of the adrenal medulla would be likely to lessen the glycogen content of the liver. The possibility must be kept in mind, however, that variation in the amount of available adrenin may affect the efficiency of the first defense.

If the first defense fails to prevent the glycemic concentration from falling, there is evidence of more widespread and more effective sympathetic discharges, characterized in cats by dilatation of the pupils, erection of hairs and salivation. As shown in figure 2, this stage is associated with spasmodically increased adrenal secretion (manifested by the marked temporary increases of the rate of the denervated heart). The culmination of this stage of agitation is the convulsive attack. Such an attack is accompanied by increased adrenal discharge (in one of our cases the denervated heart increased 33 beats per minute during the muscular spasms), and it is the occasion for further liberation of sugar into the circulation (25, p. 43). In well-fed rabbits a convulsion may be followed by temporary or permanent recovery without injection of glucose (16). It is probable, therefore, that the convulsion, and the agitation immediately preceding and attending it, form a secondary defence against the damage from an insufficient glucose supply for the needy tissues.

Still another factor which may be at work, though we have little evidence regarding the rôle it may play, is the asphyxial state (i.e., the venosity) of even arterial blood in the later stages of insulin hypoglycemia. It is quite conceivable that asphyxia, which can in itself produce powerful convulsions, either coöperates at some stage with lack of glucose, or with changes induced thereby, to excite activity in the nerve cells. This possibility has been suggested also by Olmsted and Logan (2).

The setting into operation of the sympathetic and adrenal mechanism for mobilizing sugar from the liver explains a number of conditions which have been observed in man. As already noted (p. 51), when the mechanism is started by a fall of the glycemic percentage to approximately 0.07, a patient becomes aware of a feeling of nervousness or tremulousness, or weakness, or of a sense of "goneness." These are subjective symptoms which have been described as occurring when adrenalin is injected subcutaneously (26). "The reaction may go no further than this of its own accord", so Fletcher and Campbell testify (1); this relief is precisely what would occur when the mechanism proves effective. "Or it may be cut short at this stage by the administration of carbohydrate;" the increase of available carbohydrate (sugar) is the result toward which the mechanism is operating, and which, when it is achieved, automatically cuts short the reaction. If the initial symptoms become worse, they are characterized by pallor and flushing, rapid pulse, dilated pupils and sweating, with experience of anxiety, excitement or vague emotional disturbance; again these are the severer symptoms reported by peculiarly sensitive persons when given an injection of adrenalin (26). In this connection it will be recalled that as the blood sugar falls below the critical point the denervated heart beats with increasing frequency, thus indicating both an increased adrenal secretion and a greater discharge of sympathetic nerve impulses. This relationship between hypoglycemia and adrenal output was surmised by Wilder and Boothby and their collaborators. Because of a sharp change in the rate of decrease of the glycemic percentage in a diabetic patient when the first subjective symptoms of hypoglycemia appeared, they suggested that there might be a spontaneous outpouring of adrenin at that point—a result which would tend to protect against a further decrease. And in a later paper the more rapid metabolism which set in at that point was mentioned as supporting the possibility that adrenal secretion was then called forth (26). These interesting conjectures Wilder and Boothby did not pursue further, and they have brought no evidence to support their view. The experiments presented in the foregoing pages, however, definitely confirm their insight. The use of adrenin in the treatment of insulin hypoglycemia (1) is evidently a physiological procedure; the injected adrenin is added to that normally secreted and thereby augments its efficacy, or if the adrenal medulla is more or less exhausted the injected extract has a natural replacement value. A store of glycogen in the liver is, of course, presumed.

The mechanism described in the foregoing pages acts like many others, already known, that assure stability of the organism. When the equilibrium, which normally indicates a concentration of sugar in the blood sufficient to supply needy tissues, is disturbed by serious lessening of the concentration, this compensatory mechanism is set in action to restore the equilibrium. Sugar is set free from the reserves. If the reserve station is absent, as in the experiments on liver extirpation by Mann and Magath, or if the agency reducing the sugar content of the blood is too potent, as when excessive doses of insulin are given, the mechanism is overwhelmed and the result is disastrous to the organism—convulsions and coma are followed by death. From such consequences of an inadequate sugar supply to the tissues the organism is protected by the combined activity of the sympathetic system and adrenal secretion. From the evidence adduced by Griffith, cited above, it is probable that this mechanism is normally at work well within the danger line, sustaining a requisite sugar concentration in the blood.

SUMMARY

The hypoglycemic reactions occurring when an excessive dose of insulin is given include pallor, rapid pulse, dilatation of the pupils and profuse sweating, which are indications of discharge of sympathetic impulses. The question whether adrenal secretion is involved was tested by means of the denervated heart in animals anesthetized with chloralose and in unanesthetized animals.

As the glycemic concentration falls after injection of insulin it reaches a critical point at which the rate of the denervated heart begins to be accelerated, and as the sugar percentage continues to fall the heart rate continues to rise until a maximum is reached (see figs. 1, 2 and 5).

The critical point at which the denervated heart of the animal under chloralose begins to beat faster appears to lie between 110 and 70 mgm. of glucose per 100 cc. of blood. In the unanesthetized the point lies between 70 and 80 mgm. The cardiac acceleration appears before other striking signs of sympathetic activity.

If the adrenal glands have previously been removed, or if one has been removed and the other denervated, a fall of the glycemic percentage below the critical range is not accompanied by an increased heart rate (see figs. 3 and 4). The cardiac acceleration, therefore, is not due to direct action of insulin on the heart or on the adrenal gland, but is due to increased adrenal discharge in response to nervous impulses.

If the rate of the denervated heart has been increased because of hypoglycemia, intravenous injection of glucose promptly reduces the rate (see figs. 1, A and C, and fig. 5).

As the rate of the denervated heart increases (indicating adrenal secretion), the rate of drop in the glycemic percentage decreases, i.e., the sugar curve tends to flatten (see figs. 1, 2, 5 and 7).

If the adrenal glands have previously been removed, or if one has been removed and the other denervated, and the animals are under chloralose anesthesia, the rate of drop in the glycemic percentage is usually not checked at the critical level (see figs. 3 and 6A, 6B and 6C). In unanesthetized animals without active adrenal glands the fall of blood sugar is less retarded at the critical level, and the convulsive seizures are induced sooner and with smaller doses than in animals with active glands (see table 1 and cf. figs. 2 and 4).

If not too much insulin has been given the increased rate of the denervated heart may be followed by an increase of the glycemic percentage and an attendant fall in heart rate (see fig. 7).

It is pointed out that the mechanism protecting the body from dangerous hypoglycemia probably operates in two stages—a primary stage in which sympathetic activity with adrenal secretion occurs and mobilizes sugar from the liver; and, if this proves to be inadequate, a secondary stage in which the activities of the first stage are intensified and augmented in convulsive seizures.

The mechanism here described is another remarkable example of automatic adjustments within the organism when there is a disturbance endangering its equilibrium.

BIBLIOGRAPHY

(1) Fletcher and Campbell: Journ. Metab. Res., 1922, ii, 637; also personal communications from W. D. Sansum, Bertnard Smith and Frederick M. Allen.
(2) Olmsted and Logan: This Journal, 1923, lxvi, 437.
(3) Stewart and Rogoff: This Journal, 1923, lxv, 341.
(4) See Kodama: Tohoku Journ. Exper. Med., 1923, iv, 166.
(5) Stewart and Rogoff: This Journal, 1923, lxvi, 235.
(6) Cannon and Rapport: This Journal, 1921, lviii, 308.
(7) Stewart and Rogoff: Journ. Pharm. Exper. Therap., 1917, x, 1.
(8) Stewart and Rogoff: This Journal, 1917, xliv, 149.
(9) Cannon and Carrasco-Formiguera: This Journal, 1922, lxi, 215.
(10) Cannon and Uridil: This Journal, 1921, lviii, 356.
(11) Searles: This Journal, 1923, lxvi, 408.
(12) Stewart and Rogoff: This Journal, 1923, lxiii, 436.
(13) Zunz and Govaerts: Arch. Internat. de Physiol., 1923, xxii, 87.
(14) Folin and Wu: Journ. Biol. Chem., 1920, xli, 367.
(15) Griffith: This Journal, 1923, lxvi, 618.
(16) McCormick, Macleod, Noble and O'Brien: Journ. Physiol., 1923, lvii, 235.
(17) Hepburn and Latchford: This Journal, 1922, lxii, 177.
(18) Trendelenburg: Pflüger's Arch., 1923, cci, 39.
(19) Stewart and Rogoff: This Journal, 1918, xlvi, 90.
(20) Stewart and Rogoff: This Journal, 1923, lxv, 342.
(21) Lewis: Compt. Rend. Soc. de Biol., 1923, lxxxix, 1118.
(22) Fischler: Physiologie und Pathologie der Leber, Berlin, 1916, 46–49, 96–100. See also Erdelyi: Zeitschr. f. physiol. Chem., 1914, xc, 32; and Burghold: Ibid., 1914, xc, 60.
(23) Mann and Magath: Arch. Int. Med., 1922, xxx, 73.
(24) Polonovski and Duhot: Compt. Rend. Soc. de Biol., 1921, lxxxv, 501. MacLean and Wesselow: Quart. Journ. Med., 1921, xiv, 103. Folin and Berglund: Journ. Biol. Chem., 1922, li, 213. Vladesco: Compt. Rend. Soc. de Biol., 1923, lxxxix, 910.
(25) Macleod: Physiol. Rev., 1924, iv, 21.
(26) See Goetsch: N. Y. State Med. Journ., 1918, xviii, 265; Peabody, Sturgis, Tompkins and Wearn: Amer. Journ. Med. Sci., 1921, clxi, 512; Maranon: Policlinica, 1921, no. 87, 1.
(27) Wilder, Boothby, Barborka, Kitchin and Adams: Journ. Metab. Res., 1922, ii, 722. See also Boothby and Wilder: Med. Clin. of No. Amer., 1923, vii, 55.
(28) Rogoff: This Journal, 1924, lxvii, 551.
(29) Sundberg: Compt. Rend. Soc. de Biol., 1923, lxxxix, 807.

17

SOME GENERAL FEATURES OF ENDOCRINE INFLUENCE ON METABOLISM.

W. B. CANNON, M.D.,
Department of Physiology, Harvard Medical School, Boston, Mass.

A fairly constant or steady state, maintained in many aspects of the bodily economy even when they are beset by conditions tending to disturb them, is a most remarkable characteristic of the living organism. Though the blood is the common carrier for all manner of food material and for waste, its composition varies to a surprisingly slight degree. Though warm-blooded animals are exposed to numerous and extensive alterations of surrounding temperature, they themselves undergo shifts of temperature within only a brief range. And though food is taken in as it is provided or desired and is utilized as conditions may demand, the body weight of the full-grown individual continues for long periods on a strikingly even level. Since in these, and in other instances which could be added to them, circumstances are often present which, if not controlled, would profoundly modify the constancy of the state, we must assume that controlling factors are at hand ready to act whenever the constancy is imperilled. The existence of steady states in the body and some agencies maintaining them have, of course, long been recognized. On this occasion I am laying emphasis upon these states because of an interest in some tentative general considerations with regard to physiological factors regulating them, and in the possible bearing of these considerations on problems of internal secretion. As they may have value in suggesting lines of research and bases for critical judgment, they may be worthy of attention.

Six Propositions Regarding Physiological Factors Maintaining Steady States in the Body.

The first proposition is that in an open system, such as our bodies represent, complex and subject to numberless disturbances, the very existence of a poised or steady state is in itself evidence that agencies are at hand keeping the balance, or ready to act in such a way as to keep the balance. These agencies may act antagonistically, like the nerves of the heart. A tonic action is also possible, as in arterial vasoconstriction, with constancy provided indirectly, one step back, in a nerve center subjected to opposed control. The regulation may consist of a check on the excesses of a continuous function, like the respiratory setting of the carbon-dioxide level in the blood. Again, there may be limitation by overflow, a safety-valve arrangement, as in the renal thresholds of overflow or escape for various constituents of the blood.

A second proposition is that, if the state remains steady, there is an automatic arrangement whereby any tendency towards change is effectively met by increased action of the factor or factors which resist the change. Somewhat similar statements of this proposal have been made by Spencer (1) and also by the Belgian physiologist Frédéricq (2). They did not recognize, however, failure of the state to retain its pois. because of absence or weakness of a controlling agent—a not uncommon consequence of pathological processes. For that reason I have made the statement conditional ("if the state remains steady"). As Lotka (3) has pointed out, such a reservation, which is required for constant states in living beings, and which is a recognition of their lack of permanent stability, would sharply distinguish the proposal from the strict principle of Le Chatelier which was enunciated for physico-chemical systems.

A third proposition is that any factor which operates to maintain a steady state by action in one direction does not act at the same point in the opposite direction. This declaration may appear tautological, but, as will be shown later, it has interesting bearings, and if its probable truth had been recognized, it would have led to more cautious conclusions than have been drawn.

A fourth proposition, related to the third, is that factors which may be antagonistic in one region, where they effect a balance, may be co-operative in another region. It is necessary, therefore, to define closely the field of action of opposing factors in discussing a physiological equilibrium.

A fifth proposition is that the system of checks which determines a balanced state may not be constituted of only two antagonistic factors; on either side there may be two or more, brought into action at the same time or successively. Further, the two may be of different nature, e.g., an internal secretion and a neuromuscular agency.

A sixth proposition is that when a physiological factor is known which can shift a steady state in one direction, it is reasonable to look for a physiological factor or factors having a contrary or counterbalancing effect.

In the foregoing statements the expression "ready to act" has been used as an alternative to "acting" because in certain instances there is evidence that the situation is not kept balanced by direct and immediate opposition of active agencies, but is allowed to move to and fro within limits or within critical points, possibly under control of simple physico-chemical adjustments, and only when these limits are passed are the more complex and indirect physiological agencies (i.e., those involving reactions peculiar to the organization of living beings) brought into play.

The advances in our knowledge of the relations of internal secretion to metabolism have lately been so great and so various that an attempt to present an encyclopedic survey in brief compass would be futile. Instead, may we not consider the six propositions, stated above, with regard to some recent investigations and results, and see to what degree interpretation is clarified thereby and new lines of work are indicated. In doing so I wish to disclaim any effort at an elaborate or final scheme of relationships. The concept of equilibrating factors will be used merely as a means of unifying some apparently diverse and separate observations which, in the main, have been personal experiences.

The Equilibrating Factors in the Blood-Sugar Balance.

Carbohydrate is the material most readily burned in the body. When it is provided in abundance, the combustion of fat is almost completely stopped and the combustion of protein is reduced to a minimum. In less than a half-hour after the ingestion of sugar the respiratory quotient usually rises to about 1.0 (4). The concurrent testimony of investigators supports the conclusion that carbohydrate is not only the most readily utilizable source of energy, but is essential for working the mechanism of muscular contraction. Its importance, therefore, need not be emphasized.

In the economics of the body, problems arise concerning carbohydrate demand and supply because that foodstuff is continuously being used up and is renewed only periodically. When it is taken in and is being digested it must not be permitted to increase in the blood above 170 or 180 mg. per 100 cc., for then it will be lost by escape through the kidneys. The amount available is likely to be more than is needed for current uses; the excess must be set aside in storage, in a compact and not immediately utilizable form. This is accomplished chiefly in the liver, though there are, of course, the scattered local stores. When need arises the stored carbohydrate must be released from the liver, for if the glucose in the blood falls to approximately 45 mg. convulsions are likely to occur, which may be followed by coma and death. Thus, in sum, the blood-sugar level is an equilibrium between the supply and the demand, with storage as an adjusting arrangement. The normal outer limits of the equilibrium may be regarded as at approximately 180 mg. where loss occurs, and approximately 45 mg. where convulsions occur. But the normal variations of blood sugar are ordinarily well within these extreme limits. If our earlier propositions are correct, factors are at work holding the oscillations inside the narrower confines. Some recent work, done by McIver and Bliss and myself (5), has revealed a sympathico-adrenal mechanism as one factor in the balance, protecting the body against the dangers of too low blood sugar.

As an indicator of increased adrenal secretion we have used the heart, disconnected from the central nervous system and left in the body to perform its function. If the temperature is kept normal, only changes in the blood stream can affect the rate. Alterations of blood pressure do not influence it, but a very slight increase in the adrenin content of the blood promptly makes the heart beat faster. Anrep and Daly (6) have recently found that the denervated heart is sensitive to the presence of adrenin, 1 part in 1,400 million parts of blood. All investigators now admit that medulliadrenal secretion is controlled by sympathetic impulses. Afferent stimulation and asphyxia, which are known to discharge sympathetic impulses, cause acceleration of the denervated heart—just such a change as is produced by increase of adrenin in the blood. The only agent which may confuse the response in an acute experiment is a substance given off from the liver after protein feeding—a source of trouble which can be made insignificant by study of a fasting animal or abolished by severing the hepatic nerves (7). If these precautions are taken the positive response disappears when the adrenal glands are removed or inactivated by nerve section. The faster heart rate, therefore, is then due to medulliadrenal secretion (8). The results which we have obtained by use of the denervated heart have been questioned by Stewart and Rogoff (9). These results, however, have been confirmed by Tournade and Chabrol (10) in Algiers, working with a crossed circulation; by Houssay and Molinelli (11) in Buenos Aires, also employing a crossed circulation; by Florovsky

(12) and by Baschmakoff (13) in Kasan, using the denervated salivary gland; by Hartman, McCordock and Loder (14) in Buffalo, recording the reactions of the completely denervated iris; and by Anrep (15, 16) in London, studying the denervated heart and kidney and also the isolated heart outside the body but connected with the abdominal aorta and inferior vena cava. Furthermore, Kodama (16) in Sendai, Japan, repeating the procedure of Stewart and Rogoff, has found, not what they found, but what we found. Naturally Stewart and Rogoff have criticised Kodama's results and have declared that he did not master their technique (17). A recent letter from Sendai, however, states that other workers there have refined the method, and that Kodama's data have been fully substantiated. If any investigator anywhere had given Stewart and Rogoff any support for their contention that adrenal secretion is constant, and unaffected by conditions which induce sympathetic activity, there might be reason for seriously considering it. But that support is lacking. Opposed to it is the concordant experience of other investigators, derived from a variety of methods. It justifies confidence in the result obtained by use of the denervated heart as a signal for greater discharge from the adrenal medulla.

Now to return to the question of a sympathico-adrenal factor working to keep the blood sugar normal and to protect the organism from dangerous hypoglycemia. Among the hypoglycemic reactions which occur when an excessive dose of insulin has been given are pallor, a rapid pulse, dilatation of the pupils and profuse sweating. These are phenomena resulting from sympathetic impulses. The denervated heart tells whether the adrenal medulla shares in this increased activity. When, after an injection of insulin, the glycemic concentration is falling, it reaches a critical point at which the denervated heart begins to beat faster. And as the sugar percentage continues to fall the heart rate continues to rise until a maximum is reached. In the unanesthetized test animal the critical point lies between 70 and 80 mg. of glucose per 100 c.c. of blood—a concentration at which in man also hypoglycemic symptoms first appear (18). The faster beating of the denervated heart indicates that medulliadrenal secretion has been stimulated. That inference is verified by removal of the glands, or by removal of one and denervation of the other, whereupon a fall of the glycemic percentage below the critical point is not accompanied by a more rapid heart beat. The faster beating of the heart, of course, indicates that impulses are being discharged along splanchnic fibres. These sympathetic impulses, in co-operation with increased adrenal secretion, might be expected to call forth sugar from the liver. This inference is confirmed by the fact that as the rate of the denervated heart increases (manifesting sympathico-adrenal activity), the rate of falling of the glycemic percentage is checked. And if not too much insulin has been given, the glycemic percentage may not only cease dropping, but may rise to normal. A highly significant feature of this sympathico-adrenal activity is that as the blood-sugar level rises the activity subsides. Indeed, when glucose is injected intravenously the activity has been seen to subside completely within two minutes.

Here is a remarkable mechanism having the function of keeping up the physiological percentage of blood sugar when there is danger of deficiency. After a dose of insulin the mechanism may be brought into service hours before a convulsion occurs, and if effective it may wholly protect against a convulsion. From the evidence provided by these observations we may conclude that the lower inner limit of variation of blood sugar, when the sugar is being reduced, lies in the region of 70-80 mg. Reduction below this limit calls into action a sympathico-adrenal factor working to prevent further reduction. This factor works with increasing vigor the lower the glycemic percentage, until an acme is reached in a convulsion when the denervated heart beats most rapidly.

According to Olmsted and Logan (20) the convulsive seizures do not occur in a spinal animal, even when the glycemic percentage is extremely reduced. The sensitive region appears to be in the medulla or midbrain, for in decerebrate animals typical convulsions result from hypoglycemia. Rapport and I (21) have shown that a center for medulliadrenal secretion exists in the fore part of the floor of the fourth ventricle. Quite possibly this region is especially sensitive to a reduced sugar supply and by sympathico-adrenal discharges (culminating in spasms) protects the rest of the body against the hazards of too great a reduction.

Our results have received support from Abe (22) who made use of the sensitized iris to signal a greater output from the adrenal medulla. In some instances he noted no effect, or very slight effect, during the first half-hour after giving insulin. Then there was a sudden change, marking the onset of the sympathico-adrenal function as the blood-sugar percentage fell. Houssay, Lewis and Molinelli (23) also have supported our conclusions. They employed a crossed circulation from the left adrenal vein of one dog to a jugular vein of another. When an insulin hypoglycemia was established in the donor, hyperglycemia was produced in the recipient—a result which did not occur if the left splanchnics of the donor had previously been severed. Both methods showed that adrenal secretion was increased by hypoglycemia, but unfortunately they were not used in such a way as to bring out the evidence for the critical point at which the sympathico-adrenal mechanism is set going, an observation of special interest in relation to the glycemic equilibrium.

In our second proposition we postulated that the preservation of a balanced state implies an automatic arrangement whereby a tendency to change is met by increased action of a factor or factors which resist the change. We have surveyed the evidence for a factor checking a shift of the glycemic equilibrium downwards. What agency or agencies operate in the opposite sense, to prevent a shift upwards, towards a high sugar concentration? The answer might be given promptly that it is the internal secretion of the pancreas, if competent investigators had not concluded that insulin administered to normal animals is capable of decreasing hepatic glycogen, i.e., that insulin may work in the same direction with the sympathico-adrenal mechanism. There is excellent reason for believing, however, that insulin primarily is the opposing factor and that its apparently anomalous effect in normal animals should be regarded as a secondary phenomenon. Let us examine the basis for the inference that the internal secretion of the pancreas, or insulin, normally protects the body against hyperglycemia by promoting sugar storage in the liver.

First, there is the outstanding fact that removal of the pancreas is promptly followed by hyperglycemia and a great reduction in the hepatic glycogen reserve. This phenomenon is so well known as to need no emphasis. Secondly, as Banting and his collaborators (24) have shown, the administration of insulin to depancreatized animals not only brings the blood sugar down to the normal level, but causes glycogen to accumulate again in large amounts in the hepatic cells. Indeed the hepatic glycogen of insulin-treated diabetic animals may be considerably more than in normal well-fed controls. Again, it has been found that insulin will induce a deposit of glycogen in the liver when the blood sugar is within the range of normal variations. Thus, Cori (25) has recently reported that the amount of hepatic glycogen may be quadrupled in phlorizinized rabbits by repeated doses of insulin. Finally, as bearing on the antagonism between the glycogenolytic action of adrenin and the glycogenetic action of insulin, there may be cited the experiments of McCormick, Noble and O'Brien (26). They discovered that when large doses of adrenalin were injected—doses which produced a marked reduction of hepatic glycogen in control animals—the additional injection of insulin was accompanied by a striking retention of the hepatic glycogen. The excessive doses of adrenalin used by them (e.g., 2.5 c.c. in a rabbit) allowed, they state, "enormous quantities of insulin" (e.g., 140 units) to be tolerated without convulsions. All these various observations bear witness to the importance of the internal secretion of the pancreas (insulin) as an agent operating conversely to the sympathico-adrenal mechanism in affecting the glycemic percentage by action on the liver.

The suggestion that insulin does not act in one direction only but in both directions on the liver (i.e., glycogenetically and glycogenolytically) has, to be sure, been derived from experiments on normal animals. McCormick and Macleod (27) gave insulin and carbohydrates to rabbits rendered free of glycogen and found that less glycogen was deposited in the livers of the insulin-treated animals. Dudley and Marrian (28) compared the livers of glycogen-rich mice and rabbits, some of which had received insulin and others left as controls; the hepatic glycogen of the former was very greatly reduced as compared with the latter. Babkin (29) also has reported that the glycogen content of the liver in well-nourished rabbits which had been given insulin is markedly lowered. The comments on these experiments have been somewhat equivocal. Thus Macleod (30) states that "there can be no doubt that insulin causes a rapid reduction in the amount of glycogen in the liver," but he also remarks later that at some stage after the blood sugar has become reduced by insulin "the glycogenolytic process in the liver is stimulated." A similar suggestion was offered by Dudley and Marrian. In our third proposition we intimated that a factor does not have such opposed functions with reference to the equilibrium which it protects. If we keep that idea in mind and remember that hypoglycemia calls forth sympathico-adrenal service, the results above detailed can be reasonably explained. It is significant that in the experiments of McCormick and Macleod, despite an effort to give doses of insulin "just insufficient to cause convulsions," convulsions occurred in 6 out of 15 cases; in the experiments of Dudley and Marrian the animals were killed on the appearance of convulsions; and in the experiments of Babkin, though "convulsions were not observed," the glycemic percentage was far below the critical level. In all these experiments, therefore, the conditions were such that the hepatic glycogen stores were being drawn upon, not by direct action of insulin, but by the direct action of the factor opposed to insulin—the sympathico-adrenal factor—made to operate powerfully by the insulin hypoglycemia. This conclusion is supported by the testimony of Burn (31) that a small dose of insulin, mildly effective in the normal animal, causes in an animal treated with ergotamine profound hypoglycemia, with convulsions and collapse; for ergotamine, though having no influence on the action of insulin, paralyses the protective sympathico-adrenal reaction. That the secreted adrenin plays an important part in this reaction is proved by the fact that the fall of the glycemic percentage is not checked, or is only slightly checked, at the critical point, if the adrenals have previously been inactivated (5), and also under such conditions by the appearance of convulsions sooner and after smaller doses of insulin than in animals with active glands—a phenomenon observed by Sundberg (32), by Lewis (33), and by us (5).

From all the foregoing evidence it appears that the normal level of blood sugar is preserved by the sympathico-adrenal factor acting on the low side of the equilibrium and by the pancreatic factor and the renal overflow acting on the high side. Much still remains to be learned regarding the conditions bringing the islands of Langerhans into play. From the experiments of Banting and Gairns (34) the inference may be drawn that a nervous control of insulin secretion is not necessary. That does not prove, however, that nervous control does not exist and could not be useful—the innervation of the heart is not *necessary!* The observations of De Corral (35) and of McCormick, Macleod and O'Brien (36) point to vagus impulses as causing a discharge of insulin into the blood stream. Recent (unpublished) studies by Britton at the Harvard Physiological Laboratory, as well as the investigations of Clark (37), have corroborated the earlier evidence. In harmony with these results are those announced by Lewis and Magenta (38) that after section of the abdominal vagi a given dose of insulin is less effective than before. Because in the presence of high glycemic percentages remnants of the pancreas degenerate (39) and the islet cells show signs of overwork (40), it seems probable that just as the adrenals are stimulated by hypoglycemia, so likewise the "islands" are stimulated by hyperglycemia, i.e., that each secretion is involved in a righting response which restores the normal equilibrium. If further research proves that the vagus provides a fine adjustment for insulin secretion, as the sympathetic does for adrenal secretion, it will be peculiarly interesting because of the common opposition between vagal and sympathetic functions.

A caution appears to be necessary here lest the ideas just expressed be taken as confirming Zuelzer's theory of the adrenal origin of diabetes. Indeed Stewart (42) has declared that my interpretation of the sympathico-adrenal service in maintaining the normal glycemic concentration is only an elaboration of that theory. The essence of Zuelzer's theory was that the adrenals and the pancreas are such aggressive antagonists that when the pancreas is removed diabetes follows because of the unchecked action of the adrenals. The theory has no sound basis. And our observations, that sympathico-adrenal activity, roused by hypoglycemia, is promptly checked by hyperglycemia, and that it is made to work again only when the blood sugar has *fallen* well below the usual level, should have prevented anybody from imagining that our experiments in any sense support Zuelzer's idea. Instead, they raise the question whether the high sugar level of diabetes may not be associated with a depression of sympathico-adrenal function.

It will be recalled that in our fifth proposition we suggested that when opposed factors assure an equilibrium there may be more than one on either side. Indeed, a primary action of the islands of Langerhans and the co-operative overflow arrangement in the kidneys may be regarded as a double device preventing a rise in the glycemic level. On the other arm of the balance may be placed the testimony of Olmsted and Logan (20) that the pituitary body is important. They state that typical convulsive seizures occur in decerebrate animals if the pituitary body has been removed, but that otherwise even massive doses of insulin fail to lower blood sugar to a marked degree. This result has been associated with the observation by Burn (31) that pituitrin can lessen or entirely prevent insulin hypoglycemia. The conclusion that the posterior lobe of the pituitary may collaborate with the adrenal medulla thus seems justified. In a recent investigation, however, Bulatao and I (43) have shown that in decorticated animals with intact pituitary glands insulin regularly causes a drop in the glycemic concentration, and we have called attention to the same phenomenon in three similar cases described by Olmsted and Logan. Furthermore, we showed that in decorticated and decerebrated animals sympathico-adrenal activity may be maximal and may therefore retard or lessen the effect of insulin. For Olmsted and Logan to conclude on the basis of their few cases, which are not decisive, that the high blood sugar after decerebration is maintained by pituitary influence, seems to us unwarranted. In other words, though pituitrin as a pharmacodynamic agent inhibits the action of insulin, proof is still lacking that a pituitary secretion as such has that effect, and could co-operate with adrenal secretion.

Because of the effects which hypothyroidism, hyperthyroidism and thyroid feeding have in disturbing the glycogen content of the liver and influencing sugar tolerance (44), and also because of reported hypersensitiveness of thyroidectomized animals to insulin (45), it may be that the thyroid normally collaborates with the sympathico-adrenal apparatus. The evidence on this point is still too vague and uncertain, however, to justify discussion of it in this paper.

In our fourth proposition we mentioned that factors opposed in one region of the body may be collaborators in another region. This is perhaps the case with adrenin and insulin. They are opposed with respect to glycogen storage in the liver. They may be collaborators, however, in the utilization of sugar in muscles. Various investigators have found that injected adrenalin accelerates metabolism (46), and recent Harvard studies have proved that secreted adrenin has the same effect (47), and that the effect occurs when the circulation is almost wholly confined to muscles (not yet published). Adrenalin causes in the isolated heart not only a greater consumption of oxygen per unit of time but also may cause total disappearance of the stored glycogen (48). Hepburn and Latchford (49) observed that insulin accelerates removal of glucose from the perfusing fluid by the isolated mammalian heart, and Burn and Dale (50) noted that insulin causes rapid removal of glucose in a decapitated and eviscerated preparation, i.e., when the heart and skeletal muscles are the only important metabolizing organs. The early stages of insulin action in such a preparation are characterized by increased oxygen consumption, with a respiratory quotient of unity. Thus insulin and adrenalin, antagonists in the liver, appear to be collaborators elsewhere. Insulin may promote everywhere the passage of sugar from the blood into cells. In muscle cells adrenal secretion may work with insulin to facilitate the chemical changes underlying contraction. All this, however, remains to be proved. The suggestion is mentioned here merely to illustrate the possibility of different relations of physiological agents at different places in the organism.

Some Hints at Factors Affecting Protein Equilibria.

Protein is quite as important as carbohydrate for bodily welfare and function. It is necessary for such invaluable services as building the cellular structure of the organism, repairing or restoring the worn or wasted parts, and providing the essentials for coagulation and for normal composition of the blood. It is constantly being disintegrated and lost from the body, and the possibility of securing a new supply is not always at hand. Since there is an elaborate arrangement for storing excess of carbohydrate and for liberating it as required, it would appear reasonable to look for a similar arrangement for storing indispensable protein against a time of need. The process of deaminization and the excretion of extra nitrogen indicate that accumulation of protein must be limited. The existence of a nitrogen equilibrium, capable of being altered and adapted to the requirements of the organism, implies that there prevails an orderly regulation for riddance or retention and the maintenance of a steady state.

The reality of a reserve or "deposit" protein has often been debated. Gradually there has developed a considerable body of evidence that with abundant protein feeding the hepatic cells can store protein just as they can store glycogen. More than forty years ago Afanassiew (51) noted that if "albuminates" were fed to dogs the liver became firm and resistant, and the hepatic cells increased in size and contained between their protoplasmic strands many fine sharply outlined granules, protein in nature. Seitz (52) brought further support, finding that in fed animals the total nitrogen of the liver in relation to that of the rest of the body was from 2 to 3 times as great as in fasting animals. The observations of Berg (53) have been especially significant. In 1914 he reported that in the hepatic cells of animals which are given protein there appear numerous small droplets or masses which are clearly distinguishable from the cell structure, which react as protein to Millon's reagent, which are caused to disappear by fasting, and which are restored only when protein or amino acids are fed. In 1922 he showed that these droplets can be distinguished in fresh, teased-out liver cells stained with neutral red, and that they give the ninhydrin reaction of the simpler proteins. Berg's earlier observations have been fully confirmed by Stübel (54) and by Noël (55). This histological evidence for a protein reserve has been corroborated by Tichmeneff's chemical studies (56). He starved mice for two days, then killed half of them and, after giving the others an abundance of cooked meat, killed them and compared the livers of the two groups. Expressed in percentage of body weight the livers of the meat-fed animal had increased about 20 per cent, with the hepatic nitrogen content augmented between 53 and 78 per cent.

The concept of storage of valuable material involves the idea not of miserliness but of service—the appropriate use of the material outside the storage place. There appears to be at present no proof that protein is taken from the liver for use in another part of the body which requires restoration or improvement, though the possibility that such a transfer could occur would justify an attempt to search for it. There is testimony, however, to a function which the liver performs in maintaining the protein constituents of the blood. Kerr, Hurwitz and Whipple (57) found that after depletion of blood plasma (by bleedings and injections of washed blood corpuscles in Locke's solution) the restoration of the serum protein is delayed if the liver has been damaged or has been side-tracked by an Eck fistula. Furthermore, liver injury is associated with a fall in the serum proteins. Although in the blood of the normal animal fibrinogen is rapidly reformed after being removed, Meek found that it fails to be reformed if the liver is unable to act (58). It seems, therefore, that the liver plays a fundamental rôle in the production of fibrinogen. In the production of antithrombin or antiprothrombin as well, and possibly of heparin also, there are indications that the liver is concerned and thereby might influence deeply the coagulation of the blood.

The serum proteins remain at a remarkably steady concentration, fluctuating only slightly with fasting, feeding, health or disease. And yet, as shown above, when they are reduced, the liver soon restores them. Again, the factors concerned with blood clotting are ordinarily in balance, so that coagulation takes a moderate and fairly uniform time. The relations can be disturbed, however, so that the time is longer or shorter. In experiments by Gray and myself (59) we obtained records of a shortening of the clotting time of blood by adrenalin injections if the blood is allowed to circulate through abdominal organs; the phenomenon fails if the circulation is confined above the diaphragm or if the intestines and liver have been removed. Increased adrenal secretion induced by splanchnic stimulation or excitement has a similar effect, as shown by experiments done in co-operation with Mendenhall (60). According to Grabfield (61) adrenalin increases the prothrombin element in the blood. It is significant that blood taken at the height of the hypoglycemic reaction, when sympathico-adrenal activity is maximal, "clots almost immediately it is withdrawn," as Macleod (30) has remarked, and as we have also noted. Finally, as Gray and Lunt (62) showed, a large hemorrhage hastens the coagulation of blood if the liver (and intestine) are in the circulation, but not if the blood flow is excluded from the abdomen. Since the sympathetic is stimulated by considerable hemorrhage, probably sympathico-adrenal action is normally involved in the response.

Additional evidence that the protein material of the liver can be called out by appropriate stimulation was furnished by Stübel (54) who found that the deposited droplets or small masses in the hepatic cells could be greatly reduced by injecting adrenalin subcutaneously. Moreover, in collaboration with Uridil and with Griffith (63), we have shown that after protein feeding, stimulation of the nerves of the liver causes discharge of a still undetermined substance which accelerates the heart and causes a noteworthy increase of arterial pressure. The phenomenon is reduced to insignificance in animals which have been fasting, or which have been fed carbohydrate or fat.

From the foregoing observations we may conclude that the liver is an organ of central importance not only for carbohydrate, but also for protein metabolism. It shares with other parts the power of constructing protein from amino acids, but when large amounts of protein are fed it lays by a reserve or deposit to a much greater degree than do the other parts (56). This reserve can be called out by sympathetic stimulation and by adrenin. In maintaining the normal consistency and coagulability of the blood the liver is indispensable. Here are steady states in which the liver performs quite as important a task as it performs in the carbohydrate balance of the body, and probably the factors which control the situation act with quite as great a degree of nicety. Unfortunately we are still in a twilight zone which surrounds the problem, and vision is not clear. Hints have been given as to factors affecting the protein levels, but much work must still be done before we shall be well informed regarding them or their ways of producing steady states.

An Endocrine Factor Operating to Maintain the Temperature Equilibrium.

The uniform temperature of warm-blooded animals, though they are exposed to very different environmental temperatures, has already been alluded to as an outstanding example of physiological constancy. The temperature equilibrium is maintained by opposing factors. Concerning the nature of the physical factors, vasodilation and sweating, which operate when the body temperature tends to rise, there has been no difference of opinion. On the other hand, factors which are set in action when the temperature tends to fall have been the subject of discussion for more than a third of a century. Speck (64), Loewy (65), Johansson (66) and Sjöström (67) have concluded, from experiments in which subjects were exposed to cold surroundings, that the increased metabolism which results is to be explained wholly in terms of shivering or other muscular contraction. Voit (68), Rubner (69) and recently Leonard Hill and his co-workers (70), on the contrary, have conducted experiments in which subjects were similarly exposed to cold surroundings, but usually for long periods, and have found acceleration of the metabolic rate with no demonstrable muscular movements. "It is incomprehensible," Rubner wrote, "that many physiologists can still doubt the presence of this powerful regulating mechanism. And it is just as incomprehensible that the increase of heat in the presence of cold is ascribed solely to the gross shivering of the animal." What the nature of the chemical regulator is he did not suggest.

When warm-blooded animals are exposed to cold the striking reactions are erection of hairs or ruffling of feathers, constriction of peripheral vessels and increase of blood sugar. These are signs that the sympathetic division of the autonomic system is in action. Is secretion from the adrenal medulla augmented under these conditions? The answer to this question has interesting bearings on the old controversy regarding chemical control of body temperature. Since 1906 various observers have found that adrenalin stimulates oxygen consumption in isolated organs or in the body as a whole.

This pharmacological evidence was given a physiological significance when Aub, Forman and Bright (71), at the Harvard Physiological Laboratory, found that if the adrenal glands are removed, though blood pressure and the body temperature remain unchanged, the metabolic rate may drop as much as 25 per cent below the previous normal level, and when later McIver and Bright (47) demonstrated that increased adrenal secretion accelerates the oxidative processes of the body. Clearly, if cold excites a greater outpouring of adrenin into the blood stream, it would induce a faster burning in the body just when there would be need for it.

Cramer (72) and Fujii (73) have reported reduction of the chromaffine substance of the adrenal medulla in animals exposed to low temperatures, but that condition, as Trendelenburg (74) has suggested, may not have resulted from over-discharge but from under-production. More direct evidence of an increased medulliadrenal secretion brought on by cooling was obtained by Hartman and his collaborators (75). Using the completely denervated iris of the cat to signal greater discharge of adrenin, they observed that dilatation of the pupil resulted when the cats were dipped in cold water or when they were dipped in warm water and while wet were placed in a cold room or exposed to a current of cool air. The dilatation failed to occur if the adrenal glands were removed or inactivated. The use of the iris by Hartman has been severely criticised by Stewart and Rogoff (76), who, however, have not used the completely denervated structure themselves, and who have, therefore, no good basis for judgment. Because Hartman's results have been doubted and also because we desired to eliminate quite certainly any emotional element (which wetting may have aroused in the cat), Querido and I (77) have investigated the reaction by other methods.

As is other experiments, we have employed the denervated heart as an indicator of augmented adrenal secretion. We tried several different modes of exposing animals to cold before we hit upon the one which proved most satisfactory—that of introducing into the stomach a known amount of ice-cold water. The body heat of the animal passes into the cold fluid, and, unless the temperature is to fall, heat production must be increased to compensate for the loss. We have called the situation confronting the organism under the circumstances its "heat liability." Obviously, since the weight, the temperature and the specific heat of the water are known, the number of calories which the animal must conserve or produce or both, in order to prevent a temperature fall, can be calculated and nicely adjusted to the purposes of the experiment.

When heat liabilities varying between 1,050 and 2,400 small calories per kilo were established there were invariably faster rates of the denervated heart, ranging from 15 to 64 per cent increase. When the adrenal glands were inactivated and similar heat liabilities were presented to the organism the increases ranged between 2 and 6 per cent only. And if, instead of ice-cold water, water near the body temperature was used the percentage increase was negligible. Since cold slows the rate of the isolated heart, the faster rate which follows the introduction of cold water occurs in spite of adverse conditions, a true physiological response. And since this response fails when the adrenal glands are not acting, the inference is justified that the faster heart rate is a consequence of greater adrenal secretion.

If the heat liability is more than 1,000 calories per kilo, acceleration of the denervated heart is almost uniformly attended by shivering. Thus, two calorigenic factors, secreted adrenin and muscular movements, are co-operating to protect against a shift of the temperature equilibrium towards the low side. But the shivering factor is not an essential part of the reaction, for if a heat liability of only 900 calories per kilo is to be met, shivering rarely occurs; and if it occurs, it is of short duration. The heart rate is faster, however, showing that the adrenal factor is operating.

The increase of adrenal secretion when the organism is exposed to the danger of too rapid heat loss, and the ability of naturally secreted adrenin to accelerate heat production, point to a physiological reaction the services of which to the organism ought to be demonstrable. To test the capacity of this reaction we have studied the effect of a certain heat liability on shivering when the adrenal glands were active or inactive, and we have observed the influence of heat liability on the metabolic rate in the absence of shivering.

I have already remarked that if a heat liability of 900 calories is presented to the organism (cat), shivering rarely occurs or is brief. In 15 cases shivering occurred only twice and lasted in each instance only 3 minutes. In 15 cases in which the adrenals were inactivated by removal of one gland and denervation of the other, the same heat liability per kilo caused shivering in 13, and furthermore, the shivering usually lasted considerably longer than in the control series—for 6, 7, 8, 11, 13 and even 16 and 17 minutes instead of 3 minutes. Thus the primary protection against a heat deficit in the body is found in activities of the sympathetic system (erection of hairs, vasoconstriction, liberation of easily combustible sugar, and discharge of adrenin), for these responses occur before shivering occurs and may be the only manifestation of the influence of a heat liability. If animals have been deprived of the calorigenic service of the adrenal medulla, however, greater demands are made on muscular activity as a means of maintaining the normal temperature. Under normal conditions, therefore, two lines of defense are erected against a shift of the temperature equilibrium towards the low side. The first line is the sympathico-adrenal mechanism; if that fails and the temperature tends to drop in spite of it, the second line, muscular action or shivering, is invoked to produce the needed heat. And just as the convulsive seizures in hypoglycemia are attended by a greater adrenal discharge, so likewise is shivering due to cold thus attended. In both circumstances the introduction of the muscular factors marks a climax of the action of the automatic arrangements for defense.

If the foregoing inferences are correct, it should be possible to demonstrate a faster metabolic rate in response to a bodily need for heat, quite apart from any muscular quiverings or contractions. To secure evidence on this point Querido and I, and later Bright, Britton and I, have studied the change in the metabolic rate in human beings when a known heat liability was created by taking ice-cold water or that combined with ice. We have now 19 satisfactory observations on 12 different subjects. The heat liability ranged from 360 to 512 small calories per kilo. The metabolic rate has increased in these cases as much as 35 per cent, though the usual increase was not more than 17 per cent, and this without any sign of shivering either objective or subjective. Commonly the first effect of taking the cold water is a drop in metabolism, which is followed by the rise, which reaches its maximum at periods averaging 25 minutes after the liability was established.

The observations described above have direct bearing on the old controversy regarding the nature of the chemical agencies which help to preserve the temperature equilibrium of warm-blooded animals. Careful examination of the records of the experimenters who have argued that muscular contraction alone is the source of needed heat shows that they reported substantial increases of metabolism unattended by shivering. Furthermore, they failed to recognize that shivering may co-exist with and effectively mask a true chemical calorigenesis. And finally they have inferred that because a chemical mechanism did not operate in certain short experiments the mechanism does not exist—an inference having slight value when confronted by good evidence to the contrary. The results of our experiments are definitely opposed to the idea that a single factor, the muscular factor, is relied upon to keep the temperature balance from tipping downward. They bring support to the testimony and the argument of Voit and of Rubner, and offer an explanation of the chemical regulatory device which they postulated.

The evidence which has just been presented is not to be taken as excluding still other co-operating agencies. The fact that thyroid activity accelerates metabolism, that thyroid removal lowers the metabolic rate and renders one more sensitive to cold, that the thyroid gland is probably subject to sympathetic control (78), that it enlarges in animals exposed to low temperatures (79), that in the compensatory hypertrophy of a remnant the evidences of hyperactivity are greater if the animal is exposed to frigid surroundings (80)—all these observations are highly significant of possible sympathicothyroid collaboration in the upkeep of body temperature. As Aub (81) has suggested, the sympathico-adrenal service may be employed for quick action and the thyroid may act in the same direction, but for more enduring needs. But here we must have more facts before we are warranted in giving thyroid function definitely a place among the compensatory factors evoked by a threatened shift in the temperature equilibrium.

In the foregoing discussion I have used to a large extent the results obtained in personal investigations. This emphasis will not, I trust, be interpreted as implying an over-valuation of the subjects studied or of the results achieved. It seemed best to rely on first-hand experience rather than to summarize the work of others. Furthermore, I wished to illustrate the interest which is added to a search for new facts if certain general ideas are allowed to suggest the direction of the experimenter's imagination and his efforts at securing insight.

In stating the first of the six propositions, i.e., that a steady condition in the body implies physiological agencies serving to maintain it, I mentioned several types. The type of opposed agencies has been illustrated by the operation of the sympathico-adrenal and the probable vago-insular mechanisms in keeping normal the blood-sugar level, and also by the regulation of the temperature equilibrium by physical and endocrine factors. The outflow or escape type has been illustrated by the function of the kidneys when there is too great a rise in the glycemic percentage. Questions at once appear as to other constant metabolic states, such as the storage and disposal of nitrogen, the laying by of fat, and the constancy of the organic and mineral constituents of the blood. In all these equilibria what agents are at work to limit the variations? What types do they represent? Though there are hints of answers to these questions, reliable information must still be sought.

The second proposition, it will be recalled, postulated corrective activity whenever an equilibrium is disturbed. This service was rendered, as we have seen, by the sympathico-adrenal apparatus in the presence of both hypoglycemia and hypothermia. If this second proposition is generally true, not only must we look for agents which protect a steady state from serious disturbance, but also we must search for the control which makes them act with greater vigor, the more serious the disturbance. Related to this adjusted response is the fifth proposition; that accessory protective functions may be expected to co-operate if the primary functions fail. We have seen this situation in the co-operative activity of the kidney and the pancreas in resisting hyperglycemia; we have seen it also in the vaso-constriction, the increased adrenal secretion and the shivering which constitute progressively the three lines of defense against a drop in the temperature level of the body. We should keep our minds alert, therefore, not only for regulators of steady states, but for the government of the regulators, and for the combination of regulators working towards the same end.

A general inquiry, important for clear understanding of body organization, would be one directed towards determining, so far as possible, the primary or immediate agencies, as contrasted with the secondary agencies, which induce a change of state. For example, we have seen that sympathico-adrenal stimulation is an immediate cause of glycogenolysis. Obviously any other condition inducing glycogenolysis may act, not directly on the liver, but through the sympathetic approach. Until the primary factors are known and properly recognized, we are sure to have confusion—such confusion as is implied in the idea that a corrective agent operates in both directions at the same point.

In my opinion an attempt at present to develop an elaborate and intricate conception of the interrelations of the endocrine glands would be altogether too premature and might be quite misleading. The necessary facts are not at hand. But the idea that endocrine factors play an important rôle in establishing and maintaining physiological equilibria is reasonable and moderate in its scope, and does not involve a far-reaching projection of theory into the unknown. That the idea has value not only in raising questions regarding the existence and nature of balanced states but also in suggesting ways in which endocrine agencies may affect these states has, I hope, been made evident by the instances which I have cited.

Bibliography.

1. Spencer: First Principles, New York, 1870, p. 173.
2. Frédéricq: Arch. d. Zoöl. Expér. et Gén., 1885, iii, 35.
3. Lotka: Elements of Physical Biology, Baltimore, 1925, p. 284.
4. Higgins: Am. Journ. Physiol., 1916, xli, 258.
5. Cannon, McIver and Bliss: Ibid., 1924, lxix, 46.
6. Anrep and Daly: Proc. Roy. Soc., London, 1925, B, xcvii, 454.
7. Cannon and Uridil: Am. Journ. Physiol., 1921, lviii, 353; Cannon and Griffith: Ibid., 1922, lx, 544.
8. Cannon and Rapport: Ibid., 1921. lviii, 308. Cannon and Carrasco-Formiguera: Ibid., 1922, lxi, 215.
9. Stewart and Rogoff: Ibid., 1920, lii, 304, 521.
10. Tournade and Chabrol: Compt. rend. Soc. de Biol., 1925, xcii, 418.
11. Houssay and Molinelli: Rev. d. l. Asoc. Méd. Argentina, 1924, xxxvii, 327.
12. Florovsky: Bull. Acad. Imp. d. Sci., Petrograd, 1917, ix, 119.
13. Baschmakoff: Arch. f. d. ges. Physiol., 1923, cc, 379.
14. Hartman, McCordock and Loder: Am. Journ. Physiol., 1923, lxiv, 1.
15. Anrep: Journ. Physiol., 1912, xlv, 307.
16. Kodama: Tohoku Journ. Exper. Med., 1923, iv, 166.
17. Stewart and Rogoff: Am. Journ. Physiol., 1924, lxix, 605.
18. Fletcher and Campbell: Journ. Metab. Res., 1922, ii, 637.
19. Cf. Cannon and Britton: Am. Journ. Physiol., 1925, lxxii, 283.
20. Olmsted and Logan: Ibid., 1923, lxvi, 437.
21. Cannon and Rapport: Ibid., 1921, lviii, 338.
22. Abe: Arch. f. Exper. Path. u. Pharm., 1924, ciii, 73.
23. Houssay, Molinelli and Lewis: Rev. d. l. Asoc. Méd. Argentina, 1924, xxxvii, 486.
24. Banting, Best, Collip, Macleod and Noble: Trans. Roy. Soc. Canada, 1922, xvi, Sect. v, 39.
25. Cori: Journ. Pharm. Exper. Therap., 1925, xxv, 20.
26. McCormick, Noble and O'Brien: Am. Journ. Physiol., 1924, lxviii, Trans. Am. Physiol. Soc., 144.

27. McCormick and Macleod: Trans. Roy. Soc. Canada, 1923, xvii, 63.
28. Dudley and Marrian: Biochem. Journ., 1923, xvii, 435.
29. Babkin: Brit. Journ. Exper. Path., 1923, iv, 310.
30. Macleod: Physiol. Rev., 1924, iv, 51.
31. Burn: Journ. Physiol., 1923, lvii, 326.
32. Sundberg: Compt. Rend. Soc. de Biol., 1923, lxxxix, 807.
33. Lewis: Compt. Rend Soc. de Biol., 1923, lxxxix, 1118.
34. Banting and Gairns: Am. Journ. Physiol., 1924, lxviii, 24.
35. De Corral: Ztschr. f. Biol., 1918, lxviii, 395.
36. McCormick, Macleod and O'Brien: Trans. Roy. Soc. Canada, 1923, xvii, 57.
37. Clark: Journ. Physiol., 1925, lix, 466.
38. Lewis and Magenta: Rev. d. l. Asoc. Méd. Argentina, 1924, xxxvii, 370.
39. Allen: Journ. Exper. Med., 1920, xxxi, 363, 381.
40. Howans: Journ. Med. Res., 1914, xxv, 63.
41. Zuelzer: Berliner Klin. Wchnschr., 1907, xliv, 475.
42. Stewart: Physiol. Rev., 1924, iv, 185.
43. Bulatao and Cannon: Am. Journ. Physiol., 1925, lii. 295.
44. Ringer and Baumann: Endocrinology and Metabolism, New York, 1922, iii, 260.
45. Bodansky: Proc. Soc. Exp. Biol. Med., 1923, xxi, 46. Ducheneau: Compt. Rend. Soc. d. Biol., 1924, xc, 248.
46. Boothby and Sandiford: Am. Journ. Physiol., 1923, lxvi, 93.
47. McIver and Bright: Ibid., 1924, lxviii, 622.
48. Evans and Ogawa: Journ. Physiol., 1914, xlvii, 446; and Cruickshank: Ibid., 1913, xlvii, 1.
49. Hepburn and Latchford: Am. Journ. Physiol., 1922, lxii, 177.
50. Burn and Dale: Journ. Physiol., 1924, lix, 164.
51. Afanassiew: Arch. f. d. ges. Physiol., 1883, xxx, 385.
52. Seitz: Arch. f. d. ges. Physiol., 1906, cxi, 309.
53. Berg: Biochem, Ztschr., 1914, lxi, cxciv, cxcv, 428; Münchener Med. Wchnschr., 1914, lxi, 434; Arch. f. d. ges. Physiol., 1922, 102 and 543.
54. Stübel: Arch. f. d. ges. Physiol., 1920, clxxxv, 74.
55. Noel: Presse Méd., 1923, xxxi, 158.
56. Tichmeneff: Biochem. Ztschr., 1914, lix, 326.
57. Kerr, Hurwitz and Whipple: Am. Journ. Physiol., 1918, xlvii, 379.
58. Meek: Ibid., 1912, xxx, 161.
59. Cannon and Gray: Ibid., 1914, xxxiv, 232.
60. Cannon and Mendenhall, Ibid., 1914, xxxiv, 245, 251.
61. Grabfield: Ibid., 1916, xlii, 46.
62. Gray and Lunt: Ibid., 1914, xxxiv, 332.
63. Cannon and Uridil: Ibid., 1921, lviii, 353. Cannon and Griffith: Ibid., 1922, lx, 544.
64. Speck: Deutsches Arch. f. Klin. Med., 1883, xxxiii, 388, 410.
65. Loewy: Arch. f. d. ges. Physiol., 1890, xlvi, 196, 230.
66. Johansson. Skand. Arch. f. Physiol., 1897, vii, 123.
67. Sjöström: Ibid., 1913, xxx, 1.
68. Voit: Ztschr. f. Biol., 1878, xiv, 80.

69. Rubner: Die Gesetze des Energieverbrauches bei der Ernährung, Leipzig, 1902.
70. Campbell, Hargood, Ashand and Hill: Journ. Physiol., 1921, lv, 259.
71. Aub, Bright and Forman: Am. Journ. Physiol., 1922, lxi, 349.
72. Cramer: Sixth Scientific Report, Imperial Cancer Research Fund, London, 1919, p. 1.
73. Fujii: Tohoku Journ. Exper. Med., 1921, ii, 9.
74. Trendelenburg: Ergebn. d. Physiol., 1923, xxi, 539.
75. Hartman, McCordock and Loder: Am. Journ. Physiol., 1923, lxiv, 19; Hartman and Hartman: Ibid., 1923, lxv, 612.
76. Stewart and Rogoff: Ibid., 1923, lxvi, 260.
77. Cannon and Querido: Proc. Nat. Acad. Sci., 1924, x, 245.
78. Cannon and Cattell: Am. Journ. Physiol., 1916, xli, 58; Cannon and Smith: Ibid., 1922, lx, 476.
79. Fenger: Endocrinol., 1918, ii, 98; Mills: Am. Journ. Physiol., 1918, xlvi, 329.
80. Loeb: Journ. Med. Res., 1920, xlvii, 77.
81. Aub, Forman and Bright: Am. Journ. Physiol., 1922, lxi, 342.

18

Physiological Regulation of Normal States: Some Tentative Postulates Concerning Biological Homeostatics

W. B. CANNON, M.D.

"The living being is stable. It must be so in order not to be destroyed, dissolved, or disintegrated by the colossal forces, often adverse, which surround it. . . . In a sense it is stable because it is modifiable—the slight instability is the necessary condition for the true stability of the organism." Thus wrote Richet (1) in 1900. This stability is maintained by numerous regulatory agencies which are called into action when the normal state is disturbed. As Bernard expressed it, "All the vital mechanisms, however varied they may be, have only one object, that of preserving constant the conditions of life in the internal environment" (2).

The steady states of the fluid matrix of the body are commonly preserved by physiological reactions, i.e., by more complicated processes than are involved in simple physico-chemical equilibria. Special designations, therefore, are appropriate:—"homeostasis" to designate stability of the organism; "homeostatic conditions," to indicate details of the stability; and "homeostatic reactions," to signify means for maintaining stability.

I suggest the following postulates as pertinent to homeostasis:—

1. In an open system, such as our bodies represent, composed of

unstable structure and subjected continually to disturbance, constancy is in itself evidence that agencies are acting or are ready to act to maintain this constancy. This has not been proved for all homeostatic conditions. But there are known cases which illustrate the postulate, e.g., homeothermia; and the abolition of relatively constant temperature when the homeostatic reactions are abolished, e.g., as in the effect of ether on body temperature. The homeostasis may be maintained by antagonistic agents, e.g., the cardiac nerves; or by overflow, e.g., by the kidney; or by disturbing stimuli, e.g., thirst; or by magnification of a constant process, e.g., excess of CO_2; by structural adjustments, e.g., more erythrocytes at high altitudes; and probably by other types of function.

2. If a homeostatic condition continues, it does so because any tendency towards a change is automatically met by increased effectiveness of a factor or factors which lessen the change. This postulate resembles the principle of Le Chatelier, but differs in being conditional; in biology the organism may not become more, but less, resistant to a disturbing agent, e.g., anaphylaxis. As illustrating the postulate, there is thirst, which becomes progressively more intolerable as deprivation of water continues, and which promptly disappears when water is drunk (3); and the mechanism for releasing sugar from the liver, which works more and more vigorously as hypoglycemia increases, and which stops at once when glucose is injected into a vein (4).

3. A homeostatic agent does not act in opposite directions at the same point. It may seem to do so, e.g., as when insulin is said to lessen the sugar storage in the liver in normal animals (5), and to increase the sugar storage in diabetic animals (6). But in the experiments cited as proving hepatic glycogenolysis, insulin was given until convulsions occurred, i.e., until the opposing homeostatic agent was evoked (7).

4. Homeostatic agents, antagonistic in one region of the body, may be cooperative in another region. For example, a sympathico-adrenal factor and insulin, or possibly a vagal insulin factor (8), are opposed in action on the liver, but they appear to be collaborators in the muscles, i.e., each causes acceleration of metabolism and increased utilization of sugar by active organs (9, 10). In discussing homeostatic states, therefore, the field of action of the regulating factors must be closely defined.

5. The regulating system which determines a homeostatic state may comprise a number of cooperating factors brought into action at the same time or successively. For example, when the body temperature tends to fall, vasoconstriction checks heat loss, increased adrenal secretion accelerates metabolism (11), shivering further increases heat production, and a more abundant growth of hair offers further protection; and when the temperature tends to rise, vasodilation, sweating and faster breathing oppose the change. If the oxygen delivery to the tissues is inadequate, deeper ventilation of the lungs and a faster circulation are the first reactions of the organism, and later the increased production of red corpuscles. Appetite is the first defense against a lowering of the food and when that does not meet the need, hunger (12) and thirst (3) become insistent.

6. When a factor is known which can shift a homeostatic state in one direction, it is reasonable to look for automatic control of that factor or for a factor or factors having an opposing effect.

This last postulate follows from the examples illustrating earlier postulates. It is clear that an examination of homeostatic conditions in the body and the agencies controlling them is of very great biological and medical interest and importance. It is a field which has been too little cultivated. The postulates presented above are not exhaustive and are tentative, but they may be suggestive for further investigation into the physiology of the organism.

Bibliography

1. Richet, Charles. *Dictionnaire de physiologie*. Paris, Baillière et Cie, 10 vols., 1895-1923 [1900, vol. IV, p. 721].
2. Bernard, Claude. *Leçons sur les phénomènes de la vie communs aux animaux et aux végétaux*. Paris, J. B. Baillière et Fils, 1878-79, 2 vols. [vol. I, p. 121].
3. Cannon, W. B. "Croonian Lecture. The physiological basis of thirst." *Proc. roy. Soc. B*, 1917-1919, *90*, 283-301.
4. Cannon, W. B., McIver, M. A., and Bliss, S. W. "Studies on the conditions of activity in endocrine glands. XIII. A sympathetic and adrenal mechanism for mobilizing sugar in hypoglycemia." *Amer. J. Physiol.*, 1924, *69*, 46-66.
5. McCormick, N. A. and Macleod, J. J. R. "The influence of insulin on glycogen formation in normal animals." *Trans. roy. Soc. Can., Sect. V*, 1923, 3d ser., *17*, 63-73.
6. Banting, F. G., Best, C. H., Collip, J. B., Macleod, J. J. R., and Noble,

E. C. "The effect of insulin on the percentage amounts of fat and glycogen in the liver and other organs of diabetic animals." *Trans. roy. Soc. Can., Sect. V*, 1922, 3d ser., *16*, 39-41.

7. Cannon, W. B. "Some general features of endocrine influence on metabolism." *Amer. J. med. Sci.*, 1926, *171*, 1-20.
8. Britton, S. W. "Studies on the conditions of activity in endocrine glands. XVII. The nervous control on insulin secretion." *Amer. J. Physiol.*, 1925, *74*, 291-308.
9. Hepburn, J. and Latchford, J. K. "Effect of insulin (pancreatic extract) on the sugar consumption of the isolated surviving rabbit heart." *Amer. J. Physiol.*, 1922, *62*, 177-184.
10. Burn, J. H. and Dale, H. H. "On the location and nature of the action of insulin." *J. Physiol. (Lond.)*, 1924, *59*, 164-192.
11. Cannon, W. B. and Querido, A. "The rôle of adrenal secretion in the chemical control of body temperature." *Proc. nat. Acad. Sci. (Wash.)*, 1924, *10*, 245-246.
12. Cannon, W. B. and Washburn, A. L. "An explanation of hunger." *Amer. J. Physiol.*, 1912, *29*, 441-454.

(Cannon, Walter B. "Physiological regulation of normal states: some tentative postulates concerning biological homeostatics." In *A Charles Richet: ses amis, ses collègues, ses élèves, 22 Mai 1926*. Auguste Pettit, Ed. Paris, Les Éditions Médicales, 1926. 102 pp. [pp. 91-93].)

19

Organization for Physiological Homeostasis

WALTER B. CANNON

The Laboratories of Physiology in the Harvard Medical School

Biologists have long been impressed by the ability of living beings to maintain their own stability. The idea that disease is cured by natural powers, by a *vis medicatrix naturae*, an idea which was held by Hippocrates, implies the existence of agencies ready to operate correctively when the normal state of the organism is upset. More precise modern references to self-regulatory arrangements are found in the writings of prominent physiologists. Pflüger (1877) recognized the natural adjustments leading toward the maintenance of a steady state of organisms when he laid down the dictum, "The cause of every need of a living being is also the cause of the satisfaction of the need." Similarly Fredericq (1885) declared, "The living being is an agency of such sort that each disturbing influence induces by itself the calling forth of compensatory activity to neutralize or repair the disturbance. The higher in the scale of living beings, the more numerous, the more perfect and the more complicated do these regulatory agencies become. They tend to free the organism completely from the unfavorable influences and changes occurring in the environment." Further, Richet (1900) emphasized the general phenomenon,—"The living being is stable. It must be in order not to be destroyed, dissolved or disintegrated by the colossal forces, often adverse, which surround it. By an apparent contradiction it maintains its stability only if it is excitable and capable of modifying itself according to external stimuli and adjusting its response to the stimulation. In a sense it is stable because it is modifiable—the slight instability is the necessary condition for the true stability of the organism."

To Claude Bernard (1878) belongs the credit of first giving to these general ideas a more precise analysis. He pointed out that in animals with complex organization the living parts exist in the fluids which bathe them, i.e., in the blood and lymph, which constitute the "milieu interne" or "intérieur"—the internal environment, or what we may call the *fluid matrix* of the body. This fluid matrix is made and controlled by the organism itself. And as organisms become more independent, more free from changes in the outer world, they do so by preserving uniform their own inner world in spite of shifts of outer circumstances. "It is the fixity of the 'milieu intérieur' which is the condition of free and independent life," wrote Bernard (1878, i, pp. 113 and 121), "all the vital mechanisms, however varied they may be, have only one object, that of preserving constant the conditions of life in the internal environment." "No more pregnant sentence," in Haldane's (1922) opinion, "was ever framed by a physiologist."

Definition of homeostasis. The general concept suggested in the foregoing quotations may be summarized as follows. The highly developed living being is an open system having many relations to its surroundings—in the respiratory and alimentary tracts and through surface receptors, neuromuscular organs and bony levers. Changes in the surroundings excite reactions in this system, or affect it directly, so that internal disturbances of the system are produced. Such disturbances are normally kept within narrow limits, because automatic adjustments within the system are brought into action, and thereby wide oscillations are prevented and the internal conditions are held fairly constant. The term "equilibrium" might be used to designate these constant conditions. That term, however, has come to have exact meaning as applied to relatively simple physico-chemical states in closed systems where known forces are balanced. In an exhaustive monograph L. J. Henderson (1928) has recently treated the blood from this point of view, i.e., he has defined, in relation to circumstances which affect the blood, the nice arrangements within the blood itself, which operate to keep its respiratory functions stable. Besides these arrangements, however, is the integrated coöperation of a wide range of organs—brain and nerves, heart, lungs, kidneys, spleen—which are promptly brought into action when conditions arise which might alter the blood in its respiratory services. The present discussion is concerned with the physiological rather than the physical arrangements for attaining constancy. The coördinated physiological reactions which maintain most of the steady states in the body are so complex, and are so peculiar to the living organism, that it has been suggested (Cannon, 1926) that a specific designation for these states be employed—*homeostasis.*

Objection might be offered to the use of the term *stasis*, as implying something set and immobile, a stagnation. Stasis means, however, not only that, but also a condition; it is in this sense that the term is employed. *Homeo*, the abbreviated form of *homoio*, is prefixed instead of *homo*, because the former indicates "like" or "similar" and admits some variation, whereas the latter, meaning the "same," indicates a fixed and rigid constancy. As in the branch of mechanics called "statics," the central concept is that of a steady state produced by the action of forces; *homeostatics* might therefore be regarded as preferable to homeostasis. The factors which operate in the body to maintain uniformity are often so peculiarly physiological that any hint of immediate explanation in terms of relatively simple mechanics seems misleading. For these various reasons the term homeostasis was selected. Of course, the adjectival form, *homeostatic*, would apply to the physiological reactions or agencies or to the circumstances which relate to steady states in the organism.

CLASSIFICATION OF HOMEOSTATIC CONDITIONS. According to Bernard (1878, ii, p. 7), the conditions which must be maintained constant in the fluid matrix of the body in order to favor freedom from external limitations are water, oxygen, temperature and nutriment (including salts, fat and sugar).

Naturally during the past fifty years new insight has been acquired and therefore a more ample classification than that just given should be possible. Any classification offered now, however, will probably be found to be incomplete; other materials and environmental states, whose homeostasis is essentially important for optimal activity of the organisms, are likely to be discovered in the future. Moreover, in any classification there will be cross-relations among the homeostatic states; a uniform osmotic pressure in the body fluids, for example, is dependent on constancy within them of the proportions of water, salts and protein. The classification suggested below, therefore, should not be regarded as more than a serviceable grouping of homeostatic categories; it may claim only the merit of having served as a basis for studying the means by which the organism achieves stability:

A. Material supplies for cellular needs.
   1. Material serving for the exhibition of energy, and for growth and repair—glucose, protein, fat.
   2. Water.
   3. Sodium chloride and other inorganic constituents except calcium.
   4. Calcium.
   5. Oxygen.
   6. Internal secretions having general and continuous effects.

B. Environmental factors affecting cellular activity.
  1. Osmotic pressure.
  2. Temperature.
  3. Hydrogen-ion concentration.

Each item in the foregoing list exists in a relatively uniform condition of the fluid matrix in which the living cells of the organism exist. There are variations of these conditions, but normally the variations are within narrow limits. If these limits are exceeded serious consequences may result or there may be losses from the body. A few examples will make clear these relations:

A reduction of the glucose in the blood to about 70 mgm. per cent (e.g., by insulin) induces the "hypoglycemic reaction" (Fletcher and Campbell, 1922), and a reduction below 45 mgm. per cent brings on convulsions and possibly coma and death; an increase of the percentage above 170 to 180 mgm. results in loss via the kidneys. Too much water in the body fluids results in "water intoxication," characterized by headache, nausea, dizziness, asthenia, incoördination (Rowntree, 1922); on the other hand, too little water results in lessened blood volume, greater viscosity, and the appearance of fever (Keith, 1922; Crandall, 1899). Sodium (with the attendant chloride ion) is especially important in maintaining constant the osmotic properties of the plasma; if the percentage concentration rises from 0.3 to 0.6 per cent, water is drawn from the lymph and cells, and fever may result (Freund, 1913; Cushny, 1926, p. 19); on the other hand, if the concentration is reduced, toxic symptoms appear—marked reflex irritability, followed by weakness, shivering, paresis and death (see Grünwald, 1909). The normal level of calcium in the blood is about 10 mgm. per cent; if it falls to half that concentration, twitchings and convulsions are likely to occur (MacCallum and Voegtlin, 1909); if it rises to twice that concentration, profound changes take place in the blood, which may cause death (Collip, 1926). The normal daily variations of body temperature in man range between 36.3°C. and 37.3°C.; though it may fall to 24°C. and not be fatal (Reincke, 1875), that level is much lower than is compatible with activity; and if the temperature persists at 42–43°C., it is dangerous because of the coagulation of certain proteins in nerve cells (Halliburton, 1904). The hydrogen-ion concentration of the blood may vary between approximately pH 6.95 and pH 7.7; at a pH of about 6.95 the blood becomes so acid that coma and death result (Hasselbalch and Lundsgaard, 1912); above pH 7.7 it becomes so alkaline that tetany appears (Grant and Goldman, 1920). The heart rate (of the dog) has been seen to decrease from 75 beats per minute to 50 when the pH fell from 7.4 to 7.0; and to increase from 30 per minute to about 85 when the pH rose from 7.0 to 7.8 (Andrus and Carter, 1924). The foregoing instances illustrate the importance of homeostasis in the body fluids. Ordinarily the shifts away from the mean position do not reach extremes which impair the activities of the organism or endanger its existence. Before those extremes are reached agencies are automatically called into service which act to bring back towards the mean position the disturbed state. The interest now turns to an enquiry into the character of these agencies.

An inductive unfolding of the devices employed in maintaining homeostatic conditions—an examination of each of the conditions with the object of learning how it is kept constant—would require more space than is permitted here. It will be possible, however, to define in broad terms the agencies of homeostasis and to illustrate the operation of some of those agencies by reference to the specific cases. Thus the account may be much abbreviated.

Two general types of homeostatic regulation can be distinguished dependent on whether the steady state involves *supplies* or *processes*.

HOMEOSTASIS BY REGULATING SUPPLIES. The characteristic feature of the homeostasis of supplies is provision for *A*, storage as a means of adjustment between occasional abundance and later privation and need, and for *B*, overflow or discharge from the body when there is intolerable excess. Two types of storage can be distinguished: a temporary flooding of interstices of areolar tissue by the plenteously ingested material, which may be designated *storage by inundation*; and an inclusion of the material in cells or in other relatively fixed and permanent structures—*storage by segregation*. We shall consider illustrations of these two types.

STORAGE BY INUNDATION. The analogy implied in this phrase is that of a bog or swamp into which water soaks when the supply is bountiful and from which the water seeps back into the distributing system when the supply is meager. There appears to be such an arrangement in the loose areolar connective tissue found under the skin and around and between muscles and muscle bundles, and also in other parts of the body. Connective tissue is distinguished from other kinds in being richest in extracellular colloid, in having a close relation to blood vessels—indeed, it serves as a support for the blood vessels—and in exposing an enormous surface area. In such structures chiefly do the agencies rule which hold not only mobile water but also substances dissolved in it, i.e., electrolytes and glucose. Here there are few cells, but instead "a spongy cobweb of delicate filaments," each of which is composed of minute fibrils bound together by a small amount of "cement substance" (Lewis and Bremer, 1927). Within the fine mesh of these collagenous fibres occur mucoid and small amounts of albumin and globulin. In this mesh and bound by it in some manner water and its dissolved substances appear to be held. Probably the proportions of stored water, electrolytes and glucose do not vary beyond a fairly limited range. Because there is evidence, however, that water and electrolytes, at least, may be affected somewhat independently with regard to their retention and elimination, they will be considered separately.

Water. The evidence for water storage is best demonstrated in experiments which withdraw water from its reservoirs and which permit an examination of the amount held in them. After hemorrhage all tissues lose water. By comparing one side of the body with the other in the same animal (the cat) Skelton (1927) found that most of the water which leaves the tissues after bleeding comes from the muscles and the skin—i.e., where loose areolar tissue is most abundant; the amount per 100 grams of tissue, however, is much less from the muscles than from the skin. The observations by Engels (1904) on dogs are in harmony with those of Skelton. Engels found that though 48 per cent of the total body water is in muscles, as might be expected from the great bulk of muscle tissue, about 12 per cent is in the skin, nearly half again as much as is in the fluid blood. And after injecting 0.6–0.9 per cent sodium chloride solution into a vein for an hour, he discovered that 690 grams had been retained and that the muscles and the skin had taken up the solution to almost the same per cent.

That the water stored in the tissues passes out from them as it is needed is shown by the studies of Wettendorf (1901) on the state of the blood during water deprivation. His dogs were, of course, continually losing water through respiratory surfaces and kidneys. Yet one of his animals thirsted for 3 days with no change in the freezing point of the blood, and another for 4 days with a depression of only 0.01°C. Clearly this constancy must be due to the seepage of water from the reservoirs to the blood as fast as it is lost from the body.

Just how the water is brought to the reservoirs, how it is held there, and how it is released as required for preserving the osmotic homeostasis of the blood, is not yet satisfactorily explained. Doubtless a change in the balance between filtration pressure through the capillary walls and osmotic pressure of the proteins, as expounded by Starling (1909), plays an important rôle. And naturally, conditions affecting the capillary wall (e.g., increasing its permeability), raising or lowering intracapillary blood pressure, or altering the concentration of the plasma proteins would affect the water content of the tissues. Diffusion pressure would likewise take part in the complex of active factors. Furthermore, as Adolph (1921) and Baird and Haldane (1922) have shown, the taking of sodium chloride can markedly influence the retention of water in the body. Probably other electrolytes likewise play a rôle. There is evidence also that the H- and OH-ion concentration may be important—a shift towards an alkaline reaction causing imbibition of water by connective tissue and an opposite shift resulting in release (Schade, 1925). That the thyroid gland is a determinative agent is indicated by the great increase of protein in the plasma and of albumin in the tissues in myxedema, and the disappearance of these conditions, together with a large release of water and sodium chloride, when thyroxin is administered (see Thompson, 1926). How these various factors coöperate when water and sodium chloride are needed in the circulation—after hemorrhage, for example—is not clear, and urgently calls for investigation. Krogh (1922) has written concerning the arrangement of water mobilization, "The nature of such a mechanism is entirely unknown and I should not like to venture even a guess regarding it"—and yet it is of primary significance for the organism.

Sodium chloride. There is good evidence that the sodium and chloride ions in the plasma may vary independently, and that of the two the base is much the more constant element (see Gamble and Ross, 1925; Gamble and McIver, 1925, 1928). In a study of steady conditions in the fluid matrix, therefore, the emphasis might properly be laid on the homeostasis of the fixed base. Since most of the facts now available, however, have come from experiments in which the behavior of sodium chloride has been examined, the present treatment of the subject must be concerned with that salt. The evidence for storage of sodium chloride in the body is found in retention under different conditions. With a fairly constant chloride intake abundant sweating and attendant loss of chloride through the skin are accompanied by a great reduction of the chloride output in the urine—a condition which continues although thereafter a diet rich in salt is taken; by this method of study a compensatory retention of 10 to 14 grams of sodium chloride has been observed (Cohnheim, Kreglinger, and Kreglinger, 1909). Further, the taking of concentrated sodium chloride by mouth results in the appearance in the urine of only a part of the amount ingested—most of it is retained in storage in the body; and even if thereupon enough water is drunk to produce a diuresis the urine has a low salt content, i.e., the salt is not given up readily from its storage place (Baird and Haldane, 1922).

When a search is made for the sodium chloride reserve in the body the highest percentage of chloride is found in the skin and the lowest in the muscles—indeed, on a chloride-rich diet one-third of the chloride of the body may be in the skin, and after an intravenous infusion of a sodium-chloride solution the skin may hold the stored chloride to an amount varying in different experiments between 28 and 77 per cent of the amount injected. This evidence is supported by observations on animals fed a chloride-poor diet. Under these circumstances between one and two-tenths of the chloride content of the body is lost, and of this amount between 60 and 90 per cent comes from the skin, though the skin is only 16 per cent of the total body weight (Padtberg, 1910). It is noteworthy that the blood gives up relatively little of its chloride content; again the circulating fluid is kept constant by supplies from tissue storage.

It is well to recognize that the way in which sodium chloride is held in the skin, whether by adsorption on surfaces in areolar tissue or by solution in the interstitial fluid of the areolar spaces, is not known. Probably it is osmotically inactive. That sodium chloride and water are closely related in storage, however, seems to be well established (see Adolph, 1921).

Glucose. The first, temporary depository for excessive blood sugar, as for excessive sodium chloride, is the skin. When sugar or other readily digestible carbohydrate is a large constituent of the diet the glycemic concentration rises commonly from about 100 to 170 mgm. per cent (Hansen, 1923). During this period of high percentage of sugar in the blood there is also a high percentage in the skin (Folin, Trimble and Newman, 1927). This appears to be again an example of storage by inundation. No chemical change occurs in the sugar. No special device is required either to deposit it in the temporary reservoir or to remove it therefrom. As the circulating sugar is utilized or placed in more permanent storage in the liver and in muscle cells, the glycemic level falls. Thereupon the more concentrated glucose, which has overflowed into the spaces of the skin and possibly into other regions where alveolar tissue is abundant, gradually runs back into the blood again and then follows the usual courses of the blood glucose into use or into the fixed reserves.

STORAGE BY SEGREGATION. As previously stated, this mode of storage, commonly within cells, is stable and lasting. It is seen, for example, in carbohydrate reserves as glycogen, in protein reserves as irregular masses in liver cells, in fat reserves as adipose tissue, and in calcium reserves as the trabeculae of the long bones. It differs from storage by inundation in being subject to much more complicated control. Storage by inundation may be regarded as a process of outflow from the blood stream and backflow into it according to the degree of abundance—a relatively simple process. Storage by segregation commonly involves changes of physical state or of molecular configuration and appears to be subject to nervous or neuro-endocrine government. This rather tentative statement is used because of the large gaps in our knowledge, which further consideration will reveal. We shall consider the segregated storage of carbohydrate, protein, fat and calcium.

Carbohydrate. The best example of homeostasis by means of segregation is offered by the arrangements for storage and release of carbohydrate. As is well known, when carbohydrate food is plentiful the glycogen reserves in the liver are large; in prolonged muscular work these reserves may be almost wholly discharged (Kulz, 1880); and yet, while they are being discharged, the blood sugar is maintained at concentrations which neither result in the possibility of sugar loss through the kidneys, nor in the possibility of disturbance from hypoglycemia (Campos, Cannon, Lundin and Walker, 1929). A mechanism must exist, therefore, to release sugar from the liver as it is needed.

An insight into the action of factors which prevent the fall of the blood sugar to a seriously low level may be obtained by a study of the effects of insulin. As stated above, the reduction of the glycemic concentration to about 70 mgm. per 100 cc. by insulin induces the "hypoglycemic reaction," characterized by pallor, rapid pulse, dilated pupils and profuse sweating. These are signs of sympathetic innervation. That this is part of a general display of activity by the sympathetic division of the autonomic system is shown by the involvement of the adrenal medulla. Using the denervated heart as an indicator, Cannon, McIver and Bliss (1923, 1924) found that as the blood sugar fell a critical point was reached at about 70 mgm. per cent, when the heart began to beat faster—a phenomenon which failed to appear if the adrenal glands had been inactivated. If the blood sugar continued to fall the heart rate became faster, thus indicating a greater output of adrenin; and if the blood sugar rose, either because of intravenous injection of glucose or because of a physiological reaction, the heart beat returned to its original slow rate, thus indicating a subsidence of the extra discharge of adrenin. Since medulliadrenal secretion is controlled by splanchnic impulses, and since such impulses in coöperation with secreted adrenin are highly effective in causing an increase of blood sugar (Bulatao and Cannon, 1925; Britton, 1928), it is clear that the reduction of the glycemic percentage below a critical level calls forth an agency—the sympathico-adrenal system—to correct the condition. These observations have been confirmed by Abe (1924) who used the denervated iris to signal a greater output of adrenin, and by Houssay, Lewis and Molinelli (1924) who used for that purpose an adrenal-jugular anastomosis between two dogs. If, in spite of the increasingly active service of this agency as the blood sugar falls, the fall is not checked, convulsions occur (at about 45 mgm. per cent) and each convulsion is associated with a maximal display of sympathico-adrenal activity. If the liver is well supplied with glycogen such activity can restore the blood sugar to the normal level and thus abolish the conditions which brought on the convulsive attacks (McCormick, Macleod, Noble and O'Brien, 1923).

The importance of this agency has been demonstrated by experiments on healthy non-anesthetized animals in which the adrenal glands had been inactivated. The fall of blood sugar after insulin was less retarded at the critical level in cats thus altered, and the convulsive seizures were induced sooner and with smaller doses than in animals with active glands (Cannon, McIver and Bliss, 1924). This increased sensitiveness to insulin after medulliadrenal inactivation has also been proved true of rats (Lewis, 1923), of rabbits (Sundberg, 1923) and of dogs (Lewis and Magenta, 1925; Hallion and Gayet, 1925). The evidence that a small dose of insulin, mildly effective in a normal animal, causes in an animal treated with ergotamine profound hypoglycemia, with convulsions and collapse (Burn, 1923), is in harmony with this testimony, for the drug, though without influence on the action of insulin, paralyzes the protective sympathico-adrenal mechanism. Section of the splanchnic nerves, according to Lewis and Magenta, renders animals more sensitive than does removal of one adrenal and denervation of the other by splanchnic section; under the latter circumstances, it should be noted, one splanchnic is still innervating the liver.

Operation of agencies opposed to those just considered occurs when the blood sugar tends to rise. The efficacy of these agencies is revealed when an excess of glucose is ingested. The blood sugar rises to a level close to that at which it escapes through the kidneys, but normally it does not often surpass that level (Hansen, 1923). The excess sugar, apart from that set aside by inundation, is either stored in the liver or in muscles, or is converted to fat, or is promptly utilized. There is evidence that the process of storage by segregation in hepatic and muscle cells is dependent on secretion of insulin: 1. Removal of the pancreas results in prompt appearance of hyperglycemia and a great reduction of the hepatic glycogen reserves. 2 .The administration of insulin to sugar-fed *depancreatized* dogs reduces the blood sugar to the normal percentage and causes glycogen to accumulate again in large amounts in the liver (Banting, Best, Collip and Noble, 1922). 3. Insulin in small doses causes a deposit of glycogen in the liver of phlorizinized rabbits and cats, even though no sugar is provided (Cori, 1925). 4. Insulin injected into decapitated, eviscerated cats causes a decided increase in the glycogen deposit in the muscles, especially when extra blood sugar is provided (Best, Hoet and Marks, 1926). 5. Islet cells in a remnant of the pancreas degenerate, showing signs of overwork (Allen, 1920; Homans, 1914), when carbohydrate is fed; from this evidence it appears that hyperglycemia stimulates the islet cells to secrete. This stimulation may be direct, as proved by absence of diabetes if a portion of the pancreas is transplanted under the skin and the rest of the gland is then removed and by the appearance of the disease when the engrafted piece is extirpated (Minkowski, 1908), by prompt reduction of hyperglycemia from injected glucose although the vagi have been severed (Banting and Gairns, 1924), or by reduction of diabetic hyperglycemia when a pancreas is connected with blood vessels in the neck (Gayet and Guillaumie, 1928). There is evidence also of a nerve control of insulin secretion, for stimulation of the right vagus reduces blood sugar, but not if the vessels of the pancreas are tied (Britton, 1925); and, according to Zunz and La Barre (1927), after union of the pancreatic vein of dog A to the jugular of dog B injection of glucose into A causes the blood sugar to fall in B--an effect that does not occur if the vagi of A have been cut or atropine given. That this nervous control is not necessary does not prove that it is useless when present—for example, the heart after denervation will continue beating and will maintain the circulation! It may be that the vagus provides a fine adjustment for insulin secretion as the sympathetic does for secretion of adrenin.

The general scheme which has been presented above is represented diagrammatically in figure 1. As Hansen (1923) has pointed out, there are normal oscillations in blood sugar occurring within a relatively narrow range. Possibly these ups and downs result from action of the opposing factors, depressing or elevating the glycemic level. If known elevating agencies (normally and primarily the sympathico-adrenal apparatus) are unable to bring forth sugar from storage in the liver, the glycemic level falls from about 70 to about 45 mgm. per cent, whereupon serious symptoms (convulsions and coma) may supervene. The range between 70 and 45 mgm. per cent may be regarded as the *margin of safety*. On the other hand, if the depressing agency (the insular or vago-

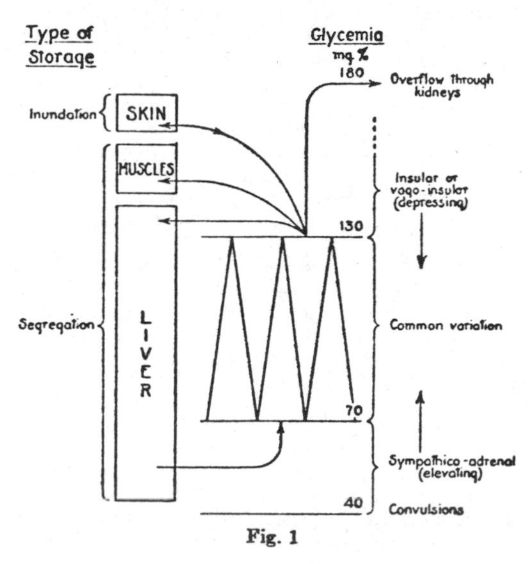

Fig. 1

insular apparatus) is ineffective, the glycemic level rises to about 180 mgm. per cent and then sugar begins to be lost through the kidneys. The range from 100 or 120 to 180 mgm. per cent may be regarded as the *margin of economy*—beyond that, homeostasis is dependent on wasting the energy contained in the sugar and the energy possibly employed by the body to bring it as glucose into the blood.[1]

[1] Evidence opposed to the foregoing views has been brought forward recently by Cori and Cori (1928). They state that "the most prominent effect of epinephrin is observed in the peripheral tissues and consists in a mobilization of muscle glycogen and in a decreased utilization of blood sugar"; and that insulin causes a rapid disappearance of the hepatic stores, due to increased use of blood

Protein. The homeostasis of protein is perhaps widely manifested in the constancy of body structure. That would include the blood, however, and since we are concerned with the conditions which keep uniform the fluid matrix of the body, we shall pay particular attention only to that.

The importance of constancy of the plasma proteins need not be

---

sugar in peripheral tissues and to "compensatory mobilization of liver glycogen." These declarations, so contrary to evidence long accepted, call for comment. First, they gave doses of adrenalin (0.2 mgm. per k.) and of insulin (7.5 units per k.) far beyond physiological limits (equivalent to 14 cc. of adrenin and 525 units of insulin in a man of 70 k.!). Pronounced physiological effects have been obtained in white rats (which they used) with a dose of adrenin *one-twentieth* of their dose. Although they argue that their adrenin doses were slowly and fairly evenly absorbed, they present no actual evidence; and the fact that the highest blood sugar in their experiments came early and was associated with the lowest glycosuria indicates both that their argument is ill-based and that the huge doses disturbed the circulation. Further, "mobilization of muscle glycogen" consists, they explain, in a change of the glycogen to lactic acid, and from this circulating lactic acid a reconstruction of glycogen by the liver. But the glycogen in muscle is there for use; to "mobilize" it without use is like withdrawing forces from the firing line and settling them in barracks! Again, in declaring that adrenin causes hyperglycemia "because the utilization of blood sugar is diminished" they neglect the evidence: 1, that intravenous injection of adrenin raises blood sugar with almost no latent period (Tatum, 1921); 2, that emotional excitement can raise blood sugar 30 per cent or more in a few minutes, but not after adrenalectomy (Britton, 1928), and that the same phenomena are seen when the splanchnics are stimulated (Macleod, 1913), and 3, that after an adrenin injection the blood sugar rises quickly in the liver veins and only later is equaled in the portal or femoral vein (Vosburgh and Richards, 1903)—all evidence against their views, because the hyperglycemia comes too soon and is too clearly of hepatic origin to be ascribed to failure of use of glucose by peripheral tissues. Moreover, their belief that adrenin causes "decreased utilization of blood sugar" is contradicted by the observation that when glucose and adrenin are supplied to the heart-lung preparation sugar consumption rises to about four times the former amount (Patterson and Starling, 1913), and that dogs exhausted by running can be made to continue (i.e., using sugar in their muscles) and will put forth from 17 to 44 per cent additional energy if they are given subcutaneously small doses of adrenin (0.02–0.04 mgm. per k., sometimes repeated) but not if a large dose (0.17 mgm. per k.—n.b., *less* than that used by the Coris) is given (Campos, Cannon, Lundin and Walker, 1929). Finally, although they mention a "compensatory mobilization of liver glycogen" as the cause of depleted hepatic stores after their enormous doses of insulin, they do not hint at the nature of the compensatory process, though they report low blood-sugar levels which would set in action the sympathico-adrenal apparatus. For these various reasons the views advanced by Cori and Cori seem not to warrant a surrender of the well established conceptions of the action of adrenin and insulin.

emphasized. Because they exert osmotic pressure and do not ordinarily escape through capillary walls, they prevent the salts dissolved in the blood from passing freely into perivascular spaces and out from the body through the renal glomeruli (Starling, 1909). When Barcroft and Straub (1910) removed much of the blood from a rabbit, separated the corpuscles, and reinjected them suspended in an equal volume of Ringer's solution instead of plasma, so that the difference was merely a reduction of the colloid, urine secretion was increased *forty times.* The development of "shock"—engorgement of liver, spleen, kidneys, intestinal mucosa and lungs, with accumulation of fluid in the intestine—noted by Whipple, Smith and Belt (1920) when the plasma proteins were reduced to 1 per cent, tells the same story. But not only does homeostasis of the plasma proteins provide for homeostasis of the blood volume; at least one of them (fibrinogen) is, in case of hemorrhage, essential for preservation of the blood itself. The very existence of the fluid matrix of the body is dependent, therefore, on constancy of the proteins in the plasma—and usually they are remarkably constant in various conditions of health and disease.

Blood is the one tissue in the body from which protein can be quantitatively removed and its restoration then studied. When by plasmapharesis the plasma proteins are reduced from about 6 to about 2 per cent there is a prompt rise in their concentration within fifteen minutes, a more gradual restoration thereafter during the first twenty-four hours to about 40 per cent, and full recovery in two to seven days. It may be that the prompt rise is relative, due to escape of salt solution from the blood vessels, although arguments have been advanced that it results from emergency discharge of the proteins from storage (Kerr, Hurwitz and Whipple, 1918). The slower recovery seems certainly dependent on the liver, for the following reasons: 1. If the liver has been injured by phosphorus or chloroform, restoration of the plasma proteins is delayed. 2. Dogs with an Eck fistula may have no restoration for the first three days after plasmapharesis. And 3, fibrinogen, which usually is completely restored within twenty-four hours, is not thus restored if the liver is unable to act (Foster and Whipple, 1922, Meek, 1912).

The evidence that the liver is important for homeostasis of proteins in the blood plasma raises the question whether protein is stored there. Results obtained by histological and biochemical methods have agreed in supporting the conclusion that hepatic cells can carry reserve protein as well as reserve carbohydrate. The early observations of Afanassiew

(1883) that the liver cells of dogs given an abundance of "albuminates" increase in size and contain protein granules between the structural strands, have been confirmed by Berg (1914, 1922), by Cahn-Bronner (1914), by Stübel (1920) and by Noël (1923). In sum, these recent histological studies show that when animals are fed protein there appear in the hepatic cells fine droplets or masses, which react to Millon's reagent, which yield the ninhydrin reaction of a simple protein, which disappear on fasting, and which reappear on feeding protein or amino acids. The biochemical analyses by Seitz (1906), who found that in fed animals the total nitrogen of the liver in relation to that in the rest of the body is from two to three times as great as in fasting animals, have been supported by the results obtained by Tichmeneff (1914). He starved mice for two days, then killed half of them and after giving the others an abundance of cooked meat killed them and compared the livers of the two groups. Expressed in percentage of body weight, the livers of the meat-fed animals increased about 20 per cent, with the hepatic nitrogen content augmented between 53 and 78 per cent.

Although the experiments on homeostasis of plasma proteins indicate that the liver is an important source of these materials in case of need, and although the testimony cited above would justify consideration of the liver as a storage place for protein, the modes of storage and release are almost wholly unknown. Stübel (1920) did, indeed, observe that the small protein droplets or masses in the hepatic cells could be greatly reduced by injecting adrenin subcutaneously. If these masses help to supply essential protein elements for blood clotting, as the dependence of fibrinogen on the liver would imply, their liberation by adrenin and by conditions which would excite the sympathico-adrenal apparatus might account for certain phenomena of faster clotting. Coagulation is more rapid after adrenin injections (see Cannon and Gray, 1914; La Barre, 1925; Hirayama, 1925), after splanchnic stimulation (Cannon and Mendenhall, 1914), or after large hemorrhage which calls the sympathetic into action (Gray and Lunt, 1914), but only if the blood is allowed to flow through the liver and intestines. In this category also is the very rapid clotting of blood taken at the height of the hypoglycemic reaction (Macleod, 1924), when sympathico-adrenal activity is maximal.

It is quite possible that protein is stored in other places than the liver, and also that the thyroid gland is an important agency for controlling both storage and release. Boothby, Sandiford and Slosse (1925) have reported that with a uniform nitrogen intake a negative nitrogen balance exists while thyroxin is establishing a new higher metabolic level. After its establishment there is a smaller deposit of nitrogen in the body. Now if thyroid dosage is stopped (while the uniform nitrogen intake continues), a positive balance obtains until a new lower metabolic level is reached, i.e., more nitrogen is deposited. These effects are much more marked in a person afflicted with myxedema than in a normal person. Indeed, as Boothby has suggested, the "edema" of myxedema may be an abnormal amount of deposit protein in and beneath the skin. The efficacy of thyroid therapy, previously noted, in reducing the increased proteins of the plasma and the albumin of the tissues, in cases of myxedema, supports the view that the thyroid gland is somehow associated with protein regulation and metabolism.

Although the foregoing review has brought out the primary importance of homeostasis of the proteins of the plasma for maintaining the volume and character of both the intravascular and extravascular fluid matrix of the organism, and for protecting the organism against loss of the essential part of the matrix—the blood—, it has revealed also how much still needs to be learned. Here, as with other useful material, constancy is attained by storage, which stands between plenty and need, and in this respect the liver plays an important rôle. The sympathico-adrenal apparatus seems to influence release from storage, and also varying activity of the thyroid gland may be determinative. Are special agencies required to manage the laying by of the reserves? We do not know.

Fat. According to Bloor (1922) the concentration of fat, cholesterol and lecithin in the blood is fairly constant in the same species of animals, but may differ greatly in different species. As is well known, ingestion of fat produces an "alimentary lipemia" which may cause the fat content of the blood to rise as high as 3 per cent in the dog and 2 per cent in man. A relatively large increase in the fat content of the blood appears to be without serious consequence. In pathological states —e.g., in diabetes—the lipemic percentage may rise to 10, 15 and even to 20 per cent without producing obvious symptoms. On the other hand the normal blood fat is remarkably persistent. Carbohydrate and protein alone may be fed for considerable periods without reduction of the lipemic level, and fasting for short periods may actually be accompanied by a rise (Schulz, 1896), although after two weeks of fasting the level may undergo a slow fall. Whether total absence of fat from the blood could be produced, and if so, whether that condition would be attended by disturbances, are questions yet to be answered.

The constancy of the lipemic level for many days in spite of relative or complete starvation implies that there is a governing agency which brings the fat from storage into the blood stream. As Lusk (1928, p. 107) has remarked, "The length of life under the condition of starvation generally depends upon the quantity of fat present in the organism at the start." Fat is stored in the liver, if carbohydrate is not fed (Rosenfeld, 1903); it is also stored under the skin, beneath serous coats (e.g., around the kidneys), in the omentum, and between and in the muscle fibres. What leads to fat storage in some individuals to a greater extent than in others is unknown. In hypothyroidism there may be a generally diffused obesity, an obesity which rapidly disappears under thyroid therapy. A slight scratch in the surface of the brain stem between the infundibular process and a mammillary body produces adiposity (Bailey and Bremer, 1921), as does a tumor or other lesion of this region. Grafe (1927) cites instances of unilateral hypertrophy or atrophy of fatty tissues and suggests that the disposal of fat is under a sympathetic control managed from the hypothalamic region. It is pertinent to note that kittens allowed to live after unilateral sympathectomy until they have doubled their weight, have no demonstrable differences in the amount or distribution of the fat on the two sides of the body (Cannon, Newton, Bright, Menkin and Moore, 1929).

If the regulation of fat storage is obscure, the regulation of its release is even more so. When fat is needed for maintaining the energies of the body it is removed from adipose tissue until the fat cells are practically empty. Yet, even when death from starvation occurs, the fat content of other tissues may not be very different from normal (Terroine, 1914). What causes the fat to move from the adipose stores is not known. Lusk's remark, that "the fasting organs attract fat from the fat deposits of the body, and it is brought to them in the circulating blood," was probably not intended to be explanatory, and it is not. Possibly the reversible reaction mediated by lipase, as described by Kastle and Loevenhart (see Loevenhart, 1902), may be an important factor in maintaining homeostasis of the lipemic level—the enzyme favoring storage when the level is raised and favoring release when the level falls. But on these points more information is desirable.

Calcium. The special and diverse uses of calcium—for the growth of the skeleton and teeth, for the repair of broken bone, for the maintenance of proper conditions of irritability of nervous and muscular tissues, for the coagulation of blood, and for the production of serviceable milk—render it a highly important element in the bodily economy. Like sugar and protein and fat, calcium may be in great demand on exceptional occasions. Under such circumstances, however, the amount in the blood must not be much reduced, for serious consequences ensue. As previously noted, there is normally a homeostasis of calcium in the blood at approximately 10 mgm. per cent. If the blood calcium is lowered to less than 7 mgm. per cent, as may be done by removal of the parathyroid glands (without change in the percentage of sodium and potassium), or by injection of sodium citrate, twitchings and tetanic convulsions occur, with a severity measured by the degree of deficit of available calcium; and these symptoms are quickly relieved by injecting a soluble calcium salt sufficient to restore the normal percentage (MacCallum and Voegtlin, 1909; MacCallum and Vogel, 1913; Trendelenburg and Goebel, 1921). On the other hand, if the blood calcium is raised above approximately 20 mgm. per cent, by injection of parathyroid extract, profound changes are produced in the blood—the viscosity is greatly increased, the osmotic pressure rises, the blood phosphates are doubled, and there are four times the normal amount of non-protein and urea nitrogen—conditions associated with vomiting, coma, and a failing circulation (Collip, 1926). Obviously homeostasis of blood calcium is of capital importance.

As in the homeostasis of other materials, that of calcium is made possible by storage, built up in times of abundance and utilized in time of need. The recent studies of Bauer, Aub and Albright (1929) have demonstrated that the trabeculae of the long bones are easily made to disappear by a persistent diet deficient in calcium and by growth, and that they are readily restored by feeding a calcium-rich diet. The trabeculae serve, therefore, as a storehouse of conveniently available calcium.

How the homeostasis of calcium is regulated has not been determined. The following evidence associates the parathyroid glands with the regulation: 1. Partial or complete removal of these glands results in a lowering of the calcium content of the blood, as mentioned above, and in a defective deposit of dentine in growing teeth and in a defective development of the callus about a bone fracture (Erdheim, 1911). 2. A diet poor in calcium induces parathyroid hyperplasia (Marine, 1913; Luce, 1923); pregnancy and lactation do likewise, without, however, reduction of the calcium percentage in the blood. 3. Diseases characterized by defects in calcification of bone—e.g., rickets and osteomalacia—are attended by hypertrophy of the parathyroid glands (see Strada, 1909; Weichselbaum, 1914). And 4, implantation of parathyroids in a parathyroidectomized rat restores the power to deposit dentine having a normal calcium content (Erdheim, 1911). But *how* the parathyroids control calcium homeostasis—whether they act directly or are stimulated by nerves, whether they act alone and by increased or decreased activity effect storage or release, or whether they coöperate with other agencies perhaps antagonistic—all this still needs investigation.

The pharmacodynamic action of thyroxin seems to involve the thyroid as well as the parathyroid glands in calcium metabolism (Aub, Bauer, Heath and Ropes, 1929). Administration of thyroxin greatly increases the calcium losses in urine and feces whether persons are normal or myxedematous. There is evidence also that in hyperthyroidism the bones show osteoporosis and that calcium excretion is much augmented. Possibly the parathyroids serve for deposit and the thyroids for release of calcium—thyroxin raises the blood calcium in a low-calcium tetany. Such hints, however, must be regarded with suspicion until put to test, for, as shown in the experiments on the action of adrenin and insulin, mentioned above, powerful pharmacodynamic agents given in doses exceeding the physiological range can produce complicated and indirect effects.

The homeostatic functions of hunger and thirst. In the foregoing discussions storage has been emphasized as a regulatory mediation between supply and demand. Back of storage, however, and assuring provisions which can be stored, are powerful motivating agencies—appetites and hunger and thirst. Because of pleasurable previous experiences with food and drink appetites invite to renewal of these experiences; thereby material for the reserves is taken in. If the reserves are not thus provided for, hunger and thirst appear as imperious stimuli. Hunger is characterized by highly disagreeable pangs which result from strong contractions of the empty stomach—pangs which disappear when food is taken (Cannon and Washburn, 1912; Carlson, 1916). Thirst is an uncomfortable sensation of dryness and stickiness in the mouth, which can be explained as due to failure of the salivary glands (which need water to make saliva) to keep the mouth moist; when water is swallowed and absorbed they, as well as the rest of the body, are provided with it and since they can consequently moisten the mouth, the thirst disappears (Cannon, 1918). By these automatic mechanisms the necessary materials for storage of food and water are assured.

Overflow. Previously, in relation to the homeostasis of blood sugar, the use of overflow as a means of checking an upward variation of constituents of the blood has been mentioned. Not only excessive sugar, but excessive water, excessive sodium and potassium and chloride ions are discharged by the kidneys. In accordance with the modern theory of urine formation (Cushny, 1926), these are all "threshold substances." They are resorbed by the kidney tubules only in such relations to one another as to preserve the normal status in the blood. All in excess of that is allowed to escape from the body.

It is interesting to note that these substances are primarily stored by flooding or inundation. When these reserve supplies are adequate, however, the ability of the overflow factor to maintain homeostasis is marvelous. The feat reported by Haldane and Priestley (1915) of drinking 5.5 liters of water in six hours—an amount exceeding by one-third the estimated volume of the blood—which was passed through the kidneys with such nicety that at no time was there appreciable reduction of the hemoglobin percentage, is a revelation not only of the efficacy of the kidney as a spillway but also of the provision in the body for maintaining a constancy of its fluid matrix.

The lungs as well as the kidneys serve for overflow. As is well known, a slight excess of carbonic acid in the arterial blood is followed by greatly increased pulmonary ventilation. Thus the extra carbon dioxide is so promptly and effectively eliminated that the alveolar air is kept nearly constant (Haldane, 1922). By this means provision is made for extra carbon dioxide to flow out from the blood over a dam set at a fixed level. In consequence, in usual circumstances, the hydrogen-ion concentration of the blood is fairly evenly maintained, and the harmful effects of an excessive shift in the alkaline or acid direction is avoided.

HOMEOSTASIS BY REGULATING PROCESSES. There are steady states in the body which do, indeed, involve the utilization of materials, but which are so much more notably dependent on altering the rate of a continuous process that they can reasonably be placed in a separate category. We shall consider two of them, the maintenance of neutrality, and the maintenance of a uniform temperature (in homeothermic animals). The physiological adjustments involved in these processes are so commonly known that a mere outline of them, without detailed description or many references, will be sufficient to illustrate the mode of regulation.

Maintenance of neutrality. The importance of confining the changes in the hydrogen-ion concentration of the blood to a narrow range has already been emphasized. This concentration is determined by the ratio, H_2CO_3:$NaHCO_3$, in the blood. On going to a high altitude the tension of carbonic acid is lessened, the ratio is lowered and the pH rises.

Under these circumstances the blood alkali also is lessened until the pH is restored. And on returning to sea level the opposite process occurs and continues until there is a normal adjustment again, due, according to Y. Henderson (1925), to "calling an increased amount of alkali into the blood," a result probably attained, however, by the passage of acid elements from the blood into the tissues or the urine.

Back of these adjustments between the blood and the tissues, however, are the preventive measures which protect the blood from danger by anticipatory action. Acid metabolites are continuously being produced in the living cells and if allowed to accumulate in them these substances interfere with or prevent further action. Elaborate arrangements are ready in the organism to forestall that contingency. To be sure, the facilities for controlling non-volatile acid are limited. But it can be dealt with in a variety of ways. The lactic acid, for example, which is developed in muscular contraction, is in part promptly neutralized—the phosphocreatine recently discovered by Fiske and Subbarow (1929) appears to be capable of functioning in an extraordinarily effective manner in neutralizing lactic acid within the muscle cells. Another part of the acid is soon oxidized; and the rest is rebuilt into neutral glycogen. For continued effectiveness of all three of these methods of disposal there must be provided an adequate supply of oxygen. Although muscles, and probably other tissues as well, go into "oxygen debt" by acting in spite of accumulating lactic acid, that state is characterized by a diminished capacity to do work, great according to the debt, by a prescription on the amount of debt allowable, and by the definite requirement of ultimate payment. When non-volatile acid is burned to volatile carbonic acid, however, it is in a form which can be carried away and disposed of to an almost unlimited amount, with only slight change in the reaction of the blood. During vigorous muscular work as much oxygen as possible must be delivered. There is practically no storage of oxygen in the body. Air-breathing animals are surrounded by an ocean of oxygen—the problem is solely that of conveyance from the boundless external supply to the exigent tissues. For that purpose circulatory and respiratory processes must be greatly accelerated. Fortunately these adjustments, required to get rid of the volatile acid, are precisely those required to bring to the tissues the oxygen which serves to make the acid volatile and readily discharged.

In vigorous muscular effort the pulmonary ventilation may be increased from 6 liters per minute to 60 or 80 liters or more, due to the effects of acid in the respiratory center. Under these circumstances the sympathico-adrenal system is active (Cannon and Britton, 1927) and it is altogether probable that thereby the bronchioles are dilated at a time when wider passageways would facilitate the to-and-fro movement of larger volumes of air. There is an ampler return of blood to the heart per minute, because of contraction of the splanchnic vessels, because of the pressures excited by the active muscles on capillaries and veins within them, and because of the pumping action of the diaphragm. Thus the heart receives a greater charge of blood and puts forth a greater amount per beat. And because of lessened vagal tone, increased tension on the venous side, and participation of the sympathetic accelerators, the heart, well charged, may beat twice as fast as it does at rest. With a much larger minute output from the heart and a constricted splanchnic area the arterial pressure is markedly raised. In the active muscles the arterioles are dilated and the closed capillaries are opened; and through these more numerous channels the high head of arterial pressure drives an abundant blood stream. Evidence indicates that the total circulation rate may be augmented as much as four times. But not only are the corpuscles utilized more effectively by being made to move faster, the *number* of corpuscles is increased by discharge from storage in the spleen (see Barcroft, 1926)—an effect which, like splanchnic constriction and cardiac acceleration, the sympathetic system helps to produce (see Izquierdo and Cannon, 1928). In the laboring muscles, where acid is being produced and where especially oxygen is needed, the excess carbon dioxide itself facilitates the unloading of oxygen from the corpuscles and also its own carriage away to the lungs. In these ways the local flow may be increased as much as 9 times and the oxygen delivery may be increased as much as 18 times what it is during rest (Bainbridge, 1923). Thus, in spite of the fact that in a short time more lactic acid by far can be produced by muscular work than could be neutralized by the buffers in the blood— condition which must inevitably cause death—the reaction of the blood is altered to only a minor degree.

No more admirable example of homeostasis can be mentioned than that of the pH of the fluid matrix of the body. It is managed by accelerating and retarding the continuous processes of pulmonary ventilation and the flow of blood. The physico-chemical changes within the blood itself, which will not be considered in this article, greatly diminish the effects of slight variations in these physiological processes. The degree of respiration is largely influenced by the hydrogen-ion concentration in the cells of the respiratory center, but they in turn are influenced by an increase of the concentration of carbonic acid in the blood. Again the disturbance brings its own cure, and as the concentration is lowered by the heavier breathing, the heavier breathing ceases. The adjustment of the circulation may be similarly managed. The faster heart, the vascular constriction (except in active areas), and the contracted spleen point to functioning of the sympathetic system. Even slight voluntary activity calls the system into service (Cannon and Britton, 1927), and asphyxia is a highly effective stimulus for it (Cannon and Carrasco-Formiguera, 1922). The centers for sympathetic control may be influenced like the respiratory center—acidity may develop in them as a consequence of oxygen-want or carbonic acid excess and the primary result may be stimulation. This suggestion is supported by the experiments of Mathison (1911), showing that asphyxia and also extra carbon dioxide in the respired air raise arterial blood pressure, and by the observation of Cannon, Linton and Linton (1924) that muscle metabolites bring into action the sympathico-adrenal system. In vigorous muscular work the remarkably close correlation between the adjustment of the respiratory and the circulatory apparatus to the needs of the organism might thus be explained,—though both systems are started into faster service by impulses incidental to a voluntary act, they might be maintained in the performance of their extra task by the increased hydrogen-ion concentration in the blood and later they would gradually return to their quiet routine functions because their extra activity had resulted in reducing the hydrogen-ion concentration to the resting level.

Maintenance of uniform temperature. The importance of uniform temperature in providing conditions favorable for a constant rate of the chemical changes in the body requires no emphasis. And the danger of a rise of temperature a relatively few degrees above the normal, as well as the depressant effect of a fall much below the normal, likewise is well recognized. For general considerations, to be discussed presently, it is pertinent, however, to mention briefly the changes which take place when the body temperature tends to rise or fall. If the change is in the direction of a rise, relaxation of peripheral vessels occurs, thus exposing warm blood to the surface where heat may escape to colder surroundings; or when that is ineffective, sweating takes place, the skin is cooled by evaporation, and the abundant blood flowing through the skin loses heat thereby. Polypnea plays a part similar to sweating, and is especially serviceable in animals not well provided with sweat glands. If, on the other hand, the change is in the direction of a fall, there is a constriction of peripheral vessels and an erection of hairs and feathers which enmesh near the skin a layer of poorly conducting air; when these means of conserving heat do not check the fall of temperature, adrenin, capable of increasing heat production, is set free in the blood stream (Cannon, Querido, Britton and Bright, 1927); and when the heat thus produced does not suffice, shivering is resorted to as the final automatic protection against a temperature drop. This highly efficient arrangement for maintaining homeostasis of body temperature involves only an acceleration or retardation of the processes of heat production and heat loss which are constantly going on. The delicate thermostat which operates the regulation appears to be located in the subthalamus (Isenschmid, 1926; Rogers, 1920), and to be influenced directly by changes in the temperature of the blood (Kahn, 1904; O'Connor, 1919; Sherrington, 1924), and also reflexly (see Hill, 1921). The noteworthy features of the total arrangement, apart from its efficiency, are the varieties of the devices for homeostasis, their appearance in a sequence of defences against change, and the close involvement of the sympathetic system in the conservation, production and dissipation of heat.

THE RÔLE OF THE AUTONOMIC NERVOUS SYSTEM IN HOMEOSTASIS. The homeostatic regulators act automatically. Although skeletal muscles and the diaphragm are, of course, under control of the cerebral cortex, their functions in the regulation of temperature (shivering) and neutrality (faster breathing) are managed low in the brain stem. And for the most part the regulators are not under voluntary government. Commonly the autonomic system, or that system in coöperation with endocrine organs, is called into action. Illustrations of these facts are seen in the vago-insular and the sympathico-adrenal influences on the glycemic level, the vagal and sympathetic effects on the heart rate and the sympathetic effects on blood vessels during vigorous muscular effort, and the sympathico-adrenal function in accelerating heat production when the body temperature tends to fall.

The facts just mentioned emphasize a distinction long recognized between the "voluntary" and the "involuntary" or "vegetative" functions of the nervous system. It is desirable to remove from physiology terms having psychological and botanical implications. The two relations of the nervous system—towards the external and towards the internal environment—naturally suggest that distinctions should be based on these opposite functions. The "voluntary" or cerebrospinal system, elaborately outfitted with exteroceptors and with muscles which operate bony levers, is arranged for altering the external environment or the position of the organism in that environment by laboring, running or fighting. These may appropriately be regarded as *exterofective* activities, and the "voluntary" system therefore as the exterofective system. The exterofective activities, however, must produce coincident changes in the internal environment—e.g., utilizing blood sugar and discharging into the blood acid waste and extra heat. Under these circumstances the "involuntary" nervous system plays its part by acting on the heart, smooth muscles and glands in such ways as will preserve the "fitness" of the internal environment for continued exterofective action. This interofective function of the "involuntary" nervous system justifies calling it the *interofective* system. Inactivity of the exterofective system establishes a basal state for the organism, because minimal functioning of that system is accompanied by minimal functioning also of the interofective system. Exterofective action is reflected in interofective action, which rises as the internal disturbance rises and subsides as the disturbance subsides.

The interofective system has been referred to thus far as if it were single instead of consisting of three divisions, with distinctive general functions. These functions were summarized by Cannon (1914) as follows—the sacral, a group of reflexes for emptying hollow organs which become filled; the cranial, a series of reflexes protective and conservative and upbuilding in their function; and the mid- or sympathetic division, a mobilizer of bodily forces for struggle. Similar views have been expressed by Hess (1926) who has recognized the "histotropic" function of the cranial division in promoting the welfare of the tissues, and the "ergotropic" function of the sympathetic division in operating to increase the facilities for doing work. The ideas developed in the foregoing discussion modify only slightly the views expressed in 1914. The emphasis is somewhat differently placed, to be sure, if the maintenance of conditions favorable for exterofective activity is regarded as the chief function of the autonomic system. Thus, although the sacral autonomic division has as one of its functions the perpetuation of the race, it is also serviceable in emptying the bladder and rectum of loads which might interfere with extreme physical effort—an interpretation which is consistent with the well-known effects of strong emotion (e.g., fear) in voiding these viscera as a preparation for struggle. Likewise, though the cranial division has other conservative functions it notably exhibits its conservative uses in providing and preserving the measures which are required to keep the fluid matrix constant when profound disturbance might occur: it gives the heart opportunity for rest and recuperation by checking its rate in quiet times, it promotes the gastro-intestinal movements and excites secretion of the digestive juices and thereby assures the reserves of energy-yielding material, and it appears to be further useful through the vago-insular system in bringing into storage some of these reserves. These divisions, the sacral and the cranial, however, operate indirectly and somewhat remotely to protect homeostasis. It is the mid- or sympathetic division which acts directly and promptly to prevent changes in the internal environment; by mobilizing reserves and by altering the rate of continuous processes; as repeatedly noted in the foregoing pages, this division works to keep constant the fluid matrix of the body and therefore may properly be regarded as the special and immediate agent of homeostasis. The idea that the sympathetic division—or the sympathico-adrenal apparatus, for the nerve impulses and adrenin coöperate—serves to assure homeostasis, that for this purpose it functions reciprocally when the exteroceptive system functions, is not a fundamental modification of the "emergency theory" (Cannon, 1914a). Recent studies have shown that if emergencies do not arise, if marked changes in the outer world or vigorous reactions to it do not occur, the sympathico-adrenal apparatus is not a necessity and can be wholly removed without consequent disorder (Cannon, Newton, Bright, Menkin and Moore, 1929). Limitations appear when circumstances alter the internal environment; it is then that the importance of the sympathico-adrenal apparatus becomes evident. As has been shown above, however, this apparatus plays its rôle not only in preserving homeostasis during grave crises which demand supreme effort, but also in the minor exterofective adjustments which might change the fluid matrix of the body.

Some postulates regarding homeostatic regulation. About four years ago Cannon (1925) advanced six tentative propositions concerned with physiological factors which maintain steady states in the body. It will be pertinent to consider them again now with reference to the foregoing discussion of homeostasis.

1. "In an open system such as our bodies represent, compounded of unstable material and subjected continually to disturbing conditions, constancy is in itself evidence that agencies are acting, or ready to act, to maintain this constancy." This is a confident inference—an inference based on some insight into the ways by which certain steady states (e.g., glycemia, body temperature, and neutrality of the blood) are regulated and a confidence that other steady states are similarly regulated. The instances cited in the previous pages have illustrated various agencies employed in the organism to that end. Although we do

not know how constancy of plasma proteins, lipemia and blood calcium, for example, is brought about, probably it results from as nice devices as those operating in the better known cases of homeostasis. Of course this realm of interest is full of problems—highly significant problems—inviting attempts at solution, and as they are solved the confidence expressed in the first postulate may be justified.

2. "If a state remains steady it does so because any tendency towards change is automatically met by increased effectiveness of the factor or factors which resist the change." Thirst, the hypoglycemic reaction, the respiratory and circulatory response to a blood shift towards acidity, the thermogenic functions, all become more intense as the disturbance of homeostasis is more pronounced, and they all subside promptly when the disturbance is relieved. Similar conditions probably prevail in other steady states. Of course, the state may not remain steady, as in pathological weakness or defect, and for that reason the postulate was made conditional. As Lotka (1925) has pointed out, this conditional statement, required for living beings and due to their lack of permanent stability, sharply distinguishes the proposal from the strict principle of Le Chatelier true for simple physical or chemical systems. Indeed, as Y. Henderson (1925) has remarked, the physiological and the chemical conceptions of equilibrium are quite different. "The one invokes energy to maintain itself, or if disturbed to recover . . . . the other in seeking balance only goes down hill dynamically."

3. "Any factor which operates to maintain a steady state by action in one direction does not also act at the same point in the opposite direction." This proposal, which should have been limited to *physiological* action, is related to the questions discussed in the footnote on p. 410. Does adrenin in physiological doses both discharge glycogen from the liver and increase glycogen storage there? Does insulin likewise act oppositely in relation to the hepatic glycogen reserves? In the footnote mentioned, reasons were given for not crediting the evidence for opposed action by a single one of these agents. An agent may exist which has an influence of a tonic type—a moderate activity—which can be varied up or down, and which can act at a given point in "high" concentration but not in a "low" concentration. The adrenal medulla, which is subject to control of opposed nervous influences (Cannon and Rapport, 1921), may be cited as an agent of that type.

4. "Homeostatic agents, antagonistic in one region of the body, may be coöperative in another region." The sympathico-adrenal and the vago-insular influences are opposed in action on the liver, but they

appear to be collaborators in their action on muscles, e.g., leading to effective use of sugar by muscle cells (Burn and Dale, 1924). Too little is known about the effects of these agents to permit this postulate to be of much significance at present.

5. "The regulating system which determines a homeostatic state may comprise a number of coöperating factors brought into action at the same time or successively." This statement is well illustrated in the arrangements for protection against a fall of temperature in which series of defences are used one after another, and also in the elaborate and complex arrangements for maintaining uniform reaction of the blood.

6. "When a factor is known which can shift a homeostatic state in one direction it is reasonable to look for automatic control of that factor or for a factor or factors having an opposing effect." This postulate is implied in earlier postulates. It is expressed as a reiteration of the confidence that homeostasis is not accidental but is a result of organized government, and that search for the governing agencies will result in their discovery.

The reader has had occasion to be impressed by the large gaps in our knowledge not only of homeostatic conditions but also of the arrangements which establish and maintain them. Repeatedly the phrase "is not known" has had to be employed. It is remarkable that features so characteristic of living beings as the steady states should have received so little attention. Innumerable questions remain to be answered. Little is known, for example, about the effective stimuli for such homeostatic reactions as are well recognized. Are there receptors which are affected in blood-sugar regulation or are the regulatory factors worked by direct action on cerebral centers? Again, there are homeostatic agencies which were not considered above, such as the extra erythrocytes produced in organisms living at high altitudes, the thicker hair growth during prolonged cold weather, and also steady states which have not been mentioned, such as the stabilization of phosphorus in relation to calcium, the evidence from constancy of basal metabolism that there is constancy in the thyroxin content of the blood, indeed the evidence from other steady states (as weight, and sex character) that other endocrine products are uniformly circulating—the questions presented by these and many other reactions which are serviceable in preserving uniformity in the fluid matrix offer a fascinating field for research.

In the two preceding sections the functions of the divisions of the autonomic system in relation to homeostasis were defined and some postulates regarding homeostasis were presented, not with the idea that the statements should be taken as conclusive but rather that they might prove suggestive for further investigation. Indeed, that point of view should be recognized as prevailing throughout this review. It is the writer's belief that the study of the particular activities of the various organs of the body has progressed to a degree which will permit to a greater extent than is generally recognized an examination of the interplay of these organs in the organism as a whole. Their relations to their internal environment seemed to offer a suggestive approach to a survey of their possible integrative functions. In such a venture errors are sure to creep in which must later be corrected, and crude ideas are sure to be projected which must later be refined. Though the present account of agencies which regulate steady states in the body is likely to prove inadequate and provisional, there is no question of the great importance of the facts of homeostasis with which it deals. This account may at least serve to rouse interest in them and their importance. The facts are significant as outstanding features of biological organization and activity. They are significant also in understanding the complex disorders of the body, for in a state normally kept regular by a group of coöperating parts, full insight into irregularity is obtained only by learning their mode of coöperation. Again, effective methods of attaining homeostasis are significant in comparison with the methods in systems where steady states are not yet well developed; the regulation of homeostasis in higher animals is probably the result of innumerable evolutionary trials, and knowledge of the stability which has finally been achieved is suggestive in relation to the less efficient arrangements operating in lower animals and also in relation to attempts at securing stability in social and economic organizations. Finally, continued analysis of biological processes in physical and chemical terms must await a full understanding of the ways in which these processes are roused to perform their service and are then returned to inactivity. Indeed, regulation in the organism is the central problem of physiology. For all these reasons further research into the operation of agencies for maintaining biological homeostasis is desirable.

BIBLIOGRAPHY

Abe, Y. 1924. Arch. f. exper. Path. u. Pharm., ciii, 73.
Adolph, E. F. 1921. Journ. Physiol., lv, 114.
Afanassiew, M. 1883. Pflüger's Arch., xxx, 385.
Allen, F. M. 1920. Journ. Exper. Med., xxxi, 363, 381.
Andrus, E. C. and E. P. Carter. 1924. Heart, xi, 106.

AUB, J. C., W. BAUER, C. HEATH AND M. ROPES. 1929. Journ. Clin. Invest., vii, 97.
BAILEY, P. AND F. BREMER. 1921. Arch. Int. Med., xxviii, 773.
BAINBRIDGE, F. A. 1923. The physiology of muscular exercise. London, p. 90.
BAIRD, M. M. AND J. B. S. HALDANE. 1922. Journ. Physiol., lvi, 259.
BANTING, F. G., C. H. BEST, J. B. COLLIP AND E. C. NOBLE. 1922. Trans. Roy. Soc. Canada, xvi, 13.
BANTING, F. G. AND S. GAIRNS. 1924. Amer. Journ. Physiol., lxviii, 24.
BARCROFT, J. AND H. STAUB. 1910. Journ. Physiol., xli, 145.
BARCROFT, J. 1926. Ergebn. d. Physiol., xxv, 818.
BAUER, W., J. C. AUB AND F. ALBRIGHT. 1929. Journ. Exper. Med., xlix, 145.
BERG, W. 1914. Biochem. Zeitschr., lxi, 428; München. med. Wochenschr., lxi, 434.
1922. Pflüger's Arch., cxciv, 102; cxcv, 543.
BERNARD, C. 1878. Les Phénomènes de la Vie. Paris, two vols.
BEST, C. H., J. P. HOET AND H. P. MARKS. 1926. Proc. Roy. Soc. (London), B, c, 32.
BLOOR, W. R. 1922. Endocrinology and metabolism. New York, iii, 294.
BOOTHBY, W. M., I. SANDIFORD AND J. SLOSSE. 1925. Ergebn. d. Physiol., xxiv, 733.
BRITTON, S. W. 1925. Amer. Journ. Physiol., lxxiv, 291.
1928. Amer. Journ. Physiol., lxxxvi, 340.
BULATAO, E. AND W. B. CANNON. 1925. Amer. Journ. Physiol., lxxii, 295.
BURN, J. H. 1923. Journ. Physiol., lvii, 318.
BURN, J. H. AND H. H. DALE. 1924. Journ. Physiol., lix, 164.
CAHN-BRONNER, C. E. 1914. Biochem. Zeitschr., lxvi, 289.
CAMPOS, F. A. DE M., W. B. CANNON, H. LUNDIN AND T. T. WALKER. 1929. Amer. Journ. Physiol., lxxxvii, 680.
CANNON, W. B. AND A. L. WASHBURN. 1912. Amer. Journ. Physiol., xxix, 441.
CANNON, W. B. 1914a. Amer. Journ. Physiol., xxxiii, 356.
1914b. Amer. Journ. Psychol., xxv, 256.
CANNON, W. B. AND H. GRAY. 1914. Amer. Journ. Physiol., xxxiv, 232.
CANNON, W. B. AND W. L. MENDENHALL. 1914. Amer. Journ. Physiol., xxxiv, 245, 251.
CANNON, W. B. 1918. Proc. Roy Soc. (London), B, xc, 283.
CANNON, W. B. AND D. RAPPORT. 1921. Amer. Journ. Physiol., lviii, 308.
CANNON, W. B. AND R. CARRASCO-FORMIGUERA. 1922. Amer. Journ. Physiol., lxi, 215.
CANNON, W. B., M. A. McIVER AND S. W. BLISS. 1923. Boston Med. and Surg. Journ., clxxxix, 141.
1924. Amer. Journ. Physiol., lxix, 46.
CANNON, W. B., J. R. LINTON AND R. R. LINTON. 1924. Amer. Journ. Physiol., lxxi, 153.
CANNON, W. B. 1925. Trans. Cong. Amer. Physicians and Surgeons, xii 31; 1926, Jubilee Volume for Charles Richet, p. 91.
CANNON, W. B. AND S. W. BRITTON. 1927. Amer. Journ. Physiol., lxxix, 433.
CANNON, W. B., A. QUERIDO, S. W. BRITTON AND E. M. BRIGHT. 1927. Amer. Journ. Physiol., lxxix, 466.
CANNON, W. B., H. F. NEWTON, E. M. BRIGHT, V. MENKIN AND R. M. MOORE. 1929. Amer. Journ. Physiol., lxxxix, 84.

CARLSON, A. J. 1916. The control of hunger in health and disease. Chicago.
COHNHEIM, O., KREGLINGER AND KREGLINGER, JR. 1909. Zeitschr. f. physiol. Chem., lxiii, 429.
COLLIP, J. B. 1926. Journ. Biol. Chem., lxiii, 395.
CORI, C. F. 1925. Journ. Pharm. Exper. Therap., xxv, 1.
CORI, C. F. AND G. T. CORI. 1928. Journ. Biol. Chem., lxxix, 309.
CRANDALL, F. M. 1899. Arch. Pediat., xvi, 174.
CUSHNY, A. R. 1926. The secretion of urine. 2nd ed., London.
ENGELS, W. 1904. Arch. f. exper. Path. u. Pharm., li, 346.
ERDHEIM, J. 1911. Zeitschr. f. Pathol., vii, 175, 238, 259.
FISKE, C. H. AND Y. SUBBAROW. 1929. Journ. Biol. Chem., lxxxi, 656.
FLETCHER, A. A. AND W. R. CAMPBELL. 1922. Journ. Metab. Res., ii, 637.
FOLIN, O., H. C. TRIMBLE AND L. H. NEWMAN. 1927. Journ. Biol. Chem., lxxv, 263.
FOSTER, D. P. AND G. H. WHIPPLE. 1922. Amer. Journ. Physiol., lviii, 393, 407.
FREDERICQ, L. 1885. Arch. de Zoöl Exper. et Gén., iii, p. xxxv.
FREUND, H. 1913. Arch. f. exper. Path. u. Pharm., lxxiv, 311.
GAMBLE, J. L. AND S. G. ROSS. 1925. Journ. Clin. Invest., i, 403.
GAMBLE, J. L. AND M. McIVER. 1925. Ibid., i, 531.
1928. Journ. Exper. Med., xlviii, 859.
GAYET, R. AND M. GUILLAUMIE. 1928. Compt. rend. Soc. de Biol., cxvii, 1613.
GRAFE, E. 1927. Oppenheimer's Handbuch der Biochemie, Jena, ix, 68.
GRANT, S. B. AND A. GOLDMAN. 1920. Amer. Journ. Physiol., lii, 209.
GRAY, H. AND L. K. LUNT. 1914. Amer. Journ. Physiol., xxxiv, 332.
GRÜNWALD, H. F. 1909. Arch. f. exper. Path. u. Pharm., lx, 360.
HALDANE, J. S. AND J. G. PRIESTLEY. 1915. Journ. Physiol., l, 296.
HALDANE, J. S. 1922. Respiration. New Haven, pp. 21–22; p. 383.
HALLIBURTON, W. D. 1904. Biochemistry of muscle and nerve. Philadelphia, p. 111.
HALLION, L. AND R. GAYET. 1925. Compt. rend. Soc. de Biol., xcii, 945.
HANSEN, K. M. 1923. Acta. Med. Scand., lviii, Suppl. iv.
HASSELBALCH, K. A. AND C. LUNDSGAARD. 1912. Skand. Arch. f. Physiol., xxvii, 13.
HENDERSON, L. J. 1928. Blood. New Haven.
HENDERSON, Y. 1925. Physiol. Rev., v, 131.
HESS, W. R. 1926. Klin. Wochenschr., v, 1353.
HILL, L. 1921. Journ. Physiol., liv, p. cxxxvi.
HIRAYAMA, S. 1925. Tohoku Journ. Exper. Med., vi, 160.
HOMANS, J. 1914. Journ. Med. Res., xxv, 63.
HOUSSAY, B. A., J. T. LEWIS AND E. A. MOLINELLI. 1924. Compt. rend. Soc. de Biol., xci, 1011.
ISENSCHMID, I. 1926. Handb. d. norm. u. path. Physiol., Berlin, xvii, 56.
IZQUIERDO, J. J. AND W. B. CANNON. 1928. Amer. Journ. Physiol., lxxxiv, 545.
KAHN, R. H. 1904. Arch. f. Physiol., Suppl. Bd., p. 81.
KEITH, N. M. 1923. Amer. Journ. Physiol., lxiii, 394.
KERR, W. J., S. H. HURWITZ AND G. H. WHIPPLE. 1918. Amer. Journ. Physiol., xlvii, 379.
KROGH, A. 1922. The anatomy and physiology of the capillaries. New Haven, p. 227.

Kulz, E. 1880. Pflüger's Arch., xxiv, 41.
LaBarre, J. 1925. Arch. Internat. de Physiol., xxv, 265.
Lewis, F. T. and J. L. Bremer. 1927. A textbook of histology. Philadelphia, p. 72.
Lewis, J. T. 1923. Compt. rend. Soc. de Biol., lxxxix, 1118.
Lewis, J. T. and M. Magenta. 1925. Compt. rend. Soc. de Biol., xcii, 821.
Loevenhart, A. S. 1902. Amer. Journ. Physiol., vi, 331.
Lotka, A. J. 1925. Elements of physical biology. Baltimore, p. 284.
Luce, E. M. 1923. Journ. Pathol. and Bact., xxvi, 200.
Lusk, G. 1928. The science of nutrition. 4th ed., Philadelphia.
MacCallum, W. G. and C. Voegtlin. 1909. Journ. Exper. Med., xi, 118.
MacCallum, W. G. and K. M. Vogel. 1913. Journ. Exper. Med., xviii, 618.
Macleod, J. J. R. 1913. Diabetes; its pathological physiology. London, p. 61.
1924. Physiol. Rev., iv, 51.
Marine, D. 1913. Proc. Soc. Exper. Biol. Med., xi, 117.
Mathison, G. C. 1911. Journ. Physiol., xlii, 283.
McCormick, N. A., J. J. R. Macleod, E. C. Noble and K. O'Brien. 1923. Journ. Physiol., lvii, 224.
Meek, W. J. 1912. Amer. Journ. Physiol., xxx, 161.
Minkowski, O. 1908. Arch. f. exper. Path. u. Pharm., Suppl. Bd., 399.
Noël, R. 1923. Presse méd., xxxi, 158.
O'Connor, J. M. 1919. Journ. Physiol., lii, 267.
Padtberg, J. H. 1910. Arch. f. exper. Path. u. Pharm., lxiii, 60.
Patterson, S. W. and E. H. Starling. 1913. Journ. Physiol., xlvii, 143.
Pflüger, E. F. W. 1877. Pflüger's Arch., xv, 57.
Reincke, J. J. 1875. Deutsch. Arch. f. Klin. Med., xvi, 12.
Richet, C. 1900. Dictionnaire de Physiologie, Paris, iv, 721.
Rogers, F. T. 1920. Arch. Neurol. und Psychiatr., iv, 148.
Rosenfeld, G. 1903. Ergebn. d. Physiol., ii, pt. 1, 86.
Rowntree, L. G. 1922. Physiol. Rev., ii, 158.
Schade, H. 1925. Wasserstoffwechsel. Oppenheimer's Handb. der Biochem., 2nd ed., Jena, p. 175.
Schulz, F. N. 1896. Pflüger's Arch., lxv, 299.
Seitz, W. 1906. Pflüger's Arch., cxi, 309.
Sherrington, C. S. 1924. Journ. Physiol., lxviii, 405.
Skelton, H. P. 1927. Arch. Int. Med., xl, 140.
Smith, H. P., A. E. Belt and G. H. Whipple. 1920. Amer. Journ. Physiol., lii, 54.
Starling, E. H. 1909. The fluids of the body. Chicago.
Strada, F. 1909. Pathologica, i, 423.
Stübel, H. 1920. Pflüger's Arch., clxxxv, 74.
Sundberg, C. G. 1923. Compt. rend. Soc. de Biol., lxxxix, 807.
Tatum, A. L. 1921. Journ. Pharm.. Exper. Therap., xviii, 121.
Terroine, E. F. 1914. Journ. de Physiol. et de Pathol. Gén., xvi, 408.
Thompson, W. O. 1926. Journ. Clin. Invest., ii, 477.
Tichmeneff, N. 1914. Biochem. Zeitschr., lix, 326.
Trendelenburg, P. and W. Goebel. 1921. Arch. exper. Path. u. Pharm., lxxxix, 171.

Vosburgh, C. H. and A. N. Richards. 1903. Amer. Journ. Physiol., ix, 35.
Weichselbaum, A. 1914. Verhandl. d. deutsch. Naturf. u. Aerzte, 85.
Wettendorf, H. 1901. Trav. du Lab. de Physiol., Inst., Solvay, iv, 353.
Whipple, G. H., H. P. Smith and A. E. Belt. 1920. Amer. Journ. Physiol., lii, 72.
Zunz, E. and J. LaBarre. 1927. Compt. rend. Soc. de Biol., xcvi, 421, 708.

20

Displacement of Equilibrium

ALFRED J. LOTKA, M.A., D.Sc.

Die Physik wird aus dem Studium des Organischen an sich noch sehr viel neue Einsichten schöpfen müssen, bevor sie auch das Organische bewältigen kann.—*E. Mach.*

In preceding pages we have passed in review some of the principal features of interest presented by systems maintained constantly at or near equilibrium, while one of more of the parameters determining such equilibrium were slowly changing, thus engendering a moving equilibrium.

One might proceed to a consideration, on a more general basis, of the changes brought about in an evolving system through changes of any kind, including rapid ones, in the parameters. In the most general case this would amount to the discussion of a system of differential equations of the form.

$$\frac{dX_i}{dt} = F_i(X_1, X_2, \ldots X_i, \ldots X_n, t)$$

in which the time t entered explicitly into the function F.

It is not proposed to take up the study of this perfectly general case; it must suffice to point to the mathematical literature regarding equations of this form.[1]

But there is another special phase of the general problem which, like the case of *slow* changes, yields with comparative ease to analytical treatment; namely, that special phase which enquires only into the ultimate effect, upon equilibrium, of a given total change in a parameter, leaving aside all questions relating to the path by which the displacement of equilibrium takes place. Such a separate consideration of this special and restricted phase of the general problem is rendered possible by the fact that, in certain cases at any rate, the displacement of the equilibrium is independent of the path of the change, and depends only on the given initial and

[1] See, for example, E. Picard, Traité d'Analyse, vol. 3, 1908, pp. 187, 188, 194, 197; E. Goursat, Cours d'Analyse, vol. 2, 1918, pp. 482, 498.

final values of the parameters whose modification provokes or is associated with the change. So, in physico-chemical transformations ("changes of state") the principle of Le Chatelier enables us to predicate, within certain limits, the ***sign*** of the displacement of equilibrium conditioned by a change in certain of the parameters upon which the equilibrium depends.

THE PRINCIPLE OF LE CHATELIER

The principle of Le Chatelier is best illustrated by a simple example. Consider the simple chemical reaction

$$2H_2 + O_2 \rightleftarrows 2H_2O + 58.3 \text{ cal.}$$

At high temperatures this reaction is reversible; that is to say, it takes place to some extent in the direction of the upper arrow, but also to some extent in the direction of the lower arrow, and an equilibrium is finally established between these two opposing reactions. Now this is what the Le Chatelier principle tells us:

If we add either H alone or O alone to the system, the equilibrium is shifted in the direction of the upper arrow, that is to say, in such direction as to absorb some of the added constituent.

Similarly, if we heat the system, the equilibrium is shifted in the direction of the lower arrow, that is to say, in the direction of the reaction which ***absorbs*** heat. The principle, as enunciated by *Le Chatelier*[2] ***himself***, is:

> Every system in chemical equilibrium, under the influence of a change of any single one of the factors of equilibrium,[3] undergoes a transformation in such direction that, if this transformation took place alone, it would produce a change in the opposite direction of the factor in question.
>
> *The factors of equilibrium are temperature, pressure, and electromotive force,* corresponding to three forms of energy—heat, electricity and mechanical energy.

The second paragraph of the principle as quoted above, requires special emphasis. It is often omitted, even by authors of the highest

[2] Recherches sur les Equilibres Chimique, 1888, pp. 48, 210; Comptes Rendus, 1884, vol. 99, p. 786; Mellor, Chemical Statics and Dynamics, 1904, pp. 435–436.

[3] It appears that some French writers employ the term "facteur d'équilibre" as synonymous with "intensity factor of an energy." (Cf. F. Michaud, Ann. de Phys., vol. 16, 1921, p. 132.)

repute,[4] with the result that a vagueness is introduced for which Le Chatelier himself cannot justly be made responsible. This vagueness is then often rendered still worse by departures from the original wording, aimed at an extension of the scope of the law to all conceivable systems and "factors," an extension which is gained with a total sacrifice of all validity of the principle. So, for example, if we seek to apply the principle as quoted above, but omitting the restriction of the second paragraph, to the water equilibrium already mentioned, and if we select as "factor" of equilibrium not pressure but volume, the principle would lead us to reason as follows: On *diminishing* the volume of the system, that transformation will take place which, did it take place alone (i.e., at constant pressure), would be accompanied by *increase* in volume; a conclusion which is false. As has been shown by Ehrenfest,[5] the error arises through failure to discriminate, in the application of the principle, between the intensity factor (e.g., pressure) and the capacity factor (e.g., volume) of an energy.

It must appear singular that so obvious a defect of the principle, *as commonly quoted*, should so generally have escaped attention, and should for example, have passed unnoted through seven editions of so excellent a work as Nernst's *Theoretische Chemie.* Ehrenfest points out that the explanation lies in the very vagueness of the principles, which permits it to be construed in each case to suit circumstances. The principle is commonly applied *ex post facto*, and its competence to predict thus escapes any serious test. This, however, is only a partial explanation. After all, the fundamental reason for the tardy recognition, and the still more tardy admission in the general literature, of the weakness of the principle, *as commonly quoted*, must be sought in an inherent weakness of the human mind: by a curious inversion of what might be expected in logical sequence, the last things to receive critical scrutiny are always the fundamental premises of our arguments. This is true both as regards the judgment of the average individual, of the people at large, and often even of the man of very superior intellect. One recalls, in this connection, MacAuley's remarks regarding Dr. Johnson: "How it chanced that a man who reasoned upon his

[4] See, for example, W. Nernst, Theoretische Chemie, 1913, p. 698.

[5] Zeitschr. f. phys. Chem., 1911, vol. 77, p. 735. Cf. also P. Duhem, Traité d'Energétique, 1911, vol. 1, p. 467.

premises so ably should assume his premises so foolishly is one of the great mysteries of human nature."

If such an outwardly slight departure from Le Chatelier's original enunciation as the omission of his second "explanatory" paragraph, thus completely destroys the validity of his principle, what is to be said of such sweepingly vague settings as in the following examples:

The broadest definition of the principle of Le Chatelier is that a system tends to change so as to minimize an external disturbance (W. D. Bancroft, Journal of the American Chemical Society, 1911, p. 92).

Every external action produces in a body or system changes in such direction, that in consequence of this change the resistance of the body or system against the external action is increased.

If we regard the faculty of adaptation of animals and plants from the point of view that the organisms undergo, under the influence of external actions, changes which render them more resistant to those actions, then the property of non-living matter which is expressed by the principle of Le Chatelier-Braun may be regarded as a sort of adaptation of such non-living matter (Chwolson, Traité de Physique, 1909, vol. 3, p. 547).

If the equilibrium of a natural complex (system of masses, organism, system of ideas) is disturbed, it adapts itself to the stimulus (Reiz) which causes the disturbance, in such manner that the said stimulus continually diminishes until finally the original or a new equilibrium is again established (J. Löwy, Kosmos, 1911, p. 331).

The last two examples are of particular interest to us here as suggesting application of the principle to biological systems. As a matter of fact, such application of the vaguely formulated principle (in a form in which it would be injustice to link it with the name of Le Chatelier) antedates by many years its enunciation by the French physicist. The following passages in Herbert Spencer's First Principles are pertinent:

Among the involved rhythmical changes constituting organic life, any disturbing force that works an excess of change in some direction is gradually diminished and finally neutralized by antagonistic forces, which thereupon work a compensating change in the opposite direction, and so, after more or less of oscillation, restore the medium condition. And this process it is which constitutes what physicians call the *vis medicatrix naturae.*

This is a conclusion which we may safely draw without knowing the special re-arrangements that effect the equilibration: If we see that a different mode of life is followed after a period of functional derangement by some altered condition of the system—if we see that this altered condition, becoming by and by established, continues without further change, we have no alternative but to say that the new forces brought to bear on the system have been compensated by the opposing forces they have evoked (First Principles, Chapter XXII, *Equilibriation*, 173).

Almost simultaneous with Le Chatelier's publication (1884) is the following pronouncement.

L'être vivant est agencé de telle manière que chaque influence perturbatrice provoque d'elle même la mise en activité de l'appareil compensateur qui doit neutraliser et reparer le dommage (Léon Frédéricq, Archives de Zoologie Exp. et Gén., ser. 2, vol. 3, 1885, p. xxxv).

Now it is not denied that such expressions as this have a certain utility, as describing with fair accuracy a goodly proportion of a class of phenomena to which they relate. But to designate such statements, as "Le Chatelier's Principle," is wholly misleading. That principle, in its exact and narrower formulation is rigorously true, as much so as the laws of thermodynamics from which it can be deduced; it has no exceptions, any more than there is any exception to the law that heat flows by simple conduction from the hotter of two bodies to the colder.

The alleged "principle," as applied to biological systems, lacks the sureness which the true Le Chatelier principle possesses, in its stricter formulations, in physical chemistry. An organism may, by exposure to a certain influence A, become *more* resistant to the influence, as in the case of acquired immunity after an attack of infectious disease, or after habituation to such a poison as arsenic. But, by exposure to another influence B it may become less resistant to B, as in the case of cumulative poisons, or of anaphylaxis. The Le Chatelier principle does not enable us here to predict in which direction the effect will take place in a new and untried case of some influence C.

Conditions of Validity of Le Chatelier's Principle. The question arises why the principle thus breaks down in its application to biological cases of the kind cited. The answer is found by examining the basis on which the proof of the principle rests. Such an examination brings out the fact that one of the necessary conditions for the applicability of the principle is stability of the equilibrium to which application is made. Now the equilibria commonly contemplated in physical chemistry are stable, so that this condition is satisfied. But it is not always satisfied in the equilibrium of the living organism. The organism is, indeed, stable with regard to many of the commonly occurring attacks of its environment. But it is of little consequence to the species whether, for example, the individual organism is stable with regard to the ingestion of a large dose of strychnine, for in nature such ingestion will occur so rarely, if at all, as to influence in no appreciable degree the life of the species. It is not necessary for the stability of the *species*, that the *individual* be stable at all times.[6] In point of fact, we know perfectly well that sooner or later each individual finds itself in a condition of instability, by "accident" or sickness, and dies. An analysis[7] of the basis of the principle of Le Chatelier reveals the fact, among others, that all demonstrations of this principle postulate, as a fundamental characteristic of the systems to which it applies, that they be in *stable* equilibrium. The principle can, therefore be applied at best only with cautious reservation to living organisms, reservation such as, for example, Le Dantec[8] makes: "In studying as closely as possible the consequences of disease in living organisms, *when they survive such diseases*, I have drawn attention . . . . to the fact that all these consequences, such as acquired immunity and the production of antitoxic sera, can be summarized in the principle of Le Chatelier." But with such reservation the principle loses its chief utility, which consists in its power to predict the course of events. Indeed, it might be accused, in such restricted form, of being little more than a tautological platitude, which tells us that if the system or unit in question is stable, then it is stable. This is not quite such a damning accusation as may at first sight appear, for the same can be brought against the principle of the survival of the fittest, which nevertheless has proved supremely fertile in biological research. In point of fact there is a close relationship between

[6] Compare what has been said in the discussion of chemical equilibrium regarding the stability of aggregates composed of individuals, themselves of limited stability, of limited life period (Chapter XII).

[7] Such an analysis, carried out in considerable detail, has been given by the writer in Proc. Am. Acad. Arts and Sci., 1922, vol. 57, pp. 21–37. The importance of the restriction to stable equilibria, in connection with biological systems, has also been pointed out by C. Benedicks, Zeitschr. f. phys. Chemie, 1922, vol. 100, pp. 42–51. A. J. Lotka, Am. Jour. Hygiene, 1923, p. 375.

[8] La Stabilité de la Vie, 1910, p. 24.

the two principles. But it is important to note that the principle of the survival of the fittest is avowedly statistical in character, and is to be applied to organisms in the gross. This is true, also, of the principle of Le Chatelier in physico-chemical systems; its field of application is to aggregates of molecules, not to the individual. But the applications that have been essayed in biology have been made to the individual; such application can at the best yield a judgment of probabilities. In physical chemistry we deal for the most part with stable equilibria. But in biology, as has already been pointed out, though the *races* that come under our observation possess stability *as races* (else they would not have survived to be our contemporaries), it does not follow at all, that each and every individual is at all times in a state of steady equilibrium.

Aside from the limitation in the applicability of the principle to *stable* systems, other limitations appear in such an analysis of its foundations as has been referred to above. So, for example, loose analogy to the physico-chemical equilibrium, as affected by the addition of a quantity of one of the reacting substances, might lead one to draw the erroneous inference that in a community infected with malaria, the introduction of additional malaria parasites would shift the equilibrium in the direction of a higher malaria rate. But there is every reason to expect, on the contrary, that the equilibrium remains unchanged by such addition. For a close analysis of the reason for this divergence in the two cases the reader must be referred to the original paper already cited. It must suffice here to state briefly that this reason is to be found in the existence in the physico-chemical case of equations of constraint, relations between certain variables, and in the absence of analogous relations in the case of malaria.

Extension of Scope of Rigorous Applicability. While the prime result of a searching analysis of the foundations of the Le Chatelier principle is to emphasize rather the restrictions of its scope, yet in certain respects such analysis does furnish a rigorous basis for a certain generalization of its applicability beyond those bounds where its warrant rests on the firm ground of thermodynamics. And this extension of the strict applicability of the principle takes place essentially in two directions. On the one hand the thermodynamic justification, at any rate in the form commonly presented, covers only true equilibria, and does not extend to steady states maintained with constant dissipation of energy. This restriction does not appear in the demonstration of the principle on the broadest grounds that suffice for its establishment,[9] Le Chatelier's principle applies, in certain cases, to steady states of the more general type, as well as to true equilibria.

The second direction in which the analysis, on general grounds, of the principle, enlarges its field of warrant, is in the matter of the kinds of "factors" to which it is properly applicable. It has already been pointed out that, in its physico-chemical application, it must be used with proper discrimination as to the distinction between the capacity and the intensity factor of an energy, as, for example, volume and pressure. It is found, upon analysis, that the applicability of the principle to the effect of a change in pressure, for example rests upon the following fundamental property of the pressure and volume of a system in stable equilibrium.

1. For every value of v, the volume of the system, there is a definite value of p_i, the pressure *which it exerts*, the internal pressure, as we may term it.

2. The volume v increases or decreases according as the internal pressure p_i is greater or less than the external pressure p_e upon the enclosure, that is to say,

$$\frac{dv}{dt} \gtreqless 0 \text{ according as } p_i \gtreqless p_e.$$

3. It can be shown that, given (1) and (2), stability demands that the curves representing the relative between p (ordinates) and v (abscissae) must slope from left to right downwards. For if such a curve slopes in the opposite direction, then the slightest displacement from equilibrium will immediately cause the system to travel with cumulative effect, avalanche-like, along the pv curve further and further away from the starting point.[10]

[9] For justification of this and other statements made in these paragraphs the reader is referred to the author's paper already cited. It may be remarked that Ehrenfest (loc. cit.) expresses the belief that such broader scope belongs to the principle, but he does not support his impression with proof.

[10] It is interesting to note that an upward slope, from left to right, occurs in the middle limb of the van der Waals' pv curve of a gas. But this limb represents an unstable state which is never realized, the gas, instead of following this part of the curve, partially condenses and traces a horizontal straight line for the pv relation.

Now these fundamental properties (1), (2) and (3), of a capacity and an intensity factor of an energy[11] are shared by certain parameters that have no direct or simple relation to energy whatsoever; and since the applicability of the principle depends upon these properties, it will extend to such other parameters possessing them. As an example may be mentioned the relation between area a occupied by a population, and the rent per unit area R_i that an (average) individual is willing to pay. If R_i is greater than R_e, the rent at market rate, the individual will move into a more spacious apartment, and a will increase, and vice versa; so that

$$\frac{da}{dt} \gtreqless 0 \text{ according as } R_i \gtreqless R_e$$

On the other hand the curves representing, in rectangular coordinates, the relation between rent and area available per head, necessarily slope from left to right downward. If it were true, as sometimes stated, that the more a man has, the more he wants, economic equilibrium would be an unstable condition.

This example is presented here with reservation. There may be various complications in practice that may form obstacles to the simple application of the principle indicated. But it will serve to show how a perfectly rigorous justification may exist for the application of the principle of Le Chatelier outside the field of plain energetics and thermodynamics. Where, and only where such justification can be clearly shown to exist, there it will be permissible and useful to apply the principle. Applications made broadcast, without prior examination of the parameters involved, perhaps without any thought at all of reasonable parameters, are of little if any worth.

One other word of caution must be said, for which the example of area and rent will furnish a suitable illustration. Before we apply

[11] Owing to the custom of counting heat *absorbed* by a system as positive, but work done *upon it* as negative, the relation analogous to that of (2) takes the form, in the case of heat energy,

$$\frac{dQ}{dt} \lesseqgtr 0 \text{ according as } \theta_1 \lesseqgtr \theta,$$

where Q is the quantity of heat absorbed by the system at a temperature from a source at the temperature θ. Here the $Q\theta$ curves slope upward from left to right. Cf. A. J. Lotka, loc. cit., p. 36.

the principle to any particular parameter, we must be sure that the contemplated change will modify this parameter alone, and not also at the same time others that are in principle, if not in physical fact, to be regarded as independent. So, for example, one reason why the example of area and rent was presented above with express reservation is that, ordinarily at any rate, it may be difficult or impossible to modify the area of a population without modifying at the same time certain other features, such as the supply of nutriments furnished in the soil, etc.

On the whole, so far, it must be said that the result of a careful analysis of the principle of Le Chatelier yields negative results, so far as practical application to biological systems is concerned. The chief conclusion is that great caution must be exercised in employing the principle. This result may be somewhat disappointing, but it is for that none the less important. Facts are stubborn things; it seems a pity to demolish the idol of a pretty generalization, but in such things we cannot permit the wish to be father to the thought. And the idol is not wholly demolished—in fact his hitherto doubtful title to certain domains has been established on a clear basis. But his province must be recognized as very definitely bounded.

DISCUSSION OF DISPLACEMENT OF EQUILIBRIUM INDEPENDENTLY OF LE CHATELIER'S PRINCIPLE

In view of the limitations in the field of strict applicability of the principle of Le Chatelier, we are in general forced to consider separately each particular case of displacement of equilibrium. How such cases may be treated may be exemplified by the following two instances.

Case 1. Displacement of Equilibrium between Food and Feeding Species. Consider a species S_2 of mass X_2, which requires for its (equilibrium) sustenance of a mass k_2X_2 of food. Let this food be derived exclusively from the slain bodies, of total mass d_1X_1, of a species S_1. Let a fraction ϵ of all the deaths in S_1 be those caused by S_2 feeding upon S_1. Then

$$k_2X_2 = \epsilon d_1X_1 \tag{1}$$

$$\frac{X_1}{X_2} = \frac{k_2}{\epsilon d_1} = \alpha \tag{2}$$

Editor's Comments on Papers 21 and 22

Closing the Feedback Loop

IV

Walter Cannon took Bernard's basic concept of "la fixité du milieu intérieur" a giant step forward both by virtue of his own research, and by his ability of conceptual synthesis. As we have seen, he coined the term "homeostasis" to designate stability of the organism; "homeostatic conditions" to indicate details of the stability; and "homeostatic reactions" to signify means for maintaining stability. Since then the concept has been broadened, additional knowledge of intricate homeostatic mechanisms gained, and a host of new terms introduced. "Feedback," "loops," "servomechanisms," "transfer functions," "cybernetics," and many other expressions are used. In addition, the basic principle is applied by engineers, sociologists, economists, ecologists, and mathematicians as well as all types of biologists. There is thus a very real problem in deciding which Benchmark articles to include in this concluding section.

Any paper published today, at least in physiology, which is worth the paper it is printed on, should further clarify a homeostatic mechanism. But they all, obviously, cannot be included here. Therefore a very simple, if inadequate decision was made to include but two publications. The first concerns temperature regulation. We started with Blagden and Hunter. In their time, body temperature could be measured fairly accurately; thus their observation of the constancy of body temperature despite wide variations in the ambient temperature constituted an important landmark. Just how far our knowledge of the homeostatic mechanisms responsible for temperature regulation has come is well illustrated in the paper by Benzinger.

Theodor Benzinger (b. 1905) was born and educated in Germany. He received both the Sc.D. and M.D. degrees from the University of Tuebingen. From 1947 until 1971 he worked in the United States Navy Medical Research Institute being concerned almost exclusively with the mechanism of human homeothermy. He is recognized as an international authority on this subject. He summarizes the current state of knowledge succinctly in the article reproduced here.

When we speak of homeostasis we speak of control. This leads to engineering and, inevitably to mathematics. Mathematics is said to be "Pure language—the language of science. It is unique among languages in its ability to provide precise expression for every thought or concept that can be formulated in its terms." In this connotation the article by Machin would seem to be the appropriate one to conclude this survey.

Kenneth E. Machin (b. 1924) received his Ph.D. in 1953 at Cambridge where he has remained. The fact that he works in the Zoology Department there obscures the fact that he is primarily a biophysicist well versed in mathematics. In the paper reproduced here, in accord with his own words, he picks his way "through the mathematical and technological jungle, collecting here and there a result or technique which may be useful in biology." He concludes by stating that, "By the time the physiologist has reached the stage of fitting the elements of the transfer function to the structure of his biological system, he is beyond the point where feedback theory can help him." That would seem to be a good point at which to end this survey.

21

THE THERMAL HOMEOSTASIS OF MAN

By T. H. BENZINGER
Bio-Energetics Laboratories, Naval Medical Research Institute,
Bethesda, Maryland

INTRODUCTION

Thermal homeostasis in the animal world is achieved by two principal systems: behavioural and autonomic regulation. In both respects, compared with animals, man holds a unique position. In behaviour, the artificial means of man for locomotion from one environment or climate into another are exceptionally fast and far-reaching. Moreover, human protective clothing, different from the furs or feathers of animals, may be applied, removed or changed at will. The shelters of man are more elaborate; the use of fire and other means of heating with external sources of energy are a privilege of the human race. Man alone bathes for comfort in artificially heated water. Less than one century ago, discoveries in physics added another dimension: cooling. This has enabled man to refrigerate or air-condition his dwellings and working spaces. Even more recently, heating or cooling, applied to fast-moving craft for land, sea, air and space travel have permitted the invasion of areas with the most adverse conditions, including polar ice-caps, ocean-depths, high altitudes and terrestrial orbits. These means will in the future permit man to inhabit remote celestial bodies. Thus not only vital present problems of survival, health and performance, but also specific requirements for the surging future of the human species are linked to temperature and to thermal physiology.

Although powerful enough to reduce almost any environmental stress-load to manageable proportions, behavioural control is crude if the maintenance of a precisely determined norm or 'set point' is the aim to be achieved. A second system, *autonomic* temperature control, is required to perform the ultimate adjustment with high precision. This system operates with sensory stimuli that are not consciously perceived and with responses that are not wilfully initiated. The autonomic system, too, appears to be more highly developed in man than in any other species.

The sweating response is singularly efficient in the absence of fur or feathers. Respiratory temperature control, a more primitive mechanism which would interfere with acid-base balance, is therefore dispensable and practically non-existent in man. In man no additional responses, such as panting or feather-raising, have to be considered. To the experimental physiologist this simplification is an advantage. In work on man—not on animals—the cortical innervation of thermoregulatory mechanisms can be eliminated by proper instruction. Therefore the autonomic responses of sweating and vasodilatation (in warm environments) and the autonomic increase of metabolism (in cold environments) may be observed without serious interference through other avenues of heat loss, or from other sources of heat production.

METHODS

The scientific investigation of human thermal homeostasis depends of course on the development of suitable methods. The essential thermoregulatory responses, sweating, vasodilatation and increased heat production, may be measured, and indeed continuously recorded, with speed and high precision using modern principles of calorimetry. Gradient calorimetry was used extensively in the investigations described in this paper. The method has been thoroughly described elsewhere (Benzinger & Kitzinger, 1949, 1963). For indirect calorimetry, continuously recording methods of gas analysis are now available (Rein, 1937; Benzinger 1938; Pratt, 1958).

Between the thermoregulatory responses and the stimuli by which they are elicited, various connexions are conceivable. For example, links between rate of exercise and sweating have been hypothetically adduced (Nielsen, 1938; van Beaumont & Bullard, 1963). Andersson (1962) has established, by sound experimental evidence, hormonal connexions between thermal stimuli and metabolic responses. However, slow hormonal mechanisms cannot explain the minute-by-minute adjustments of internal body temperature to the norm, near 37° C. Neural connexions between stimulation by temperature and responses by heat loss or production, are the basis of mammalian and human thermo-regulation. The experimental proof of this postulate had already been obtained for heat *loss* in 1885, with the destruction of the *anterior* hypothalamic heat loss centre, by Aronsohn & Sachs; and for heat *production* in 1912, with the destruction of the *posterior* hypothalamic heat maintenance centre, by Isenschmid & Krehl. On the following pages an attempt will be made to describe in quantitative terms of causes and effects (stimuli and responses) the working mechanisms of human thermo-regulation in which these two 'centres' play their crucial and basically different roles (Benzinger, 1959, 1961, 1963; Benzinger, Kitzinger & Pratt, 1961, 1963).

Such an analysis presupposes not only the measurement of the *responses* by direct or indirect calorimetry, but also depends on the reliable measurement of temperature as a *stimulus* at the site of its physiological reception and not elsewhere. The rectum, time-honoured site of temperature measurement in physiology and medicine, is not a site of neural reception of temperature. The alternative presupposition, that by coincidence the rectum and whatever sites of central thermoreception may exist have similar patterns of temperature fluctuation, was tacitly accepted during a period of 100 years of experimentation and practical use. This hypothetical presupposition has not stood up to experimental testing (see Figs. 1 and 2).

Cranial measurements of body temperature

In the experiment in Fig. 1, immersion of the body in warm water excluded the head but included the pelvis. Nevertheless, rectal temperature lagged far behind the rapid rise of cranial internal temperature, measured at

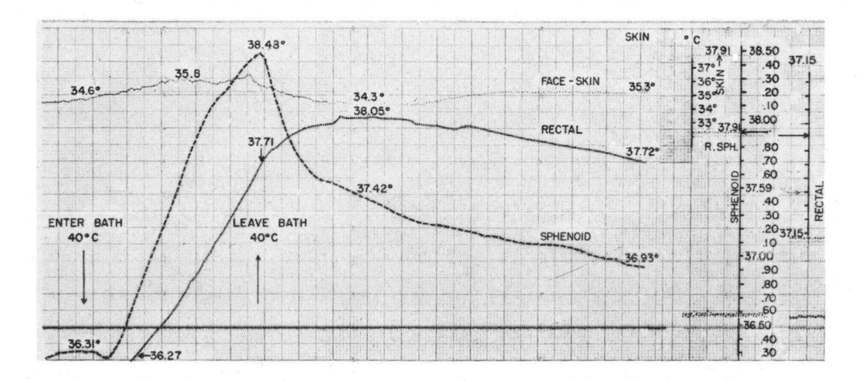

Fig. 1. Upon entering or leaving a warm bath, with the head kept out of the water, rectal temperature is lagging behind cranial temperature, taken at the anterior wall of the sphenoid sinus, with nostrils occluded. After leaving the warm water the relationship of the two temperatures is inverted. Note that the combined deviations $-0{\cdot}77$ and $+0{\cdot}79$ are as wide as the physiological range over which autonomic temperature regulation is effective (see Fig. 10).

the anterior wall of the sphenoid sinus. On leaving the bath rectal temperature kept rising and stayed elevated for more than 1 hr. While a substantial part of these discrepancies might be described as an inertial delay of rectal temperature versus cranial, it is doubtful if the time-courses actually observed in Fig. 1 could be resolved or predicted by any mathematical analysis as applied to inertial distortions. Similarly, other relations between cranial and rectal temperatures seem to defy attempts at a reduction of rectal temperature to a 'true' body temperature. While cranial temperature undergoes drastic alterations upon putting both arms into warm water, the accompanying changes of rectal temperature are hardly noticeable. Even applying a stimulus from the *inside* the body, by eating ice, seems to have little effect on pelvic internal temperature, but a great effect on cranial temperature. A different situation again exists during and after muscular exertion (Fig. 2) which has a particularly late, strong and lasting effect on temperature in the pelvis, where venous influx from the legs and pelvic muscles is predominant. None of the discrepancies between the temperature of neurological interest (in the cranium) and the temperature conventionally measured (in the rectum) appears to be predictable or suited for analysis and correction. In the work that is reported here, rectal measurements were therefore abandoned and replaced.

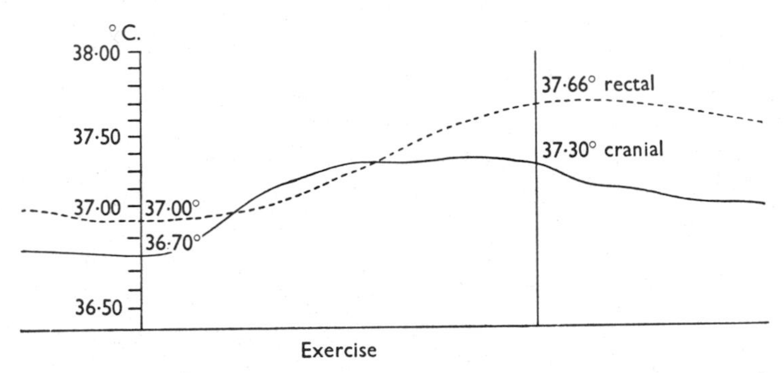

Fig. 2. Rectal temperature shows unacceptable time lag in reflecting changes of central temperature, induced by internal heat production through exercise of 50 min. duration.

Desirable substitutes for the unobtainable hypothalamic temperature would have been sphenoidal measurements, or temperatures taken in Rosenmueller's fossa near the stem of the internal carotid artery, under cocaine anaesthesia. For practical considerations the measurement at the tympanic membrane of the ear was introduced instead. It does not require surface anaesthesia. Although it is more inert than other cranial measurements it shows the same essential patterns following external or internal stimulation (Fig. 3).

Besides the cranial measurement of temperature, consideration must be given in any study of thermoregulation to the established existence of multiple, superficial warm and cold receptors at the skin. Skin temperatures were therefore recorded in ten conventionally selected places, and integrated into one summary measurement of 'average' skin temperature.

CONTINUOUS CALORIMETRIC RECORDINGS OF THE SWEATING RESPONSE

Using the calorimetric and thermometric methods described, an attempt was made to uncover reproducible relations between the stimulus of temperature at the skin or in the cranium, and the responses in loss or production of heat. For these trials it was important to select properly not only the methods, but also the experimental conditions. If the investigator wants to see which one of two stimuli, cutaneous or cranial temperature, produces

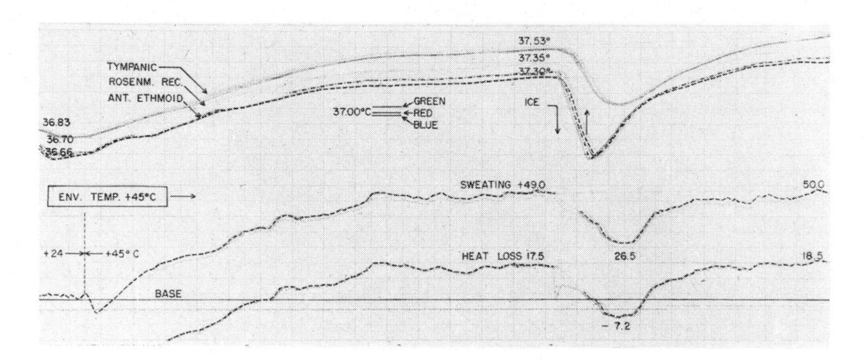

Fig. 3. In response to (1) a change of environmental temperature from 24 to 45° C., and (2) eating 1 pint of sherbet ice, the course of tympanic temperature parallels the course of deep cranial temperature in the anterior ethmoidal region and in Rosenmueller's fossa. (The narrowed difference between tympanic and deep cranial temperature at the left end of the recording is an artefact caused by lack of insulation on the ear.) Note the otherwise parallel course of all three cranial temperatures and also the rate of sweating (response to the central warm stimulation). Tympanic temperature is higher by 0·2° C. It is not, like the temperatures at mucous membranes, depressed by small rates of water evaporation.

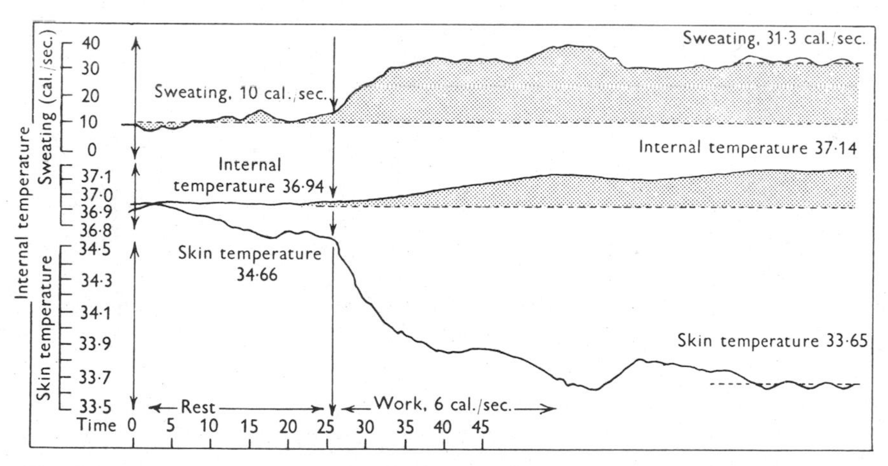

Fig. 4. Sweating and skin temperature are paradoxically related. The relation of sweating and tympanic temperature is concordant. A relative inertia of the tympanic measurement is apparent here and in Fig. 3. Deep cranial measurements (Fig. 6) respond more rapidly.

a response observed, the two stimuli must be subjected to experimental alterations with the largest possible degree of mutual independence. External thermal stimulation of the skin induces changes of internal temperature in the same direction. It does not, therefore, always give unequivocal answers. Internal change of temperature by heat-producing exercise (see Fig. 6) proved to be more suitable than external stimulation; unexpectedly, it did not result in an increase of temperature at the skin.

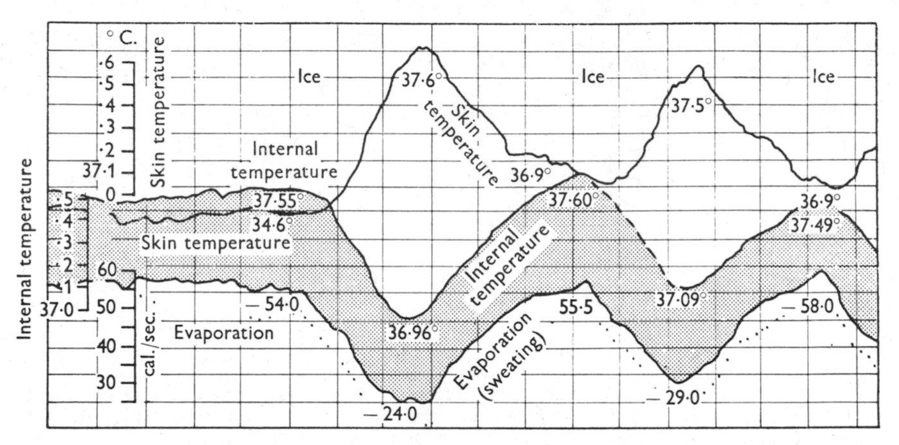

Fig. 5. Upon eating ice three times (1 pint of sherbet each time), tympanic temperature falls sharply. It returns each time spontaneously to normal. The rate of sweating (response) closely parallels tympanic temperature, the central stimulus. Skin temperature moves in the opposite direction. While the internal cranial temperature determines sweating rate by *physiological* (stimulus-response) relationship, sweating rate determines skin temperature by *physical* relationship (cooling effect of water evaporation). Environmental temperature in the gradient calorimeter was 45° C.

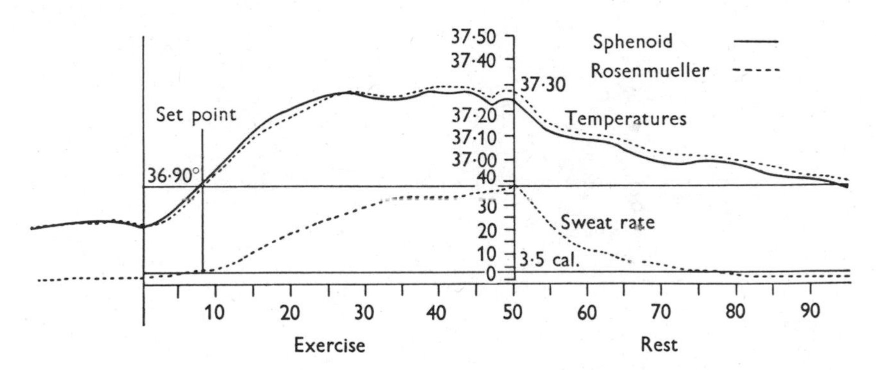

Fig. 6. When measured at the fast-following deep cranial sites, central temperature rises rapidly after the initiation of exercise. Sweating rate, the physiological response, appears to lag behind the stimulus-pattern after the onset, yet to precede it after cessation (explanation in text). In spite of rising central temperature sweating does not begin until set point is reached.

One of the basic tenets of classical thermoregulatory theory was that warm reception at the skin is the origin of physical heat regulation. Impulses from cutaneous warm receptors were thought to ascend into the anterior hypothalamic heat loss centre, where they were modified by some influence of central temperature, and then relayed into descending pathways for sweat gland action. This concept is not confirmed when subjected to experimental test (Figs. 4, 5, 6). In warm environments—where sweating is the effector mechanism of temperature control—gradient calorimetry reveals that skin temperature and sweating are paradoxically related whenever the stimulus is applied *internally*. On the other hand when the stimulus is *externally* applied there is *no* paradoxical relation between internal cranial temperature and sweating rate. Associated with the onset or *increase* of sweating with exercise is a *decreasing, not a rising*, skin temperature. For purely physical, not physiological, reasons this must be so, because the evaporation of water on the skin has a drastic cooling effect which is the very basis of the thermoregulatory effect of sweating. Our conclusion is, therefore, that sweating largely determines temperature of the skin, not vice versa, and that the relation between skin temperature and sweating is of physical, not physiological, causality. In Figs. 4 and 5, an opposed, mirror-image relation between skin temperature and sweating rate is demonstrated under quite different experimental conditions: in Fig. 4, central temperature was raised by metabolic action in muscular exercise; in Fig. 5, it was sharply depressed by ingestion of ice into the stomach. In both cases the patterns of skin temperature are paradoxically related to the patterns of sweating intensity.

On the other hand, from Figs. 4 and 5—and also from Figs. 6 and 7—it appears that the relation between cranial internal temperature and sweating is entirely different; it is a matching, not an opposed course, regardless of external or internal application of the stimulus. It is the relation expected when sweating is the response to cranial internal temperature as the stimulus. In Fig. 5 the relation appears as an almost perfect imitation of the stimulus-pattern by the pattern of response. In Fig. 4 the stimulus, measured this time at the eardrum, appears to be delayed *behind* the response. However, using a more rapidly responding method (sphenoid thermometry) for the stimulus in Fig. 6, the stimulus appears to *precede* the response. In this way, depending on the specific inertia of the methods of measurements, deviations from ideally matched time-courses may be observed in either direction. Since these deviations are well within the inertial delays of the methods applied, we must conclude that within our limits of error the response is explained as a result of the stimulus. Unknown, additional mechanisms of temperature regulation, as postulated by others (Belding & Hertig, 1962; van Beaumont, 1963), cannot be expected to be very significant quantitatively. Whether, in a body of 75,000 cal./deg. heat capacity, a response to exercise of doubled heat loss (50 instead of 25 cal./sec.) comes three seconds sooner or later would make a difference of only $\frac{1}{1000}$ °C. in average body temperature.

While different time relations must be expected when methods of different inertia are applied in two experiments, there is no similar explanation for a different time-course in one experiment, between the onset and the discontinuation of the central warm stimulation. In Figs. 6 and 7 and also, to a lesser degree, in Fig. 5, the change in response appears delayed after the onset, and accelerated after the cessation, of the stimulus. This relation was observed regardless of the way in which the control stimulus was altered, for example, by cooling with ingestion of ice, by warming

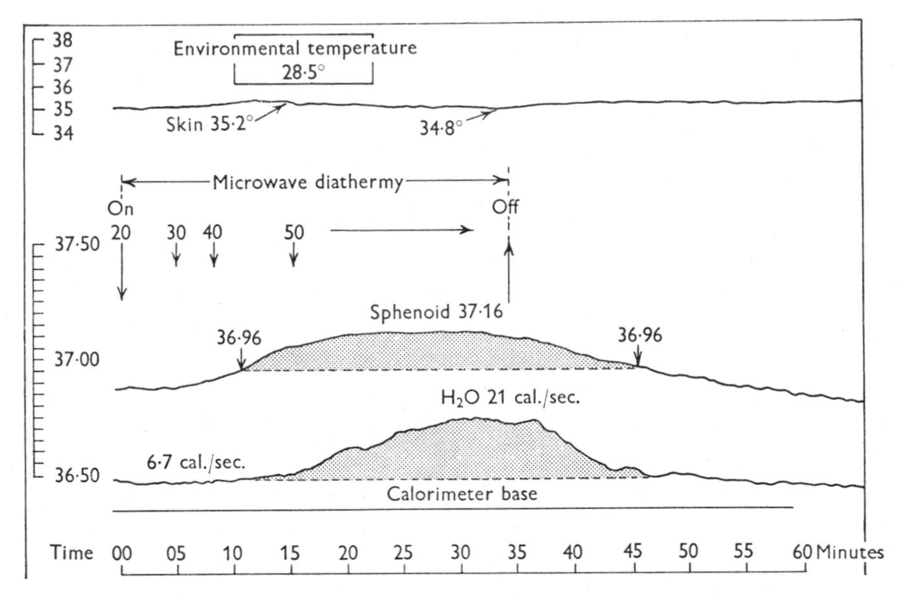

Fig. 7. When central temperature rises beyond the set point by heating with bilateral application of microwave diathermy to the carotid arteries, sweating rate rises with some delay. When the heating is discontinued, sweating rate falls more swiftly than the diminishing stimulus.

through physical exercise, or by microwave-diathermy applied to the carotid arteries. It seems therefore to be inherent in the physiological mechanism and thermoregulatory sweating as a characteristic of the neuronal networks on which the ultimate response depends. It seems, therefore, that whatever discrepancies in time-course are found between central warm stimulation and the response of sweating, they can be satisfactorily explained by one of two reasons: the physiological, inertial characteristics of the neural mechanism, or simply the inertial factors of the measurements. Speed as a physiological factor in thermo-regulation is frequently overrated. With an inertial body heat capacity of 80,000 cal./sec.° C. and a maximal sweating response of 80 cal./sec., deviations due to belated responses are small. A delay in response by as much as 100 sec. would be required to produce a transient increase of internal temperature by 0·1° C. This difference would tend to disappear in the steady state. Therefore the stimulus-response relations in the steady state deserve primary consideration. Fortunately, experiments in steady states are not subject to inertial errors which may obscure relations during transients. Steady states are best for seeing whether or not central temperature is only one of two stimuli, or indeed the dominant stimulus; whether the effect of *skin* warm reception on sweating is actually zero; and whether as a consequence, a different nature of the essential neurones in the anterior hypothalamus must be postulated, namely that they are terminal sensory receptors of a 'temperature eye', and not transmission-neurones of a synaptic chain, with an undefined sensory quality of the synapses to satisfy the classical assumptions concerning their temperature sensitivity.

PHYSICAL HEAT REGULATION IN THE STEADY STATE

The absence of driving warm impulses from the skin may be demonstrated by observations on the same individual during exercise or rest, and in a wide variety of environmental conditions. As shown in Fig. 4, muscular exercise in a warm environment with the associated overproduction of heat produces an *increase of internal* temperature while it usually results (because of intense sweating) in a *decrease of skin* temperature. Therefore, when skin temperatures of the same individual in a steady state of exercise and in a steady state of rest are compared—in two environments so chosen that internal temperature happens to be the same during the two observations—the skin temperature is much lower while the man is working than while he is at rest. If, in spite of the differences in skin temperature, the rates of sweating at the given cranial internal temperature were found to be the same in both observations, then there was no influence of skin temperature upon the intensity of sweating, and no warm impulses from the skin influenced the autonomic sudomotor system. Such pairs, or near pairs, of observations may be obtained by taking the human subject on various days through steady states of rest or exercise at many different environmental temperatures. Subsequently, after weeks or months, the observed rates of sweating may be plotted against skin temperature and cranial internal temperature (see Fig. 10). With this technique and evaluation, skin plots were found to be widely scattered. On the other hand, the plots against cranial internal temperature reveal (Fig. 8) that thermoregulatory sweating in a warm environment is a reproducible function of central temperature, regardless of skin temperature. Skin temperature was much lower—1½° C. on the average—in the work observations as compared with the observations at rest. In spite of this large difference, measurements fell upon one regression line within experimental errors. If skin temperature had any influence upon the rates of sweating in warm environments, all observations during exercise would have been shifted to the right or downward. The fact that there is no detectable, systematic right-shift indicates that warm reception at the skin plays, if any, an imperceptible role. On the other hand, the unusual power of the driving *central* warm

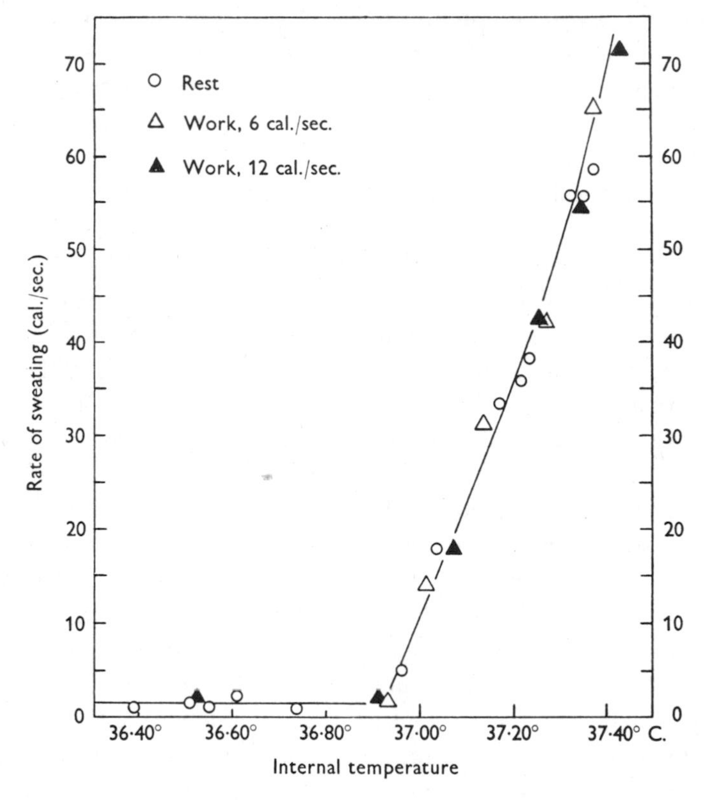

Fig. 8. Steady states of sweating rate were attained at rest (circles) and during exercise (triangles) in the gradient calorimeter in a wide variety of environmental temperatures (nude, 10–45° C.). Plotting of the sweating rates against tympanic temperatures reveals no 'right- or downward-shift' of the working observations (cooler skin) compared with resting observations (warmer skin). Sweating is determined directly by *central* warm stimulation, independent of warm impulses from the skin.

impulses appears from the graph in Fig. 8. To produce a change in the sweating heat loss rate equivalent to the heat produced by 10% of a normal metabolism, a temperature change of 1/100° C. suffices. The regression line meets the zero base of sweating at a sharply determined intersect, the 'set point of the human thermostat'.

These experimental facts lead to a far-reaching conclusion: the anterior hypothalamic 'heat-loss centre' from which, by general agreement, the efferent impulses for sweating and vasodilatation originate, is not what it was assumed to be in classical theory, that is, a synaptic relay station for peripheral drives from the skin. It can operate *without* afferent, sensory impulses. Therefore its essential neurones must be terminal sensory cells for temperature reception comparable to the first neurones of the retina for

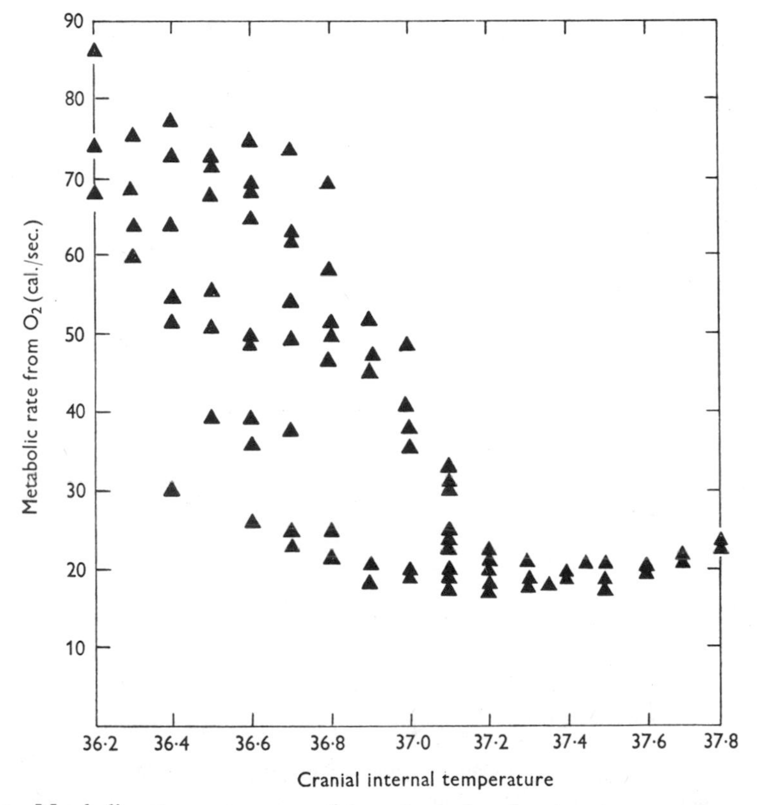

Fig. 9. Metabolic rates were measured in water baths of various temperatures, ranging from 36 to 14° C., where a large variety of combinations between central and cutaneous temperatures can be obtained. As opposed to *sweating* rates (Fig. 8) the *metabolic* rates observed are *not* a unique function of cranial internal temperature. Resolution of this scattered plot is shown in Fig. 10. It became possible by considering cold reception at the skin as the driving stimulus.

the reception of light. This 'temperature eye' is only one of several internal sense organs originating from, or located within, the brain stem: the optical eye, the osmoreceptors of the hypothalamus, the chemoreceptors for hunger or satiation, and the particularly strong receptors for pH or CO_2 in the medulla from which unconscious or conscious reactions of unbearable urge arise.

The experimental proof of these conclusions on the temperature sensor by human calorimetry has been accepted with considerable reluctance by workers in the field. Apart from numerous unrecorded discussions of the subject at meetings of the American Physiological Society there are, fortunately, several pertinent publications with *experimental* contributions. A quantitative study was undertaken by Belding & Hertig (1962) with

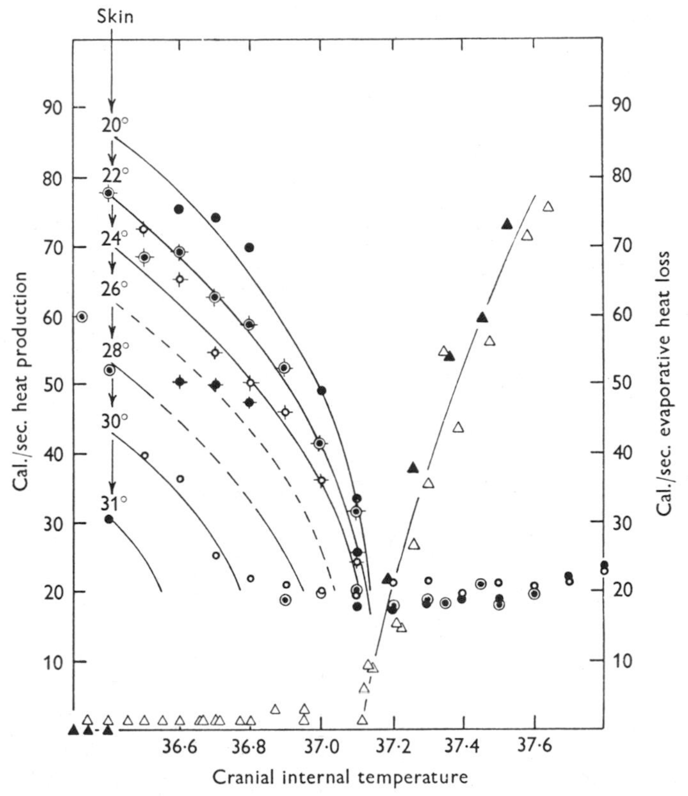

Fig. 10. Rates of metabolic heat production in response to cold (circles), and rates of sweating in response to warming (triangles) were plotted against cranial internal temperatures. The function of the 'human thermostat' with set point at 37·1° C. is presented in terms of the chemical and physical responses to peripheral cold and central warm stimulation. (Healthy young man, age 20, height 179 cm., weight 69·1 kg.)

determinations of sweating from body weight. They found, as we did, no systematic differences in sweating at widely different skin temperatures during the states of rest or exercise, but at equal cranial internal temperatures. Belding & Hertig observed that 'the correlations are good though not as perfect as those reported by Benzinger'. Another, similar, study was made by Minard (1963). He also found considerable scattering and, unexpectedly, a higher sweating rate in the working individual with a cooler skin (which is the opposite of what might have been anticipated). What the experiments by Minard as well as Belding suggest, within the limits of their errors, is a tentative conclusion that warm impulses from the skin have no demonstrable effect on sweating. The wider scattering of their measurements may be the result of a less-defined experimental situation. Convection and other environmental factors are not as tightly controlled in open warm rooms, as they are in the narrow space of a gradient calorimeter with a forced air circulation of 900 l./min. The accumulation of water on the skin has been reported by Buettner (personal communication) to impede the secretion of sweat. The increased convection during exercise may be one explanation for the elevated sweating rate of working individuals in Minard's observations and also for scattering in open room experiments in general. It is possible that gradient calorimetry will be needed as a method of investigation if the results of Figs. 8, 10 and 15 are to be confirmed in other laboratories with similar reproducibility. Most recently, however, Jeremy Crocker (1964, personal communication) has obtained undisturbed regression-lines and sharply determined, individually different set points for sweating rates plotted against tympanic temperatures, in measurements with *local* sweating of one arm in a separate enclosure.

Randall *et al.* (1963) have attempted to disprove the conclusions of Benzinger (1959) in a different way. With a highly sensitive, qualitative test for water on the skin (iodine–starch–paper) they observed, in the warm chamber, the appearance of water on the skin of paraplegic patients with complete transections of the spinal cord and on healthy individuals contralaterally, when an arm was immersed in water of 47° C. No estimate was given of the possible quantitative significance of this phenomenon for the loss of calories. Therefore as long as quantitative evidence is not produced, it must be concluded from series of quantitative experiments such as those in Figs. 8, 10 and 15, that for a man with an intact nervous system there is no measurable contribution from skin warm reception to the sweating response under physiological stress. This does not mean that observations on paraplegics, or on healthy individuals, sweating under contralateral stimulation with 47° C. water (an excessive temperature), do not promise interesting results. Such observations point to important spinal connexions between certain segmental sensory and effector pathways or even to abortive spinal mechanisms of control. The results must be applied with caution to problems of thermoregulation as they operate under physiological conditions: a skin temperature of 47° C., as in Randall's tests, is far out of the range in which the physiological temperature regulation is operative in warm environments. Not even in environments of 50° C.

(122° F.) have skin temperatures beyond 38° C. ever been observed in this laboratory. At 47° C. skin temperature, pain and vascular effects are active. These may for various reasons result in a qualitatively discernible emergence of water from sweat glands.

Similarly, Randall *et al.* contested the validity of tympanic observations of temperature with animal experiments during which the carotid arteries were ligated, and discrepancies between hypothalamic and tympanic temperatures were observed. Obviously, it must be expected that the temperature of the heat-losing tympanic membrane would fall, whereas the temperature of the heat-producing hypothalamic tissues must rise when circulation is interrupted. Yet Randall considered the observed differences as a valid argument against the method of tympanic thermometry. However as a parallel the accepted method of rectal thermometry was not subjected to a test by ligation of the abdominal aorta.

The favourable characteristics of tympanic measurements were recently confirmed, on primates by Rawson & Hammel (1963) and on man by Piironen (1963). Their usefulness must be assessed against the fallacy of rectal temperature, against the discomfort associated with oesophageal thermometry, against the requirements of surface anaesthesia for sphenoid sinus or Rosenmueller fossa measurements and against the impossibility of taking measurements in the hypothalamus of humans. Rawson & Abrams (1964) have recently observed a tight homeostatic control of hypothalamic temperature in freely-moving dogs over long periods by telemetering devices.

Appropriately, this discussion should close with some remarks on the results of Fig. 8. Sensory warm impulses from the skin exert no measurable influence on sweating. In retrospect it appears that warm receptors could not possibly exert such an influence without upsetting the thermoregulatory system and defeating its purpose. If sweating depended on warm-impulses from the skin, then the declining temperature of the skin with every onset of strenuous exercise would curtail or eliminate the response of sweating at the time when it is most needed. (It is not in the state of rest but in the state of exercise that a hot environment is dangerous to survival.) When the skin is the area of attack for a vigorous cooling mechanism it is not simultaneously a suitable location for the controlled temperature variable in a delicate system of regulation. Temperature regulation in a home would not be improved if the warm sensor of the thermostat were sprinkled with water and cooled during periods of overheating. For these and similar considerations Fig. 8 may be considered as a clarifying, principal result of the experiments reported here, and a starting point from which the other thermoregulatory mechanisms may be resolved by similar techniques and arguments.

CHEMICAL HEAT REGULATION

For an investigation of the 'chemical' control of human body temperature in cold environments, it is an advantage to replace the direct with an indirect method of calorimetry, where the response is an increased rate of chemical change, reflected quantitatively in oxygen consumption. It is not, as in *physical* temperature regulation, a change in loss of metabolic heat first stored in the enormous heat capacity of the body. The absence of a requirement for *direct* calorimetry makes it possible to bring about the necessary changes simply through water baths of chosen temperatures. The artificial introduction or withdrawal of heat for stimulation—which would make direct calorimetry almost impossible—does not interfere with indirect calorimetry by measurement of oxygen consumption. Observations in water baths also eliminate immediately the problems of manipulation and measurement of skin temperature: following an old tradition, the water temperature may be taken as the surface temperature of the skin, when the water is vigorously stirred. For oxygen consumption, the method of continuous gas analysis by Pratt (1958) and a method for direct recording of metabolic rate (Benzinger, 1938) were applied. Using these methods, an attempt was made to find reproducible relations between internal cranial temperature and metabolism.

Measurements taken under a great variety of conditions were expected to align themselves, like the measurements of sweating, in a mirror image of Fig. 8, with oxygen consumption rising from a set point as internal temperatures decline. Instead, the measurements were found to be widely scattered (Fig. 9). Nevertheless, the rates of oxygen consumption appeared to obey one rule: above the set point of the human thermostat for the onset of sweating (37·1° C. in this individual), oxygen consumption was *not* elevated regardless of how low the temperature of the skin happened to be. This clearly indicated an inhibitory role of internal warm reception in the chemical heat regulation of man.

Meanwhile a drastic influence of cold reception at the skin was found during short immersions in cold water in which there was not enough time for cranial internal temperature to be substantially lowered. It became apparent that both temperatures, cutaneous as well as cranial internal, were involved in the determination of the metabolic rate. To resolve the quantitative roles of the two factors graphical analysis was again applied.

In Fig. 10, a plot of metabolic rates (and also sweating) against *internal cranial* temperatures, 'best lines' were drawn through points of equal skin temperature. Any of the lines in the graphic resolution of Fig. 10 therefore shows the relation of metabolic rate to *central* thermo-reception while skin temperature is constant. In Fig. 11, a plot of the same metabolic rates against *skin* temperatures, 'best lines' were drawn connecting points of equal cranial internal temperature. Therefore, the lines in Fig. 11 illustrate how cold reception at the *skin* affects the metabolic rate, while cranial internal temperature is constant. This leads in both representations to a quantitative resolution of 'chemical' heat regulation in terms of one effect—the metabolic response—and two causes: the central and peripheral temperature stimuli.

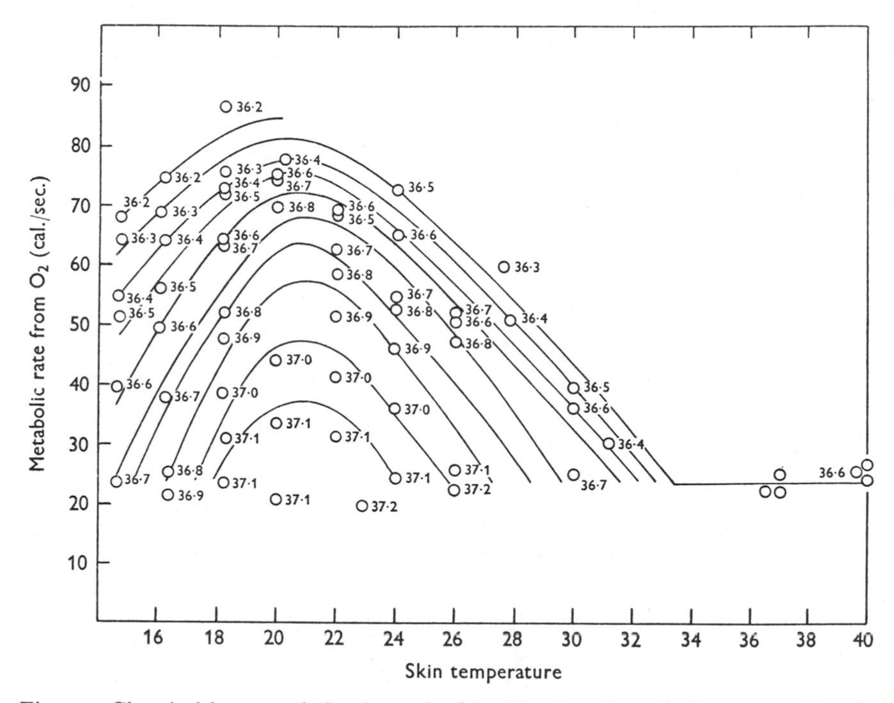

Fig. 11. Chemical heat regulation is resolved in this plot of metabolic rates against *skin* temperatures with lines drawn through points of equal *central* temperature. The maximal rates of O_2-consumption are observed at 20° C., the temperature of maximal firing rates of human skin cold receptors (Hensel & Boman, 1960). As central temperature increases, the metabolic response is progressively inhibited. Inhibition becomes complete at the set point for sweating, 37·1° C. With skin temperatures above the cold threshold metabolism is a resting level, regardless of central temperature.

Besides the metabolic rates in response to cold, the rates of sweating in response to central warm reception of this individual have been represented in Fig. 10. With their undisturbed regression line, the measurements of sweating demonstrate the basic difference between the sudomotor and the metabolic systems. The *sweating* rate in a warm environment is uniquely determined by an *internal warm* receptor, without demonstrable influences from the skin. In contrast the *metabolic* response to cold, while elicited by *cutaneous* cold reception, is sharply and reproducibly restrained by increasing central temperature. The central warm inhibition of the metabolic response to cold becomes complete at the 'set point' of the human thermostat for sweating, 37·1° C. in this individual.

THE NEUROLOGICAL BASIS OF HEAT REGULATION

As shown in Fig. 10, peripheral cold and central warm reception, the primary causes of the physiological process of thermoregulation, are inseparably linked with their ultimate effects (excessive losses or overproduction of heat). Extending between these terminal points of the causal physiological chain there must exist a network of *anatomical* structures such as receptors, afferent pathways, synapses and efferent pathways. Moreover, physiological activity must be demonstrable by action potentials in these anatomical components of the thermoregulatory system, whenever they are in action. Such individual structures and activities had been discovered long before it became possible to resolve the mechanisms of thermoregulation as a whole in quantitative terms by calorimetry. The neurosurgical observations are documented in an extensive literature to which reference must now be made. This will lead not only to a more thorough understanding, but also to a mutually independent confirmation of the results of one approach through the results obtained by another in the long scientific history of this problem.

Electrical characteristics of skin cold receptors

The neurological considerations begin, appropriately, with cold receptors, by reference to the classical work of Hensel & Zotterman (1951). The dome-shaped curves in Fig. 11—representing the metabolic response of man to cold reception at the skin—appear like a reproduction of Hensel & Zotterman's plots, representing firing rates from single cold receptors of the cat in response to cold stimulation (Fig. 18). Thresholds and maxima are naturally different in the mouth of the cat from thermoreceptor thresholds and maxima in the skin of man. However, when Hensel & Boman (1960) extended their observations to nerve fibres exposed in their own radial nerves, they found for maximal response and for the threshold the same receptor temperatures, where direct calorimetry in man reveals the maximum—or the extrapolated threshold—of oxygen consumption (see Fig. 17). There could be no doubt that Rubner's 'chemical' heat regulation, the metabolic response to cold, is indeed produced by elevated firing-rates from the combined cutaneous cold receptors. Further analogies confirmed the tentative conclusion that the firing rates of isolated neurones and the oxygen consumption of man respond in the same way, namely with

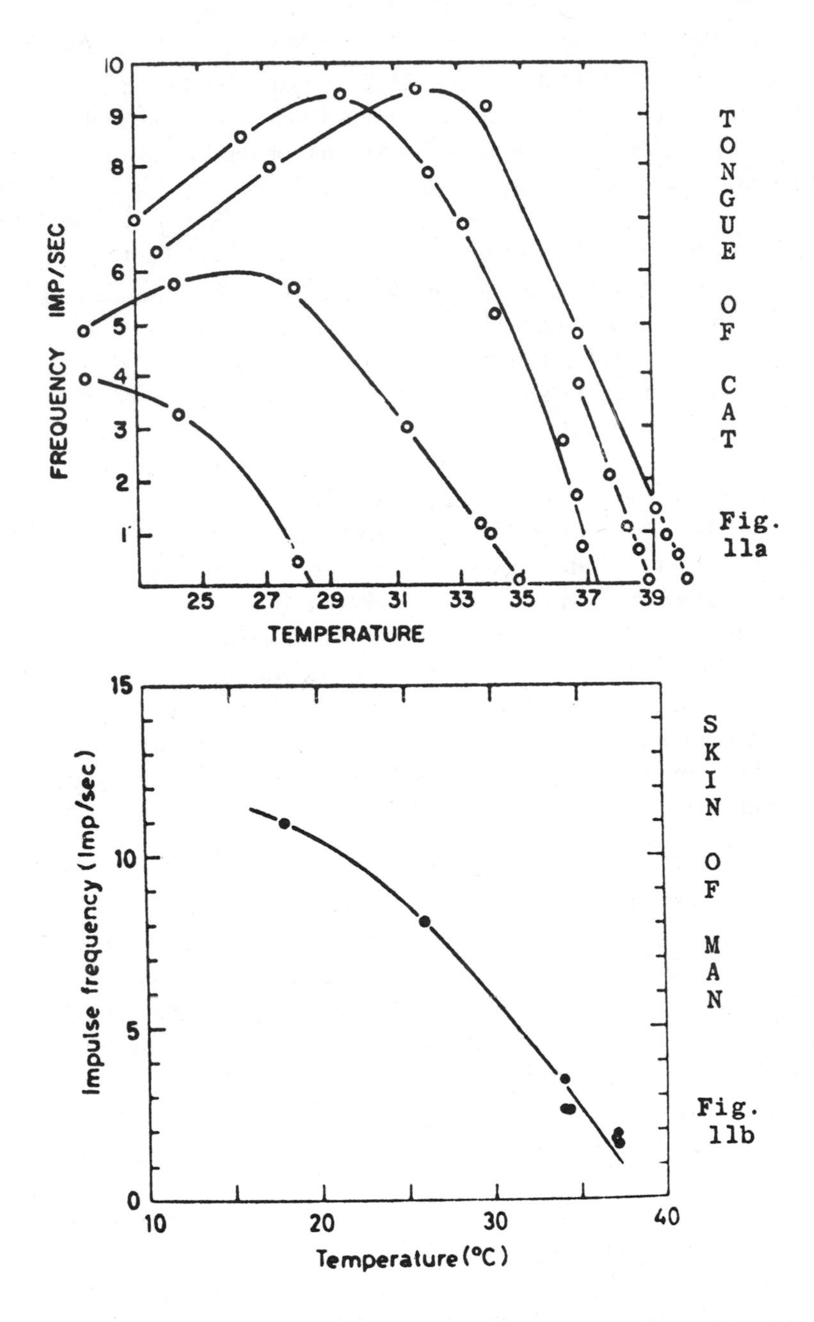

FIRING RATES OF COLD RECEPTORS IN RESPONSE TO TEMPERATURE
REF. HENSEL AND BOMAN(1960) AND HENSEL AND ZOTTERMAN(1951)
NOTE SIMILARITY WITH OXYGEN CONSUMPTION IN MAN (FIG.11)

overshooting intensity to sudden cooling and with overshooting inhibition to sudden rewarming. When the initial 'spike' subsides, both the firing rates of thermoreceptors and the oxygen consumption of man settle at new, appropriate levels corresponding to the temperature of a steady, or a slowly changing state (see Figs. 14 and 15). A principal function of the cold receptors in the skin is to elicit and to drive chemical heat regulation, the metabolic response to cold.

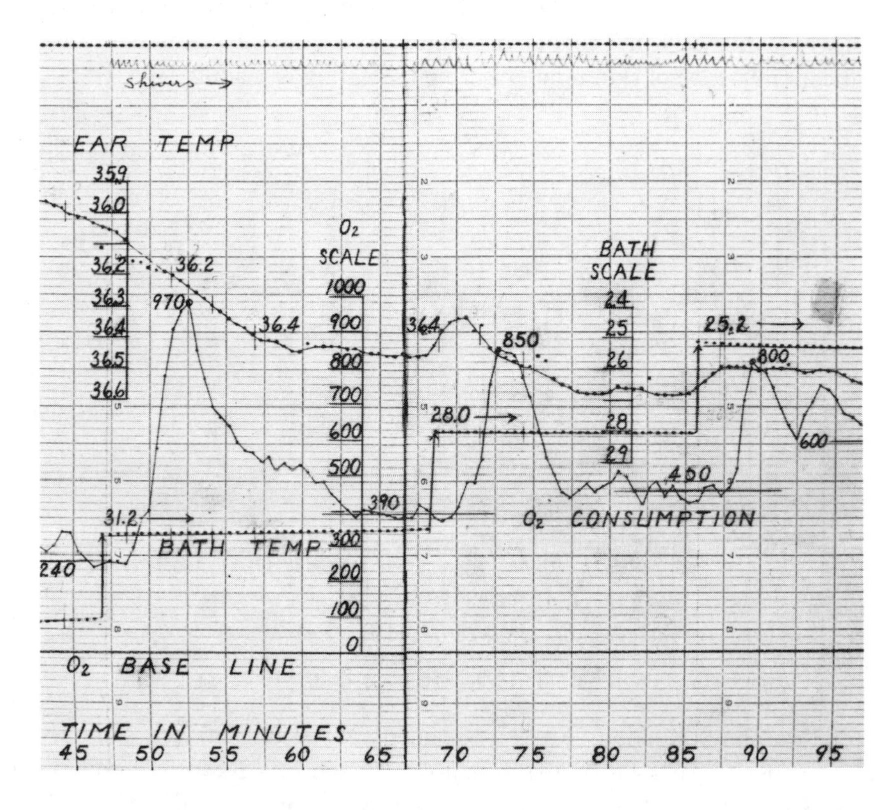

Fig. 12. Response by metabolic heat production to cold stimulation of the skin. Each time upon the sudden lowering of temperature an overshooting, transient response by oxygen consumption is observed. Then a new and elevated level is attained depending on the new, lower level of skin temperature. (See Fig. 13 for corresponding temperature-dependent patterns of firing rates of cold receptors.)

The inhibition of sweating by cold impulses from the skin

Growing experience with the threshold and with other characteristics of the cold receptors of the skin has made it possible to demonstrate another, an inhibiting, influence of these receptors, on sweating (Benzinger, 1961). A seeming contradiction between experiments made by this and other laboratories was thus resolved. Kuno (1934), Robinson (1949) and others had observed that sweating may be diminished when the skin is cooled. Their interpretation was that cooling of the skin abolished cutaneous warm impulses, which they assumed were the origin of the sweating response. From Fig. 14 it appears instead that sweating rates are reduced only when skin temperature is lowered into the range below 33° C., where cold receptors fire at increased rates and where metabolic heat production rises when no central warm inhibition takes place. This right and downward shift of sweating rates is shown with an obese subject as he works intensely in a cool environment. The graph explains how any significant influence of skin temperature on sweating is readily detected by

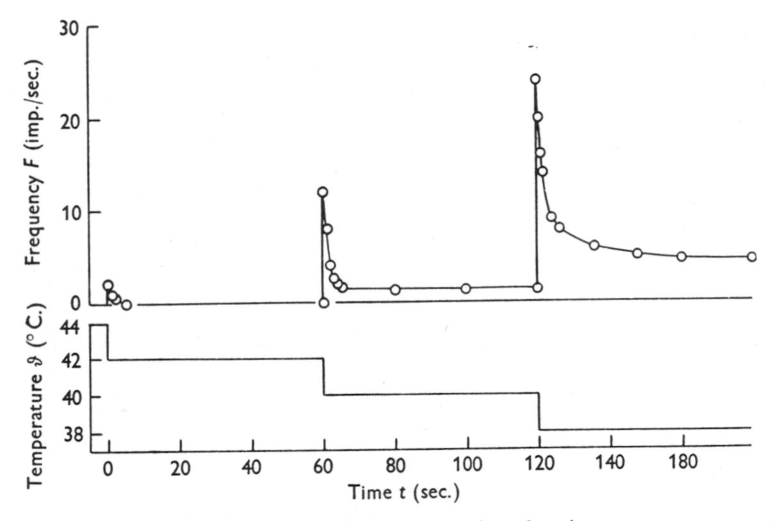

Fig. 13. Firing rates of cold receptors of the cat are plotted against receptor temperatures. Note the correspondence of these patterns with skin-temperature-dependent patterns of human oxygen consumption in Fig. 12. (After Hensel & Zotterman, 1951.)

this method and evaluation. Fig. 15, obtained on the same subject, demonstrates again that such an influence is absent when *cold* receptors are not active. In short, while sweating is inhibited by peripheral cold reception, the central warm drive for sweating operates *without* peripheral sensory support. It is a *central* sensory function.

The inhibition of sweating from the skin is 'anti-homeostatic'. During strenuous exercise in a cool environment it must inevitably lead to internal temperatures which exceed the closely guarded norm. This phenomenon has long interested physiologists. Nielsen (1938) has given an excellent description of the facts. He advanced the hypothesis that an elevated internal temperature is favourable to the performance of exercise and that a special, unknown mechanism of thermoregulation achieves this adjustment. The explanation of human thermoregulation as a whole did not become easier through this additional requirement at the time when the postulate was made.

The anti-homeostatic effect of inhibited sweating lasts no longer than a paradoxical condition of the body with a cold shell and an overheated core. Such a discrepancy cannot endure, as heat will be conveyed from the

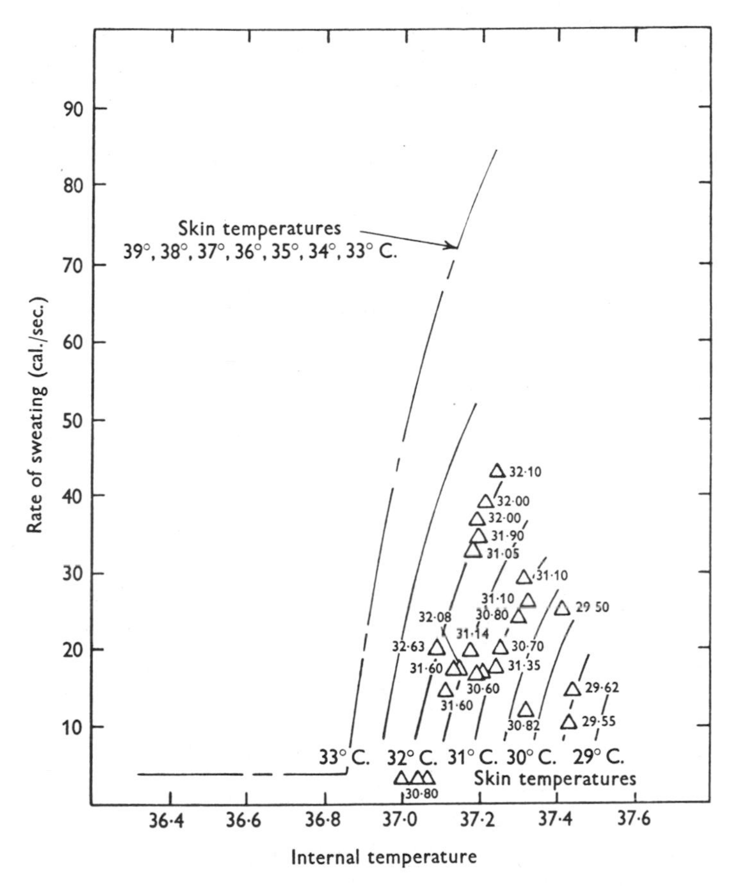

Fig. 14. Cold inhibition of sweating is demonstrated in the gradient calorimeter with an obese subject working strenuously at cool environmental temperatures. Sweating rates at skin temperatures below 33° C. are 'right-shifted' in comparison with similar observations in the warm range of skin temperature (Fig. 15).

heated core into the chilled shell with the blood stream. Once the internal norm of central temperature is restored, excess sweat on the skin would be of no advantage. At cool skin temperatures water would only evaporate slowly. Yet it would lead to uncomfortable conscious sensations of cold. Through longer lasting effects it could result in central cooling below the norm. The inhibition of sweating during cold reception at the skin, though anti-homeostatic, prevents such overshooting action. A simultaneous activation of shivering and sweating mechanisms is prevented by double safeguards: internal warm reception prevents shivering; sweating is inhibited when cold reception at the skin occurs. No regulatory activity takes place while central and peripheral messages on temperature are in contradiction.

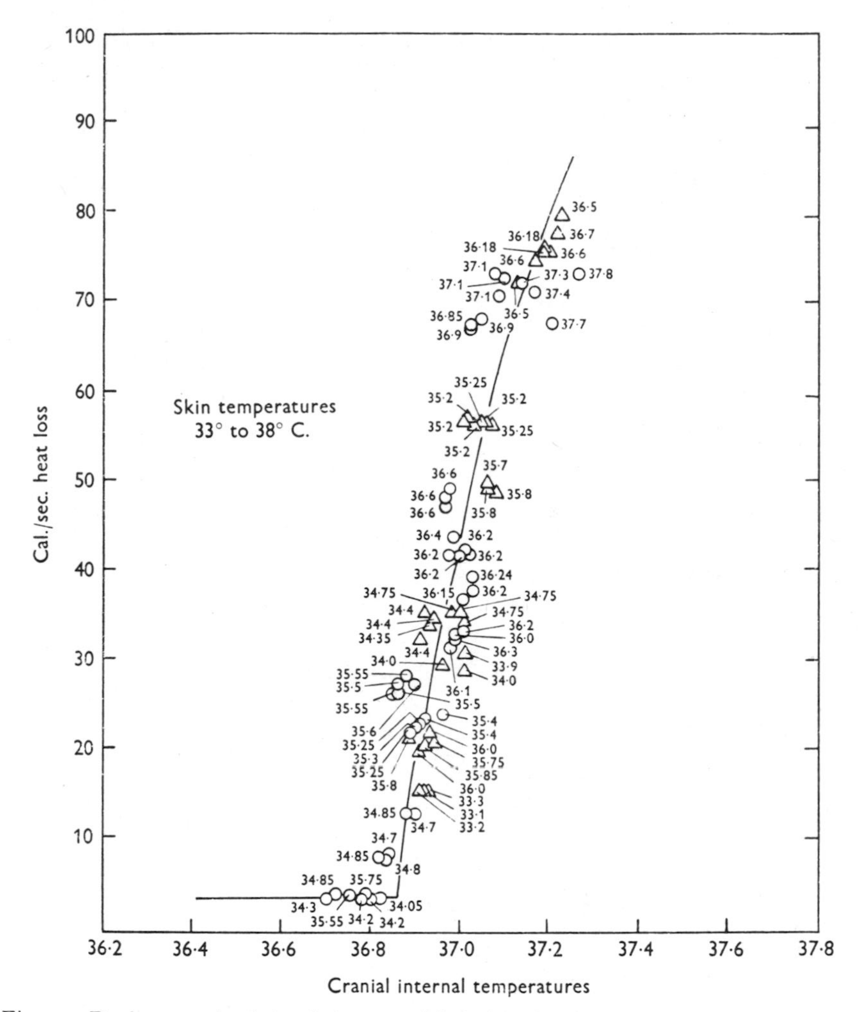

Fig. 15. During exercise (triangles) or rest (circles) in the gradient calorimeter at various environmental temperatures, sweating rates were measured and plotted against cranial internal temperatures. At skin temperatures above 33° C. there is no right shift of working observations (cool skin) compared with resting observations (warmer skin), as observed in the cold range on the same individual (Fig. 14).

The opposite of this condition—warm skin with subnormal central temperatures—must also be briefly considered. In our experiments this condition was frequently encountered and found to be the most pleasant of all. Thermoregulatory action was found to be absent in this state. Subjects quite often fell asleep. Neither peripheral nor central drives, neither conscious nor autonomic receptions seemed to occur under such conditions in the range of normal thermoregulation. Only in hypothermia, on animals, have central cold effects on metabolic rates been observed by R. Thauer (personal communication) and referred to cold effects upon spinal neurones. Synaptic cold block of descending impulses for central inhibition in the spinal cord has not yet been ruled out as a cause of this phenomenon in hypothermia, which is not pertinent to homeostasis, the maintenance of the norm.

Centres and pathways of thermoregulation

The temperature-dependent patterns of human oxygen consumption in response to cold and the electrical characteristics of skin cold receptors form a link of knowledge between the terminal sensory stimulus and the firing rates of the first neurones. This connexion opens the way for the present discussion to enter into pathways that are already established and accepted.

After ascending in the lateral spino-thalamic (Gower's) tract, sensory fibres have been traced toward main central synapses. In studies by Anderson & Berry (1959) with degeneration methods, fibres from the tract were seen to ascend into and to terminate within the subthalamus and postero-lateral (not anterior) hypothalamic regions. Accordingly the afferent pathway of peripheral cold reception leads into the posterior hypothalamic 'heat maintenance centre'.

The Krehl–Isenschmid centre of chemical temperature regulation

The calorimetric observations on the metabolic response to cold have left no doubt that sensory impulses with firing rates controlled by the temperature of cutaneous cold receptors are ultimately relayed to effectors in metabolizing tissues. These are, at least predominantly, muscles, as demonstrated with the parallel phenomenon of shivering. The site (Fig. 16) at which the afferent and efferent pathways for shivering are connected in the central nervous system, *P*, was discovered by Isenschmid & Krehl (1912). Destruction of this 'heat maintenance centre' in the posterior hypothalamus eliminates the metabolic response to cold, while physical responses to overheating are left intact (Keller & Hare, 1932). On the other hand, extensive destruction of tissue surrounding the Krehl–Isenschmid centre does *not* eliminate the shivering response (Isenschmid & Schnitzler, 1914). A destruction of the anterior hypothalamic 'heat-loss centre' may even lead to more intense responses against cold by shivering (Pinkston, Bard & Rioch, 1934).

It is important to note that the shivering centre does not itself react to local cooling. Hemingway, Rasmussen, Wickoff & Rasmussen (1940) showed, and Freeman & Davis (1954) have confirmed, that cooling of the area in dogs does not abolish the shivering response. This proves the purely synaptic nature of this central function. The Krehl–Isenschmid centre is 'temperature blind'. It operates with afferent cold impulses from the

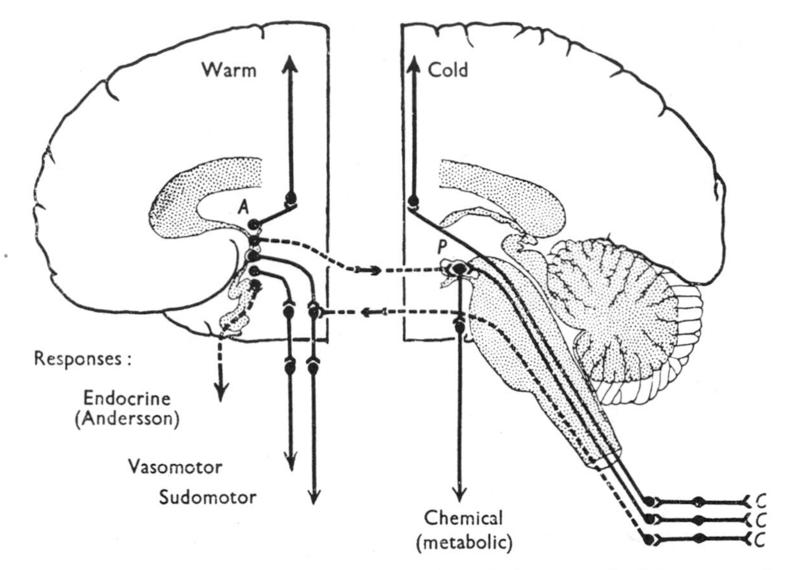

Fig. 16. Centres and pathways of thermoregulation. Pathways and minimum numbers of synaptic stations are shown as postulated from observations by indirect and direct calorimetry. Background drawing of brain locates 'centres' *A* and *P* (according to the classical observations of experimental surgery) in the anterior and posterior hypothalamic centres. Function of *P* is merely synaptic, for transmission of driving cold impulses from the skin. *A* is a terminal warm sensor, emitting driving warm impulses for sweating, for vasodilatation and for conscious, unpleasant sensations of warmth (see Fig. 18). The only connexions between the anterior and posterior thermoregulatory systems are two inhibiting pathways (dotted). The upper pathway inhibits metabolic overproduction of heat when *A* is warmed, the lower pathway inhibits sweating when the skin is chilled. Driving impulses for the conscious sensation of cold arrive at centres of higher nervous activity *without* inhibition at 'centre' *P* (see Fig. 17).

periphery and is not capable of producing thermosensory impulses of its own. It is a structure fundamentally different from the anterior hypothalamic 'heat-loss centre' to be discussed in the following section.

In Fig. 16 the Krehl–Isenschmid centre receives an afferent pathway for driving impulses from the skin through spinal ganglia. The efferent pathway to metabolizing effectors is shown only schematically with one synapse as a minimum, instead of a chain of neurones. The descending pathways for shivering have been thoroughly established by Hemingway,

Forgrave & Birzis (1954) with lesions and with the electrical recording of impulses transmitted through them, as well as with observations of shivering on dogs. With termination at the 'centre' *P* an afferent, descending pathway is drawn in Fig. 16 which will soon be discussed, together with its origin, the anterior hypothalamic 'heat-loss centre' *A*. Without such an afferent pathway from above, the 'shivering centre' *P* would receive no information on central temperature. Since it has been proved to be itself indifferent to thermal stimulation, it can only react to cold stimulation of the *skin*. It cannot by itself fulfil a homeostatic function with respect to the *internal* temperature of the body. The homeostatic, 'set-point-oriented' function of human chemical thermoregulation must have a different origin.

The 'temperature eye' or 'human thermostat'

In 1885 two medical students, Aronsohn & Sachs, working with the physiologist N. Zuntz, destroyed experimentally in rats a certain area of the brain stem. They observed that these animals were no longer capable of maintaining a normal temperature in a warm environment. Through this discovery physical heat regulation became known to be a neural homeostatic mechanism. Not until much later was another decisive observation made by an American medical student, Barbour, working with Meyer in Vienna (Barbour, 1912). *Thermal* stimulation applied to the anterior hypothalamus of dogs elicited in his experiments thermoregulatory changes of internal temperature. This discovery was not exploited further until the area was described more clearly and the responses which it elicits were directly observed (Magoun, Harrison, Brobeck & Ranson, 1938). After confirmation of their findings by several others a most thorough delineation of the structure was carried out by Andersson, Grant & Larsson (1956), who also succeeded in producing deep hypothermia in goats by electrical stimulation of the region. Andersson (1962) also first demonstrated the release of hormonal responses upon cooling of the 'centre', given by the thyroid, through mediation of the hypophysis.

Action potentials from the preoptic region in response to temperature changes were first observed by Curt von Euler (1950) who saw steady levels of potential rising and falling with experimentally induced temperature changes. In 1958 the terminal sensory nature of central warm reception was established by calorimetric experiments on man. Spike potentials in response to *warm* stimulation were first recorded by Nakayama *et al.* (1961) on 500 single units in the preoptic region. From no other region could such potentials be elicited. No units were observed that increased their firing rates on cooling.

Thus 'centre' *A* in Fig. 16 and its function as a warm sensor appear to be established by independent evidence from many different sources. The individual neurological findings are not only consistent with one another but also with the evidence obtained by gradient calorimetry, by recording oxygen consumption and cranial thermometry in man. The peculiar co-ordination (Fig. 10) by which sweating begins at the temperature where the inhibition of thermoregulatory metabolism becomes complete, strongly suggests that the two functions are exerted by the same structure. The independent, neurosurgical evidence for this postulate is strong. Hemingway *et al.* (1940) eliminated shivering in dogs by warming 'centre' *A* from below with diathermy applied to the base of the brain, anterior to the optic nerve crossing. Freeman & Davis (1954) confirmed this finding with electrodes for heating, placed inside the preoptic centre. Similar heating applied to the Krehl-Isenschmid centre was ineffective in both of these investigations. Moreover, Hemingway (1957) identified a pathway leading from 'centre' *A* to 'centre' *P* by action currents recorded from this pathway during electrical stimulation of 'centre' *A*. The function of inhibiting shivering and heat production must therefore be ascribed to the anterior hypothalamic 'centre' *A* from which the responses of sweating and vasodilatation originate. All central thermosensory functions seem to be vested and co-ordinated in one single organ.

Between the two 'centres' *A* and *P* there seem to be no other proved connexions except the two represented in Fig. 16 by dotted lines. Both of these pathways convey inhibiting, not driving, impulses. The lower of the two inhibiting pathways, dotted in Fig. 16, is not known in detail. Yet the results of calorimetry with the inhibition of sweating leave no doubt that cold receptor impulses from the skin gain access to the neural pathway of sweating somewhere between its origin and termination. In Fig. 16 the connexion for this inhibition was arbitrarily entered at the only synapse which represents the network schematically. The actual inhibiting synapse may be located at one of many conceivable levels, perhaps as low as in the spinal cord.

This leaves for further discussion only two of the pathways in Fig. 16, namely the ones leading upward into the cerebral cortex for conscious sensations of warm or cold, from where the voluntary motor actions of behavioural temperature regulation arise.

THE MAINSPRING OF THERMAL DISCOMFORT AND THERMAL BEHAVIOUR

Thermal behaviour utilizes as effector mechanisms the combined resources of human motility, experience and intelligence. These effectors are not different from those required in any other type of intelligent behaviour.

Nevertheless, the sensory mechanisms are specific and unique to thermoregulation. They must consist of first neurones for temperature sensation, and of afferent pathways, which ultimately gain access to the pre-motor neurones of the cerebral cortex for behavioural action. For automatic thermoregulation *central warm* and *peripheral cold* reception were found to be the two main sources of essential impulses. For the *conscious* sensations of warm or cold on which behavioural control depends the sources of sensory impulses had to be uncovered.

Sensation of cold

Human subjects can satisfactorily reproduce the subjective threshold, where overall cold sensation begins, in a water bath when temperature is very gradually lowered. From a warm, through an indifferent, into a cool range the temperature was lowered at a rate of one degree per hour. A wide variety of cranial internal temperatures coinciding with certain skin temperatures was attained by preconditioning in various water baths. Subjective thresholds of cold sensation were then observed and plotted against cranial internal temperature in Fig. 17. The threshold of subjective cold sensation does not appear to be dependent on cranial internal temperature. There is only a slight decline with higher cranial temperatures. It is presumably an artefact, characteristic for a lean individual. With an obese individual the declining tendency was hardly noticeable. The temperatures of the receptors proper are higher than their substitutes—the surface temperatures of the skin which had been measured and plotted in Fig. 17—only when the cutaneous receptors are not isolated by body fat from the core and when internal temperature is appreciably above the temperature of the skin. With appropriate consideration for this error, no visible dependence of cold sensation on *internal* cranial temperature is found in this range.

This observation seems essential and quite unexpected. The *metabolic* response to the excitation of cold receptors in the skin (as shown previously with Fig. 10) is *not* independent of internal temperature. It is dramatically inhibited when cranial internal temperature rises toward the set point. This appears again from Fig. 17, where the thresholds of the *metabolic* drive in response to cold at the skin were plotted against internal cranial temperature together with the thresholds of the *conscious* cold sensation. At the set point of the human thermostat the threshold for metabolic action approaches infinity. In contrast the threshold for conscious cold sensation seems unaffected. Two major conclusions follow from this plot:

(1) Central temperature seems *not* to contribute driving impulses to the conscious sensation of 'being cold' in this physiological range.

(2) Unlike those impulses that drive the metabolic system the impulses that make us *feel* cold are *not* subject to central warm inhibition.

It follows that there must be two different kinds of afferent pathways for cold impulses from the skin: one kind that possesses synaptic connexions in the posterior hypothalamic centre, subject to inhibition when central warm reception occurs, and another kind that bypasses the hypothalamic inhibition on its way through the thalamus opticus into cortical areas.

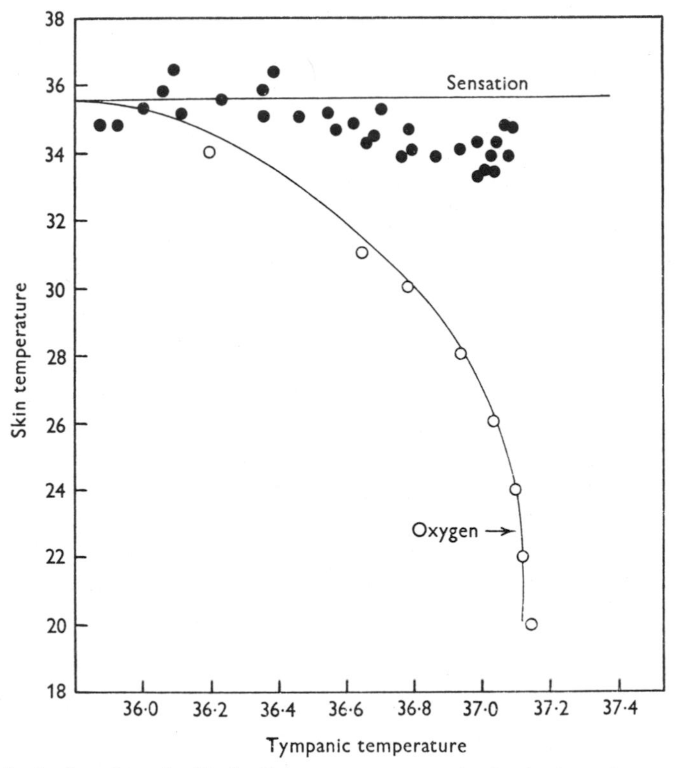

Fig. 17. In baths of gradually declining temperature the beginning of conscious cold *sensation* was observed and plotted (black dots) against the tympanic temperature simultaneously measured. The open rings are observations of threshold skin temperatures at which *oxygen consumption* begins to rise in response to cold. While the metabolic response is dramatically inhibited by central warm-reception, the *sensation* of cold appears unaffected. (Slight decline with increasing central temperature is an artefact described in the text.)

Fig. 16 describes these conditions. Arrows in Fig. 16 pointing toward the cortex do not mean to exclude the participation of lower levels in the subjective awareness of internal temperature. Yet, for intelligent behaviour, synapses with cortical pre-motor neurones are ultimately required.

Other than in their pathways, cold receptors for conscious sensations do not seem to differ from the ones that elicit the metabolic response: (1) the thresholds extrapolate to the same level, as shown in Fig. 17;

(2) the maxima seem to be near 20° C., for the conscious as well as for the metabolic reception and for the firing rates; and (3) overshooting response to sudden cooling and overshooting inhibition upon sudden rewarming are observed in subjective cold sensations as well as in the metabolic action and in the firing rates of the cold receptors.

The block imposed by central warm reception upon the metabolic response, and the absence of such inhibition in the conscious sensation of cold, are factors of extraordinary significance. They determine, in a cold environment, the relative roles of behavioural and autonomic regulation. The uninhibited sensation for behavioural action arises first. It is accentuated through the overshooting discharges of cold receptors with the onset of cooling. If the conscious early warning is wilfully answered the autonomic response becomes unnecessary. It is relegated to the role of an emergency function. Moreover, it appears from Fig. 12 that skin temperatures must be as low as 31° C. for shivering to be elicited with an internal temperature as low as 36·6° C. Under such conditions more heat is likely to be withdrawn from the body than the shivering response may be able to replace. Shivering is therefore less suited than the response of sweating to maintain a long-term steady state by thermostatic action. Behavioural responses to cold tend to correct this inadequacy. They keep conditions in a state where the human thermostat can exert its control by imperceptible sweating, near the setpoint.

In long-term adaptation to cold a third mechanism gains importance: hormonal changes provide mounting 'non-shivering thermogenesis' at the expense of the response to cold by shivering (Hart 1963).

Sensation of warmth

As opposed to the distinctly unpleasant sensation of cold the sensation of warmth on the skin is *pleasant* in character. Pain arises from cutaneous warm stimulation only at temperatures above 42° C.—far out of the range of skin temperature in which human physical heat regulation operates with maxima always below 38° C., even in 50° C. environment. Pain is unrelated to thermal homeostasis and will therefore not be discussed in this context. The *pleasant* warm sensation from the skin, however, is pertinent. It may act as a pleasure-seeking drive that reinforces the intentions to escape *cold* environmental conditions, but for human behaviour intended to avoid environments that are too *warm* pleasant warm sensations from the skin cannot possibly be a driving force. The main sensation of being overheated is of different origin. This can be simply and convincingly demonstrated.

Human subjects were accommodated in a bath of 38·5° C., a skin temperature so high that it is hardly ever encountered at hot environmental temperatures in desert or jungle climates. Surprisingly, this bath temperature is not at all unpleasant, as long as the internal cranial temperature of the subject is normal, or low, upon entering. Inevitably in a bath of 38·5° C., internal temperature will slowly rise, because no metabolic heat can be released when skin temperature is higher than the core-temperature of the body. With increasing internal temperature the unpleasant sensation of being overheated develops in a steady manner. To describe this gradual

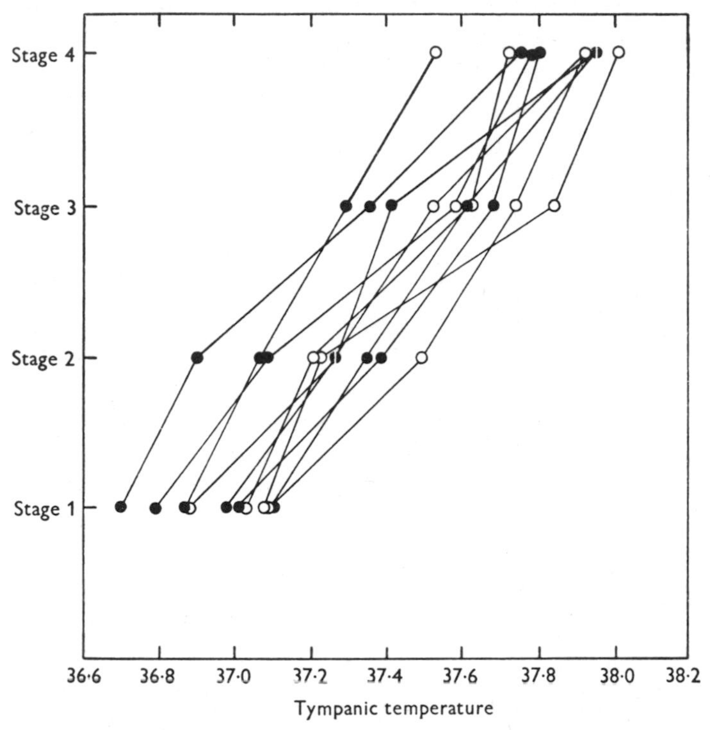

Fig. 18. In baths of *one* temperature, 38·5° C., four subjective stages of feeling warm or overwarmed (see below) were observed and plotted against tympanic temperatures. These stages of subjective comfort or discomfort extend from the mere recognition of pleasant warmth (stage 1) to almost unbearable heat (stage 4).

change, four subjective stages have been arbitrarily discerned and plotted in Fig. 18. In each of these four stages the subject would answer in the affirmative the relevant question of those given below:

(1) Do you feel warm rather than cold or indifferent? (stage 1.)
(2) Would you be more comfortable in cooler water? (stage 2.)
(3) Do you feel oppressed or restless now? (stage 3.)
(4) Do you definitely want to leave the warm bath now? (stage 4.)

Thus at one and the same skin temperature, 38·5° C., one can experience four different stages of feeling warm. These range from pleasant indifference to almost unbearable heat. The factor which produces these mounting sensations is obviously not the temperature of the skin, which had not been changed at all. The principal factor in the conscious sensation of being warm or too warm, is *central*, not peripheral temperature reception. Occasionally this observation is obscured, namely when cold receptors of the skin are active while internal temperature is elevated. Chilling the skin has a dramatic quenching effect upon the subjective sensation of being overheated. This peripheral sensation of cold overrules the central sensation of warmth. But a peripheral sensation of warmth cannot by itself produce serious discomfort.

The specific site where the *internal* warm reception takes place cannot be concluded from Fig. 18 without additional evidence. Two alternatives are open: (*a*) to assume that there exists for the reception of warmth (with pathways to the sensory cortex) a certain internal structure yet unknown and to be newly discovered; or (*b*) to assume that the known warm sensor in the anterior hypothalamus is the site from which the unpleasant, conscious sensation of warmth arises. The latter assumption seems far more conservative and attractive. It is proved at least on animals, with an observation by Hardy, Hammel & Nakayama (1962): upon local heating of the preoptic region dogs exhibit 'basking behaviour'. They act as they do when exposed to a hot environment.

It is usual in temperature physiology to accept neurosurgical evidence obtained on animals as valid also for man. In the present instance acceptance of animal behaviour as proof of the elicitation of conscious temperature sensations from centre *A* would attribute an additional capability and function to the human thermostat: notwithstanding a somewhat blurred vision (see scattering of the observations in Fig. 18) the thermostat can 'see' the internal temperature of man.

ACKNOWLEDGEMENTS

This work was supported under Contract No. R-38 by the National Aeronautics and Space Administration.

REFERENCES

Anderson, F. D. & Berry, C. M. (1959). Degeneration studies of long ascending fiber systems in the cat brain stem. *J. comp. Neurol.* **111**, 195–229.

Andersson, B. (1962). Activation of the thyroid gland by cooling of the pre-optic area in the goat. *Acta physiol. scand.* **54**, 191–2.

Andersson, B., Grant, R. & Larsson, S. (1956). Central control of heat loss mechanisms in the goat. *Acta physiol. scand.* **37**, 261–80.

Aronsohn, E. & Sachs, J. (1885). Die Beziehungen des Gehirns zur Körperwärme und zum Fieber. *Pflüg. Arch. ges. Physiol.* **37**, 232–301.

BARBOUR, H. G. (1912). Die Wirkung unmittelbarer Erwärmung und Abkühlung der Wärmezentra auf die Körpertemperatur. *Arch. exp. Path. Pharmak.* **70**, 1–26.

BEAUMONT, W. VAN & BULLARD, L. W. (1963). Sweating: its rapid response to muscular work. *Science*, **141**, 643–6.

BELDING, H. S. & HERTIG, B. A. (1962). Sweating and body temperatures following abrupt changes in environmental temperature. *J. appl. Physiol.* **17**, 103–6.

BENZINGER, T. H. (1938). Untersuchungen über die Atmung und den Gasstoffwechsel, insbesondere bei Sauerstoffmangel und Unterdruck, mit fortlaufend unmittelbar aufzeichnenden Methoden. *Ergebn. Physiol.* **40**, 1–52.

BENZINGER, T. H. (1959). On physical heat regulation and the sense of temperature in man. *Proc. nat. Acad. Sci., Wash.* **45**, 645–59.

BENZINGER, T. H. (1961). The diminution of thermoregulatory sweating during cold-reception at the skin. *Proc. nat. Acad. Sci., Wash.* **47**, 1683–8.

BENZINGER, T. H. (1963). Peripheral cold and central warm reception, main origins of human thermal discomfort. *Proc. nat. Acad. Sci., Wash.* **49**, 832–9.

BENZINGER, T. H. & KITZINGER, C. (1949). Direct calorimetry by means of the gradient principle. *Rev. sci. Instrum.* **20**, 849–60.

BENZINGER, T. H. & KITZINGER, C. (1963). Gradient layer calorimetry and human calorimetry. In *Temperature: its Measurement and Control in Science and Industry*, **3**, 87– . New York: Reinhold.

BENZINGER, T. H., KITZINGER, C. & PRATT, A. W. (1963). The human thermostat. In *Temperature: its Measurement and Control in Science and Industry*, **3**, 637– . New York: Reinhold.

BENZINGER, T. H., PRATT, A. W. & KITZINGER, C. (1961). The thermostatic control of human metabolic heat production. *Proc. nat. Acad. Sci., Wash.* **47**, 730–9.

BENZINGER, T. H. and TAYLOR, G. W. (1963). Cranial measurements of internal temperature in man. In *Temperature: its Measurement and Control in Science and Industry*, **3**, 111–000. New York: Reinhold.

EULER, C. VON (1950). Slow 'temperature potentials' in the hypothalamus. *J. cell. comp. Physiol.* **36**, 333–50.

FREEMAN, W. J. & DAVIS, D. D. (1959). Effects on cats of conductive hypothalamic cooling. *Amer. J. Physiol.* **197**, 145–8.

HARDY, J. D., HAMMEL, H. T. & NAKAYAMA, T. (1962). Observations on the physiological thermostat in homoiotherms. *Science*, **136**, 326.

HART, J. S. (1963). Insulative and metabolic adaptation to cold in vertebrates. *Symp. Soc. exp. Biol.* **18**, 31–48.

HEMINGWAY, A. (1957). Nervous control of shivering. *Alaskan Air Command Technical Note AAL-TN-57-40.*

HEMINGWAY, A., FORGRAVE, P. & BIRZIS, L. (1954). Shivering suppression by hypothalamic stimulation. *J. Neurophysiol.* **17**, 375–86.

HEMINGWAY, A., RASMUSSEN, T., WICKOFF, H. & RASMUSSEN, A. T. (1940). Effects of heating hypothalamus of dogs by diathermy. *J. Neurophysiol.* **3**, 329–38.

HENSEL, H. & BOMAN, K. K. A. (1960). Afferent impulses in cutaneous sensory nerves in human subjects. *J. Neurophysiol.* **23**, 564–78.

HENSEL, H. & ZOTTERMAN, Y. (1951). Quantitative Beziehungen zwischen der Entladung einzelner Kältefasern und der Temperatur. *Acta physiol. scand.* **23**, 291–319.

ISENSCHMID, R. & KREHL, L. (1912). Über den Einfluss des Gehirns auf die Wärmeregulation. *Arch. exp. Path. Pharmak.* **70**, 109–34.

ISENSCHMID, R. & SCHNITZLER, W. (1914). Beitrag zur Lokalisation des der Wärmeregulation vorstehenden Zentralapparates im Zwischenhirn. *Arch. exp Path. Pharmak.* **76**, 202–23.

KELLER, A. D. & HARE, W. K. (1932). Heat regulation in medullary and midbrain preparations. *Proc. Soc. exp. Biol., N.Y.* **29**, 1067–8.

KUNO, Y. (1934). *The Physiology of Human Perspiration.* London: Churchill.

MAGOUN, H. W., HARRISON, F., BROBECK, J. R. & RANSON, S. W. (1938). Activation of heat loss mechanisms by local heating of the brain. *J. Neurophysiol.* **1**, 101–14.

MINARD, D. (1963). Sweat rate during work and rest at elevated internal temperatures. *Fed. Proc.* **22**, 177.

NAKAYAMA, T., EISENMAN, J. S. & HARDY, J. D. (1961). Single unit activity of anterior hypothalamus during local heating. *Science,* **134**, 560–1.

NIELSEN, M. (1938). Die Regulation der Körpertemperatur bei Muskelarbeit. *Skand. Arch. Physiol.* **79**, 193–230.

PIIRONEN, P. (1963). The effects of exposures to extremely hot environments on the temperatures measured at the tympanic membrane, in the oesophagus and in the rectum of men. *AF EOAR* 62–31 *TR Biosciences, Inst. Occupational Health, Helsinki, Finland.*

PINKSTON, J. O., BARD, P. & RIOCH, D.McK. (1934). The responses to changes in environmental temperatures after removal of portions of the forebrain. *Amer. J. Physiol.* **109**, 515–31.

PRATT, A. W. (1958). A small-animal analytical respirometer. *J. nat. Cancer Inst.* **20**, 161–72.

RANDALL, W. C., RAWSON, R. O., MCCOOK, R. D. & PEISS, C. N. (1963). Central and peripheral factors in dynamic thermoregulation. *J. appl. Physiol.* **18**, 61–4.

RAWSON, R. & ABRAMS, R. (1964). Hypothalmic temperatures of unrestrained cold-acclimatized and unacclimatized dogs in cold and neutral environments. *Fed. Proc.* **23**, 566.

RAWSON, R. O. & HAMMEL, H. T. (1963). Hypothalamic and tympanic membrane temperatures in rhesus monkey. *Fed. Proc.* **22**, 283.

REIN, H. (1937). Ein Gaswechselschreiber. *Abderhaldens Handbuch d. biol. Arbeits methoden Abt. IV Teil* 13, 754.

ROBINSON, S. (1949). Physiology and heat regulation and the science of clothing. In *Physiological Adjustments to Heat* (ed. Newsburg). Philadelphia: Saunders.

22

FEEDBACK THEORY AND ITS APPLICATION TO BIOLOGICAL SYSTEMS

By K. E. MACHIN
Department of Zoology, University of Cambridge

INTRODUCTION

Formal feedback theory is a happy hunting ground for mathematicians. They enter the field bearing such exotic weapons as contour integration, Fourier, Laplace and Hilbert transforms, and operational calculus. Their work is embodied in books more notable for rigour and completeness than for comprehensibility and immediate usefulness. The practising engineer, as is his wont, has found methods of avoiding the mathematics; graphical techniques of analysis and other approximate methods are in common use in the places where feedback systems are actually built. In this paper I shall try to pick my way through the mathematical and technological jungle, collecting here and there a result or technique which may be useful in biology.

The problems of the engineer and the biologist are largely complementary; the engineer's work is synthetic, the biologist's analytical. Engineers have to meet specifications with the hardware available: 'Design a system which will point a gun with an accuracy of such-and-such, under the control of information coming out of this radar set. Cost and/or weight and/or unreliability must be minimized'. With living material this work has already been done before the biologist appeared on the scene. In engineering terms his problem might read: 'Find out how the other man has designed this gun-laying system. You are allowed to watch it working; you may push the gun, or jam the radar set, and see what happens. You are even allowed to investigate the mechanism of the internal computer by pushing in a crowbar and observing the waveforms picked up on it.'

In relating properties and structure, the biologist will frequently make use of models, whether they be conceptual, mathematical, schematic or even actual. Feedback theory can often help him to relate properties and model, but never model and structure. The dangers of comparing model and structure were fully discussed by this Society in its 1959 Symposium, and I shall not in general consider this stage of biological analysis. How, then, can feedback theory guide the process of relating properties and model? Most engineering literature on feedback and control systems is concerned with meeting specifications and optimizing performance. This synthetic approach may answer the question 'Can you think of *any* way, however unrealistic, in which an animal with this performance could be constructed?' In other words, it may help in the setting-up of what Williams (1960) has called a 'first-order model'. But the physiologist is really interested in models of a higher order; he needs to reduce the system under study to a set of interconnected simple 'black boxes'. His problem is rather similar to that of the economist who can observe trade cycles and the like, and wishes to trace their causes to simple interactions. The physiologist is perhaps more fortunate than the economist: he can impose perturbations on his system, and compare their results with his predictions. The perturbation of an economic system in the name of pure research is likely to be both expensive and unpopular. This paper will be mainly concerned with interpreting the results of perturbation experiments in terms of high-order models. Before doing this, it will be necessary to outline how a black box can be described quantitatively, and to show what will happen when a number of black boxes are interconnected to form a feedback system.

THE TRANSFER FUNCTION

The essential feature of a black box is that its contents are both unobservable and irrelevant; only the relation between input and output is of interest. In the common case of one input and one output, this relation is called the *transfer function*. Pringle (1962) has urged the more widespread use of the transfer function in the description of biological systems, having set a good example (Pringle & Wilson, 1952) by describing a receptor in this way. The use of the transfer function (either explicitly or implicitly) in the analysis of feedback mechanisms is almost essential. The majority of feedback theory deals with *linear transfer functions*, even though practically every physical or biological system is to some extent non-linear. The input Q_i and output Q_o of a linear black box are related by a linear differential equation:

$$a_0 Q_i + a_1 \frac{dQ_i}{dt} + a_2 \frac{d^2Q_i}{dt^2} \ldots + a_n \frac{d^nQ_i}{dt^n} = b_0 Q_o + b_1 \frac{dQ_o}{dt} + b_2 \frac{d^2Q_o}{dt^2} \ldots + b_m \frac{d^mQ_o}{dt^m}. \tag{1}$$

The relation between input and output can be put into a much more usable form by writing the operator† s for d/dt; the transfer function Φ is then written as the quotient of two polynomials in s:

$$\Phi = \frac{Q_o}{Q_i} = \frac{a_0 + a_1 s + a_2 s^2 \ldots + a_n s^n}{b_0 + b_1 s + b_2 s^2 \ldots + b_m s^m}. \tag{2}$$

† The symbols p and D are also used for this operator.

By the type of schizophrenia which comes easily to mathematicians, but which most of us have to take on trust, we now treat s as an ordinary variable; the transfer function can then be factorized:

$$\Phi = \frac{(s-z_1)(s-z_2)\dots(s-z_n)}{(s-p_1)(s-p_2)\dots(s-p_m)}. \qquad (3)$$

The quantities z_1, z_2, etc., are called *zeros* of the transfer function; when $s = z_i$ the ratio Q_o/Q_i is zero. p_1, p_2, etc. are the *poles* of the functions; when s takes any of these values the ratio is infinite. It may not be possible to factorize the transfer function into real factors; some of the poles and zeros may thus have complex values such as $\sigma+j\omega$ (where $j = \sqrt{-1}$).

The poles and zeros cannot take any values; the transfer function must represent a piece of hardware which could in principle exist, and this imposes limitations on the values. Several conditions of physical realizability of varying degrees of severity have to be met. The first of these conditions demands that for a real input, the black box must give a real output. An output which is a function of time such as $Q_o = F(t)+jG(t)$ is not permissible; the term involving j is a mathematical fiction, and cannot come out of a piece of hardware. From this condition it follows that the poles and zeros of the transfer function must either occur in complex conjugate pairs, such as $\sigma \pm j\omega$, or else must be purely real. This criterion is the most severe one, and is quite inviolable.

The second condition is concerned with the number of poles and zeros of the transfer function. Circuit theory requires that the order of the numerator of the transfer function be not greater than that of its denominator, i.e. $m \geqslant n$. This criterion again is absolute, and must be met even for idealized devices which can only be designed on paper. The output of any physical device will tend to zero as the frequency of its input signal tends to infinity; inertia in a mechanical system, and stray capacities in an electrical circuit, will see to this. This property of the real world rules out the case $m = n$, leaving us with a criterion which must always be satisfied by the transfer function of a real device: $m > n$.

The third condition is concerned with the stability of the black box. In the absence of an input, it must give rise to no outputs which increase with time. An unstable linear black box will give an output which increases indefinitely with time; the operation of any real device with these properties (such as an oscillator) will ultimately become non-linear. The device thereby removes itself from consideration by linear circuit theory. This stability condition demands that all the poles have negative real parts. By some authors the principle of causality is invoked to attain the same result; the black box must not give an output before it has received an input, otherwise it will be behaving as an anticipating device. I believe that this attractive argument is bogus, and rests on two logical fallacies.

To what extent are we to be worried if we find in a biological system an element whose transfer function violates one of these rules? If the first criterion is violated, the answer is very simple: we have made a mistake in the algebra. If, however, we find $m \leqslant n$, then our experiments are inadequate. Either we have not measured the properties of our element over a sufficiently wide band of frequencies or, what amounts to the same thing, our time resolution is not sufficiently good. If, now, we find that the element under investigation is apparently unstable, we must examine how it is interconnected within the system. It is possible to incorporate into a negative feedback loop an element which, if isolated, would be unstable. The system can then be stable. I shall return to this possibility later.

Much of circuit theory is concerned with the properties of 'passive' elements, i.e. those which contain no source of energy. For these all the above rules are automatically obeyed; in addition certain considerations of energy have to be met. In biological systems, on the other hand, nearly every element contains a source of chemical energy; the theorems for passive circuits are almost universally inapplicable.

If several black boxes are connected in cascade, the output for one forms the input for the next. From equation (2), the transfer function of the whole system will then be the product of the transfer functions of its elements. Conversely, a complicated transfer function can be decomposed into the product of a number of simple ones; the black box it describes is equivalent to a number of simpler elements in cascade. This decomposition of a complicated system into elements is the ultimate aim of any analysis of the system.

The criterion $m > n$ mentioned above has an interesting consequence: it is impossible perfectly to cancel the effect of one black box by adding another in cascade. Such a cancellation would imply that $\Phi_2 = 1/\Phi_1$, where Φ_1 and Φ_2 are the transfer functions of the two boxes. Clearly if $m > n$ for Φ_1 then this criterion is violated by Φ_2. In the limiting ideal case of $m = n$ this cancellation would be possible, but only for certain forms of Φ_1. Since the stability condition must apply both to Φ_1 and to Φ_2, it is apparent that the real parts of both poles and zeros must be negative. Transfer functions which obey this criterion are called *minimum phase* type. When any input is applied at a certain time to a minimum phase device, it will always give an output which starts at the same moment as the input. Non-minimum phase devices have either (*a*) two or more separate paths from input to output, so that for certain inputs the signals on the two paths cancel, giving zero output; or alternatively (*b*) a *finite delay* characteristic due to the propagation within the device of information over a distance with a definite velocity. Nearly all complete biological systems are of the non-minimum phase type, due to the finite transport velocity of nervous or chemical information. Strictly every physical system is also of the non-minimum phase type, since it will be of finite size, and information cannot be passed through it at a velocity greater than that of sound (for a mechanical system) or of light (for an electrical system). But often in the normal operation of all these systems the finite delays are very much smaller than the time-scale of interest, and the systems may without undue error be regarded as of the minimum phase type.

The transfer function of a black box can in principle be found experimentally by supplying an arbitrary input to the box, and observing its output. But the subsequent computation is greatly reduced by choosing inputs which fall into one of two classes. The first class includes transients, such as delta-functions and step-functions. The delta-function (also called the impulse or the spike) is theoretically a pulse of zero width and infinite height, but of finite area. In practice the width of the pulse only needs to be small compared with the time-scale of the response. The step-function, a discontinuous jump from one constant level (often zero) to another, is a transient much used in the analysis of both biological and physical systems. Another transient which occasionally proves useful is the ramp-function: a signal whose rate of change suddenly changes from zero to a constant value. The second class of useful inputs includes sinusoids of various frequencies.

If the analytical form of the transfer function of a black box is known, it is a matter of straightforward (if tedious) computation to predict its response to a step function or an impulse. Similarly, the response to a sinusoid of any frequency can be derived. The biologist's problem is the inverse one: how to derive the transfer function from the response. It matters little for a linear box whether the input is a transient or a set of sinusoids; the information is in principle equally available from either response. The choice is largely a matter of experimental convenience. The broad outlines of the transient response give information about the terms of the transfer function with low values of m and n; the fine detail about those with high values. Similarly, the response to low-frequency sinusoids is mainly governed by the low values of m and n, while the high-frequency response is the concern of the high-value terms. There is no universally applicable procedure for going from the response to the transfer function; the experimenter must work largely by 'feel' and experience, matching the response with that of known functions. In subsequent sections I shall give examples of this process.

OPEN LOOP AND CLOSED LOOP

So far we have been concerned only with the properties of a single black box, but this Symposium's primary concern is with assemblages of black boxes in closed rings, called variously feedback loops, reflex arcs, closed-cycles, etc., according to the discipline of the writer. One point of the ring is designed 'input' and one 'output' (Fig. 1). If both input and output are

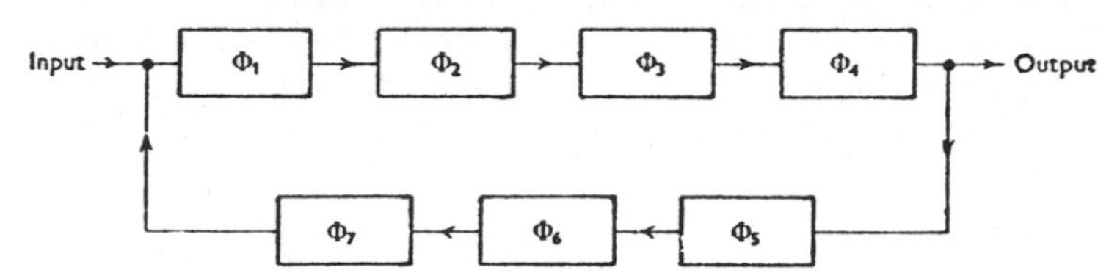

Fig. 1. A closed ring of black boxes forming a feedback loop.

accessible to the environment, the system is called a servomechanism, control system or feedback mechanism; if the input terminal is concealed within the system we have a regulator, stabilizer or homeostatic mechanism. This distinction clearly depends on where one chooses to draw the boundary of the environment, and is not of fundamental importance.

If we are constrained to make all our measurements at the input and output only, and are not allowed to stick probes into the system, we are unable to say where one black box ends and the next begins. The many boxes of Fig. 1 can be compressed into two: the forward path with transfer function Φ_F and the feedback path with transfer function Φ_B. The result of this compression is shown in Fig. 2; the 'subtracting' box has been intro-

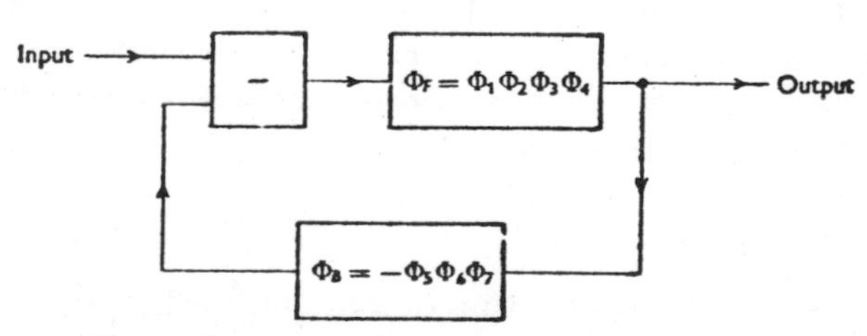

Fig. 2. The system of Fig. 1 reduced to two black boxes.

duced to make it clear that the feedback is negative, and a change of sign has been made in the feedback path to compensate for its introduction. The complete feedback system has an input terminal and an output terminal, and can therefore be regarded as a new black box with a transfer function which we may call Φ_C. It is straightforward to ..ow 'hat these various transfer functions are related by the equation

$$\Phi_C = \frac{\Phi_F}{1+\Phi_F\Phi_B}. \tag{4}$$

If the feedback loop were broken and the ends of the break treated as input and output, the transfer function between them would be $-\Phi_F \Phi_B$. The quantity $\Phi_o = \Phi_F\Phi_B$ (the minus sign is omitted by convention; the reasons for this are historical) is called the *open-loop transfer function*, to distinguish it from Φ_C, the *closed-loop transfer function.*

Equation (4) can be rewritten

$$\Phi_C = \frac{\Phi_o}{1+\Phi_o}\frac{1}{\Phi_B}. \tag{5}$$

When $\Phi_o \gg 1$, Φ_C approximates to $1/\Phi_B$; in other words, when the open-loop transfer function is large, the closed-loop transfer function becomes

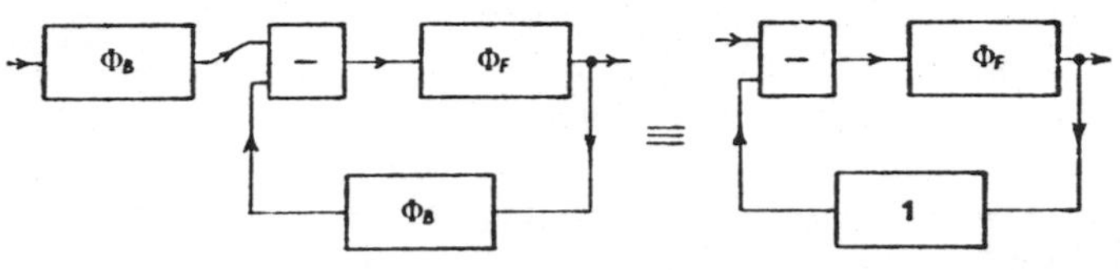

Fig. 3. The compensation of a feedback system with frequency-dependent feedback path.

the inverse of that of the feedback path. In servomechanism literature it is usual to consider only the case when $\Phi_B = 1$: the feedback path is not usually frequency-dependent within the range of interest, and a servomechanism is commonly designed so that it gives an output equal to its input. In most biological feedback systems in which both input and output terminals are accessible (such as an optomotor reflex mechanism) the condition $\Phi_B = 1$ obtains. For these systems the relation between the closed- and open-loop transfer functions is

$$\Phi_C = \frac{\Phi_o}{1+\Phi_o}. \tag{6}$$

In certain biological systems it is clear that the feedback path cannot be frequency-independent. In a postural reflex, for example, the response of the proprioceptors may be both phasic and tonic. It is possible to compensate for the characteristic of the feedback path by including an element with a similar transfer function in cascade with the input terminal, as shown in Fig. 3. Most systems with frequency-dependent feedback derive their input from the central nervous system, and if compensation were desirable it could readily be provided centrally. Since the true input terminal is probably inaccessible, and indeed usually unidentifiable, the closed-loop characteristic cannot be measured and its relation to the open-loop transfer function is thus of academic interest only.

METHODS OF ANALYSIS

The first aim of the analysis of a feedback mechanism is to determine Φ_F and, where appropriate, Φ_B. Ideally these should be available separately, but one often has to be content with a knowledge of their product Φ_o. The obvious method of measuring Φ_o, as mentioned in the last section, involves opening the loop and passing a suitable signal through it. The loop may be opened, for example, by section of a nerve, by blocking an information pathway with drugs, by rigidly constraining a limb or by a number of more subtle methods (e.g. Stark & Sherman, 1957).

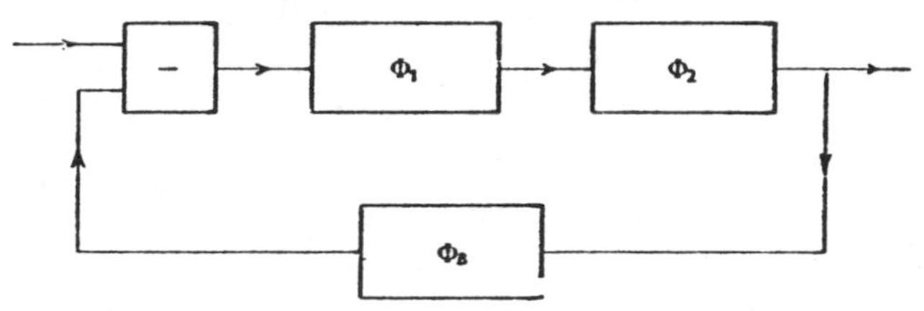

Fig. 4. A feedback system which includes a simple terminal element.

If it is not possible to open the loop, and both input and output terminals are accessible, a test signal may be fed through the intact feedback system. In this way the closed-loop transfer function can be found, and from it the open-loop transfer function inferred. If the input terminal is not accessible, such as in a homeostatic system, a process analogous to measuring the output impedance of an electronic circuit can be used. Between the controlled quantity and the effector that controls it there is often a relatively inert element with a clearly definable transfer function. Thus if Fig. 4 represents a device for controlling the concentration of a chemical in a fixed volume of fluid, the rate of supply of the chemical can be regarded as the input to the ultimate element, while the concentration represents the output. The transfer function Φ_2 is then simply V/s, where V is the volume of the fluid. Alternatively if Fig. 4 represents a feedback system controlling the position of a limb of mass m, the input and output of the ultimate element will be force and position respectively. Φ_2 is then $1/ms^2$. A perturbing signal can be fed into the input of the ultimate element, and its output observed. In the first example a change of rate of efflux from the fluid can be arranged by changing, say, the concentration of the environment; this is equivalent to adding a constant quantity to the input of the ultimate element. The corresponding change of concentration is then observed. With the limb, a force may be applied directly to it, and the resulting displacement noted. The transfer function Φ_T between the point of application of the test signal and the output can be shown to be

$$\Phi_T = \frac{\Phi_2}{1+\Phi_B\Phi_1}. \tag{7}$$

Knowing Φ_T and Φ_2, the quantity $\Phi_B\Phi_1$, which is the remaining part of the open-loop transfer function, can be found.

It has been assumed hitherto that it is possible and permissible to inject perturbing signals into the system. In some cases, such as ecological systems, this is not practicable; indeed it is virtually impossible to define the 'input' and 'output' terminals. It is sometimes possible in these circumstances to use the fluctuations arising from the natural activity of the system as 'signals'. Apart from this case the usual test signals are sinusoids or transients; their use is considered in more detail in the next three sections.

The Nyquist plot and the Bode plot

When a sinusoid of frequency $\omega/2\pi$ is fed into a linear black box, the output will be a sinusoid of the same frequency. The differential operator s operating on such a sinusoid will multiply it by ω and shift it in phase by 90°; thus for sinusoids s is equivalent to $j\omega$. If in the transfer function of a black box $j\omega$ is written for s everywhere, the resulting expression gives, in amplitude and phase, the ratio between output and input for a sinusoid of frequency $\omega/2\pi$. The amplitude ratio will be called the *gain*, G, and the phase shift will be called θ. The variation of G and θ with ω is called the frequency response, and defines the transfer function of the black box. It is usual to present this information vectorially. From the origin of a set of axes a line of length G is drawn at an angle θ to the abscissa, the poin at the end of this line is labelled with the appropriate value of ω. The curve joining all these points in order of increasing ω is then characteristic of the transfer function; it is known as a Nyquist plot (Fig. 5).

The response of the box to any transient may be calculated from its sinusoidal response by a Fourier transform process. But the principal feature of the transient response can often be guessed by inspection of the Nyquist plot. Thus a bulge in the Nyquist plot, with widely spaced frequency points, indicates a sharp resonance; the transient response may be expected to show lightly damped oscillations at this frequency. The high- and low-frequency ends of the Nyquist plot describe the transient response after short and long times respectively; the frequency above which G becomes small gives information about the rise-time of the transient response.

In a feedback system which is governed by equation (6) there is a simple geometrical method of deriving the closed-loop Nyquist plot from the open-loop plot, and vice versa. This method is shown in Fig. 6. It is usual

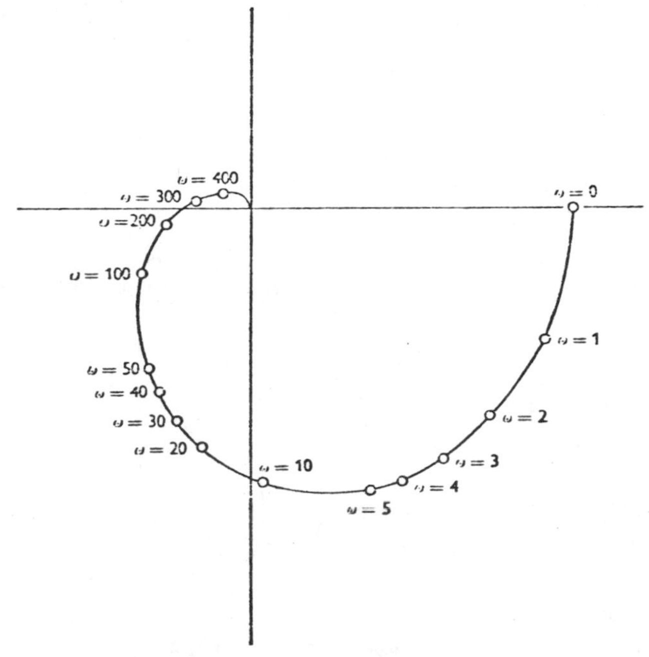

Fig. 5. A typical Nyquist plot.

in the analysis of feedback systems to work from the open-loop Nyquist plot, but if the only measurements possible are those on the closed loop, the open-loop plot can be derived, in principle, by using this construction. I use 'in principle' because the very merit of a feedback system is that its behaviour is relatively insensitive to the properties of most of its components; it is a device which conspires, as it were, to conceal its construction from the experimenter. Any inaccuracies in the measurement of closed-loop characteristics are greatly magnified in the transition to the open-loop Nyquist plot. This is illustrated in Fig. 7, where three different open-loop plots and their corresponding closed-loop plots are shown. An inverse moral may be drawn: to achieve a given performance, a biological system can be constructed in a large number of widely different ways.

Certain important features of the behaviour of a feedback system can be

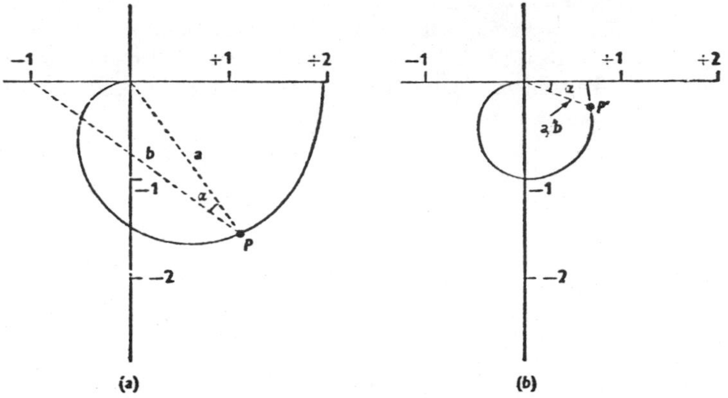

Fig. 6. The relation between (*a*) the closed-loop, and (*b*) the open-loop Nyquist plots.

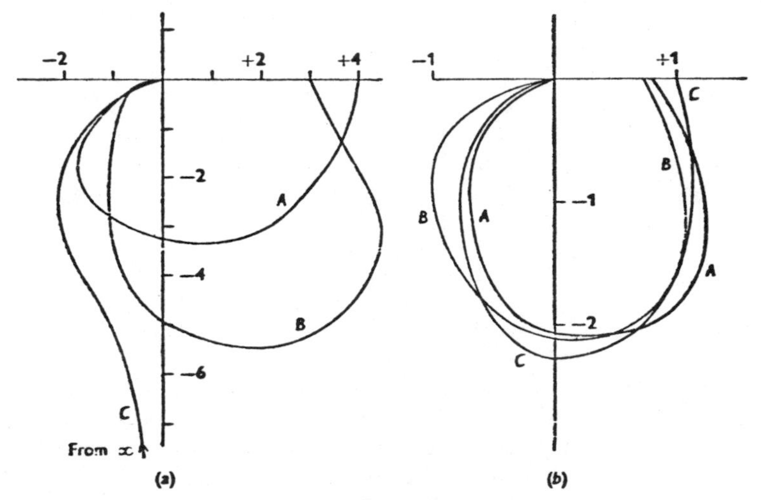

Fig. 7. (*a*) Three open-loop Nyquist plots with (*b*) their corresponding closed-loop plots

inferred immediately from its open-loop Nyquist plot. Considering first the low-frequency (or long time-scale) part of the behaviour: the gain at zero frequency determines the 'static error' of the system when it has reached a steady state. If the low-frequency end of the open-loop Nyquist plot terminates on the real axis, then the corresponding end of the closed-loop plot will not pass exactly through the point (1,0). In other words, however long we wait for transients in the output to die away, the output of the system will never exactly equal its input. This in itself is no great disadvantage; it is usually possible to allow for this error elsewhere. But the system will also have become load-sensitive; if the environment at the output end of the system changes, the output will change, even though the input be unchanged. Only by making the zero-frequency gain infinite can

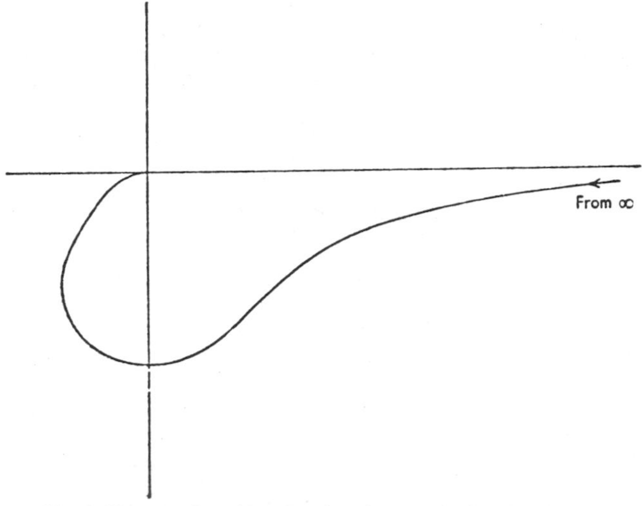

Fig. 8. The open-loop Nyquist plot of a system with 'just infinite' low-frequency gain.

this effect be avoided. Infinite low-frequency gain can be achieved in two ways. In the first, positive feedback is used within the forward path of the negative feedback system, so as to increase its gain 'just to infinity'. But what happens if this positive feedback is too large—if the gain is increased to 'just over infinity'? The system does in fact remain stable, held under control by the overall negative feedback. But the static error, and with it the dependence on load, changes sign. If, for instance, the position of a limb were controlled by such a system, and a downward force were applied to the end of the limb, it would react by an *upward* movement. A servo-mechanism with this paradoxical behaviour was christened by an eminent engineer 'the perfect civil servant'. The open-loop response of a system with 'just infinite' gain (which is, of course, unmeasurable—as soon as the loop is opened, the system would become unstable) is shown in Fig. 8.

Fig. 9 shows another approach to infinite gain. Here a single integrator has been included in the loop. This could occur, for instance, in a postural reflex in which the *velocity* of the effector was the controlled quantity, or in a homeostatic mechanism stabilizing some chemical concentration if the *rate* of production of the chemical were controlled. The response to a change of load or environment would then not have a static error term.

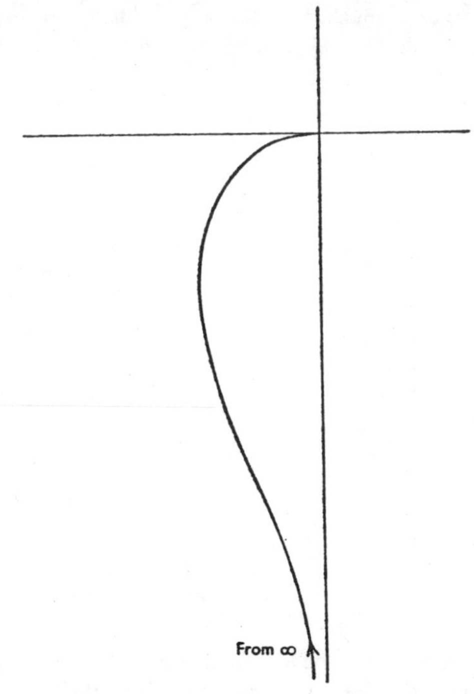

Fig. 9. The open-loop Nyquist plot of a system which includes an integrator.

The region of the open loop Nyquist plot near the point $(-1, 0)$ exercises a profound control over the closed-loop response of the system. If, for instance, the open-loop plot passes through $(-1, 0)$ as in Fig. 10, then at the appropriate frequency the closed-loop gain will be infinite. The response to any input whatsoever will then contain a sinusoid of constant amplitude. To understand what will happen if the Nyquist plot passes near, but not through, $(-1, 0)$ it is necessary to introduce the concept of *imaginary frequency*. A sinusoid of frequency $\omega/2\pi$ can be represented by $\frac{1}{2}\{e^{j\omega t} + (e^{j\omega t})^*\}$ where the asterisk signifies the complex conjugate. If now ω is purely imaginary, say equal to $-j\sigma$, the 'sinusoid' becomes $e^{\sigma t}$, a growing exponential. A complex frequency, such as $\omega - j\sigma$, represents $e^{\sigma t} \cos \omega t$, a sinusoid growing or decaying exponentially according to the sign of σ. The Nyquist plot describes the behaviour of the system at real frequencies only; it can be extended to include complex frequencies as shown in Fig. 11. Following the reasoning used above, it is clear that the closed-loop plot will become infinite at the *complex* frequency $\omega - j\sigma$ where the open-loop plot passes through $(-1,0)$. If σ is negative, the response to a general input will contain a damped sinusoid: if σ is positive,

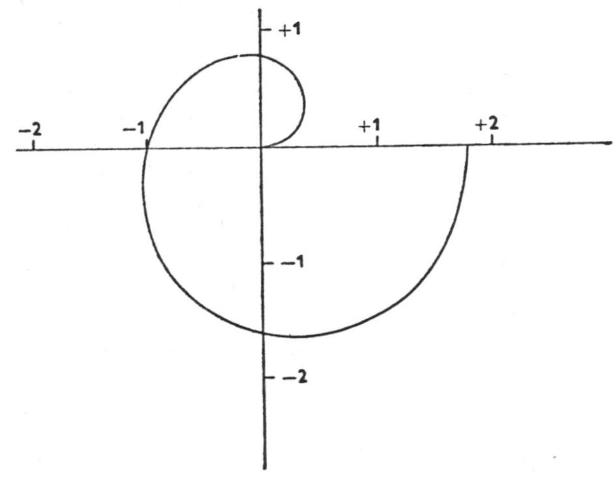

Fig. 10. The open-loop Nyquist plot of an unstable system.

an exponentially growing sinusoid will be present. The closed-loop system will then be unstable. Those familiar with the literature of feedback systems will recognize this as an application of the Nyquist stability criterion. It also provides, for the connoisseur of oddities, a clue to the behaviour of that bizarre device, the conditionally stable amplifier.

If the Nyquist plot for complex frequencies is drawn for a complicated transfer function, overlapping of various parts of the plot can occur. The plot can then pass through the point $(-1,0)$ at a number of different complex frequencies. Each such frequency represents (if the system is stable) a damped sinusoid in the output of the closed loop; the output for a particular input will be compounded of these damped sinusoids in a manner determined by the form of the input. Some of the frequencies at $(-1,0)$ may be purely imaginary; these represent decaying exponentials in the output.

Bode (1945) introduced a modified method of plotting the frequency response of a black box; its utility will become apparent in the next section. The radius vector to a point on the Nyquist plot can be expressed in complex number notation as $G\,e^{j\theta}$; the logarithm of this expression is $\log G+j\theta$. The real part is the gain expressed on a logarithmic scale (such as by quoting it in decibels) while the imaginary part is the phase angle. These two, plotted separately against frequency f, convey the same information as the Nyquist plot in a more convenient form. If the Nyquist

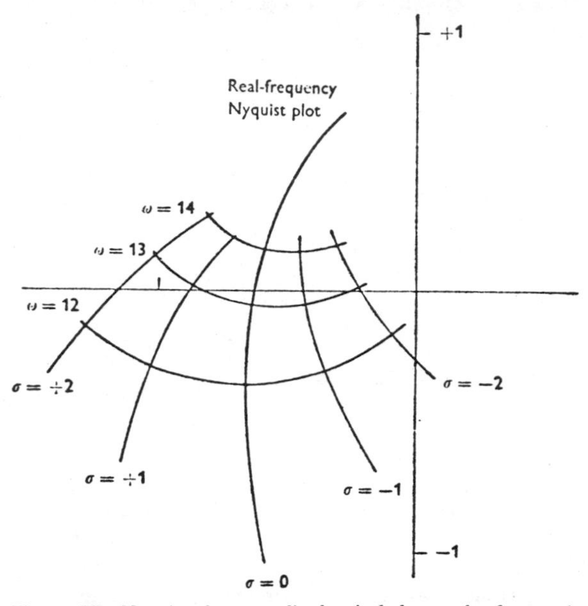

Fig. 11. The Nyquist plot generalized to include complex frequencies.

plots of two black boxes are known, the computation of the plot for the two in cascade involves multiplication of vectors, while the same process on the logarithmic plot involves mere addition of two pairs of curves. In a true Bode plot $\log G$ and θ are plotted, not against f but against $\log f$. Transfer functions which are merely powers of s then appear as straight lines on the $\log G/\log f$ plot.

Sinusoidal analysis

Suppose that the open-loop frequency response has been determined for some biological system, either by opening the loop or by inference from the closed-loop behaviour. How are we to derive the form of the transfer function—how are we to start the process of model-building? In principle

28-2

the experimental data may be fed into a computer, with instructions to find values for the poles and zeros of the transfer function which give the best (e.g. least squares) fit to the data. Such a method owes more to brute force than to elegance, and is not even very efficient. Unless the programmer specifies the number of terms in the transfer function, the machine will obediently churn out a transfer function with as many terms as there are points in the data, and will announce that it has fitted the data perfectly. But if the programmer artificially limits the number of terms, he is applying unjustifiable constraints to the results. If he has guessed too high, he will be presented with too many terms, some of which are meaningless and confusing. If his guess is too low, some essential feature of the system will have been ignored. Furthermore, the data may not be of equal value at all frequencies; the experimenter will know how much reliance he can place on the various parts of the data, but he may have difficulty in formalizing his feelings into machine instructions.

One can, instead, compile a 'library' of Nyquist plots of common elements (Fig. 12) and proceed by a process of successive approximations to match the experimental data, which is plotted in the same way. A prominent feature of the experimental plot is identified with a known element; the experimental Nyquist plot is then divided, point by point, by the plot for the element. The resulting plot forms the starting point for the next approximation. This process requires some practice; it is only too easy to concentrate on the shape of the plot and not to pay sufficient attention to the distribution of frequency points on the curve. There is in fact rather too much information for convenience on the Nyquist plot, and for this reason the Bode plot is easier to use.

For minimum phase systems, Bode also demonstrated that the phase plot could be derived from the amplitude plot, and vice versa. One of the two curves of the Bode plot is therefore redundant. Without the restriction to minimum phase systems, a given gain plot can correspond with a number of different phase plots; for minimum phase systems, all the possible phase plots are identical, and include less phase shift than all those for non-minimum phase systems.† In biology non-minimum phase systems of the double-path type are rare, but those involving a finite delay are common. The finite delay contributes to the phase plot a phase shift which increases linearly with frequency, and to the gain plot precisely nothing. It is usual, therefore, to fit the gain plot with minimum phase elements, and to calculate what phase plot would result. The difference between this and the observed phase plot, the 'excess phase', can then usually be fitted by a finite delay.

† A somewhat esoteric exception to this rule will be mentioned in a later section.

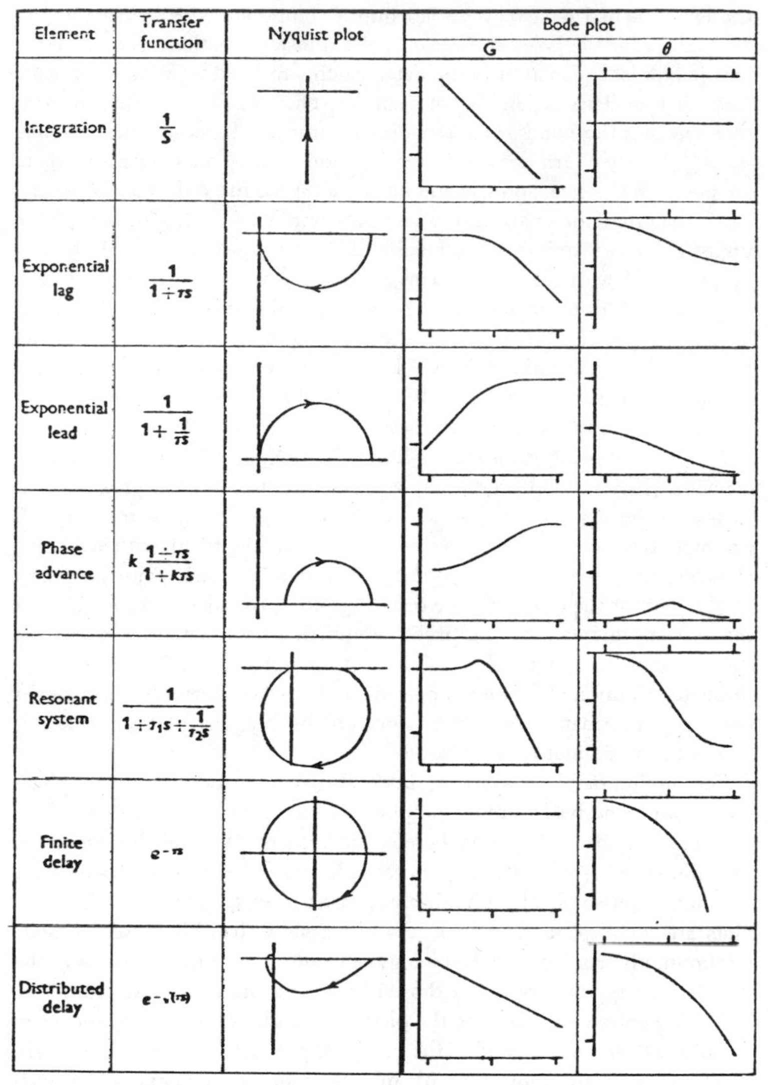

Fig. 12. The transfer function and frequency response of some typical common elements. (The arrows on the Nyquist plots give the direction of increasing frequency. The calibrations on the axes of the Bode plots are provided at the following intervals: gain and frequency, every decade; phase, every 90°.)

The fitting of the gain plot is expedited by the use of a library of the characteristics of common elements (Fig. 12). As mentioned above, Bode plots are additive, so that only subtraction is involved in the process of fitting. The gain plots of most common elements become linear at low and high frequencies, and have characteristic asymptotic slopes. The principal elements of the transfer function can therefore often be identified by inspection of the asymptotic slope of the low- and high-frequency ends of the experimental gain plot.

Since the frequency scale of the Bode plot is logarithmic, elements of the same form but different time-constants will have gain plots of the same shape, but merely shifted along the frequency axis. Standard templates can therefore be prepared for common elements, and moved around on the experimental gain plot as required. A detailed description of the fitting of Bode plots is given in many text-books on control theory (e.g. Brown & Campbell, 1948).

Transient analysis

If a transient input, such as a step-function, is fed to a stable system, its output will be the sum of a number of exponentials of various time-constants and a number of damped sinusoids of various frequencies and damping coefficients. Each exponential corresponds to a purely real pole of the transfer function, while each damped sinusoid represents a pair of poles such as $-\sigma \pm j\omega$. In principle, then, inspection of the output will immediately reveal all the poles of the system's transfer function. In practice, if one is lucky, it is possible to identify one exponential and one damped sinusoid in a step-function response. If these identified wave-forms are subtracted from the observed response, the remainder is usually so 'noisy' that attempts to extract further wave-forms are hopeless. From the two dominant modes a first approximation to the transfer function can be derived by the use of the Laplace transform (Evans, 1954).

The step-function response is better used as a rough guide to the general form of the transfer function, rather than as a tool for complete formal analysis. Thus the static error of a system may be found from a closed-loop step-function response; this gives some indication of the gain of the system when open-loop. The rise-time of the step-function response provides information about the high-frequency cut-off of the closed-loop transfer function. The presence of a damped sinusoid in an open-loop step-function response always indicates a pair of poles of the type $-\sigma \pm j\omega$; such a sinusoid in a closed-loop response implies that the open-loop transfer function has at least two poles, though they may be purely real.

Non-linearities

The above techniques of analysis apply only to linear systems, while all practical systems are non-linear. Are these techniques, then, useless? The answer to this question depends on the nature of the non-linearity. If the relation between input and output of the non-linear element can be represented by a smooth curve, free of discontinuities of ordinate or slope, then linear analysis can be useful. A very small test signal, applied to such a device, will sweep it over only a small part of this curved characteristic, and provided that the signal is small enough this part of the curve can be considered as a straight line. Ordinary linear analysis can then be used, but will give information only about the system at that particular working point. The full evaluation of the properties of the system involves the repetition of the analysis at many working points throughout the range of interest; the complete characteristic can then be synthesized.

It is rarely necessary to investigate a system as thoroughly as this. One usually needs to know the form of the transfer function only, the non-linearities being considered as complications of a non-fundamental nature. It is then only necessary to ensure that the test signal is sufficiently small for linear analysis to be applicable. If sinusoidal signals are used, the harmonic distortion in the output provides a useful criterion (Machin & Pringle, 1960). As the signal is reduced, the proportion of harmonics should decrease steadily to zero; when only a few percent of harmonic distortion are present, linear analysis is usually legitimate. If transient signals are used, the amplitude of the signal is reduced until the ratio output/input has become sensibly constant.

Some non-linear systems have characteristics which include discontinuities; useful analysis is then extremely difficult. As examples of such systems, the following may be quoted: (*a*) mechanical devices including 'click' mechanisms, such as the wing articulation of insects; (*b*) a single vertebrate muscle, in which the force/velocity relation has a discontinuity of slope at zero velocity; (*c*) sense organs which have a 'threshold'. Many systems have discontinuities which fortunately lie outside the range of interest, and can therefore be ignored, so long as their effect in extreme cases is recognized. Such systems include ecological systems, which have a discontinuity at zero population (negative animals not being permissible) and effector systems with 'end stops'.

Analysis without perturbation

A feedback system in its natural habitat may receive a large variety of inputs and will give out corresponding outputs. By inspection of these inputs and outputs, it is possible to derive the transfer function of the closed-loop system (Westcott, 1960). The techniques of sinusoidal analysis can be used by resolving a sample of the input into Fourier components, which are then compared with the Fourier components of the output over the same time. Alternatively, analysis in the time domain rather than the frequency domain can be used; the coefficients of the differential equation of the transfer function can be found by, in effect, solving a large number of finite difference equations based on samples of the input and output at nearby time-intervals. Both these procedures are only possible if a computer is available.

Two special-purpose analogue computers have been built which analyse systems under their normal operating conditions. One of them (Donaldson, 1960) was constructed to simulate the learning of manual skill, while the other (Gabor, Wilby & Woodcock, 1961) carried the ambitious title 'Universal non-linear filter, predictor and simulator'. But both were effectively model-building automata, which studied a system by way of its input and output, and gradually constructed a model whose transfer function was the same as the system under study. Donaldson's machine was designed as an aid to thought about learning processes, and was never intended to be used for the analysis of complex systems. Gabor's machine, on the other hand, appears to have been conceived as an analytical tool. It has been maintained (Tizard, 1961), with some justification, that its function can be carried out equally well by a general-purpose digital computer. Arguments of this type have raged over every analogue computer built since the war; I do not propose to discuss what is largely a matter of economics and opinion.

All analyses of this type suffer from a great disadvantage: the properties of the system can only be found over that part of the spectrum which is represented in the input. One cannot expect, for example, a learner driver to have evolved a perfectly stabilized control system until he has had some experience of step-function changes of track induced by the random behaviour of the local fauna. Donaldson's machine in its present form builds a model of a human operator balancing an inverted pendulum; in early experiments the machine would converge rapidly to a model, but now that the operator has become more skilled at balancing, and therefore makes fewer sharp movements, the machine's 'learning' is slower. It is, in fact, of little use studying a system by this technique unless the system is 'busy'. It may be possible to increase the activity of the system under study, without unduly disturbing its habits, by applying random signals covering a wide frequency band to some point of the system.

A limited amount of information about a feedback system can be obtained by studying the random fluctuations of its output only. In a number of systems the natural 'tremor' of the output has been identified with random noise filtered by the closed-loop transfer function (e.g. Robson, 1959, 1962). In others (e.g. Deutsch & Clarkson, 1959) such tremors have been attributed to the instability of the feedback loop. In both cases the frequency of tremor could be altered by modifying the transfer function of that part of the loop which was accessible. Alternatively, oscillations can be induced in an otherwise stable system by modifying the external part of the loop (Stark, 1959; Fender & Nye, 1961; Robson, 1962). Such experiments can usually assist in the identification of a feedback system, or can confirm the results of more detailed studies on the transfer function of a feedback loop. But one should not expect to learn from them very much about the detailed form of the transfer function.

FEEDBACK THEORY AND BIOLOGICAL SYSTEMS

Once it has been recognized that some biological mechanism depends for its action on feedback, its analysis would ideally proceed along the following lines. A method should be found whereby the loop can be opened, and the characteristics of the open loop determined. The closed-loop characteristics can then be predicted and compared with experiment. A transfer function should then be fitted to the open-loop characteristic, the elements of which form the models of the elements of the biological system. The process of analysis is complete when a one-to-one identification has been made of the model elements with the elements of the system. The details of these steps—whether sinusoidal or transient analysis is used, the investigation of the linearity of the system, the technique for loop opening, the method of fitting the transfer function—are all trivial, but will occupy most of the time of the investigation.

This complete sequence has not yet been carried out for any biological system. For few, indeed, has progress been made beyond the stage of recognition that feedback is occurring. Even this first stage of the process is not without its hazards; the constancy of a given parameter does not necessarily mean that it is subject to feedback control. The feedback enthusiast who sees the constancy of a natural population as the result of ecological feedback should not apply his ideas to the constancy of population of a laboratory culture; the animals may be long-lived and all of one sex. To take a less trivial example: the wing-beat frequency of, say, a pigeon is variable, while that of a bee is remarkably constant. This constancy is not due to feedback, but to the fact that in the bee the wing-beat frequency is determined by the mechanical resonant frequency of the wing system, while in the pigeon it is under central control. Even if a control mechanism is operating, it may be of the open-loop (or feed-forward) rather than the closed-loop or (feedback) type. Thus the compensatory rolling movements of the human eye, at first sight due to a classical optomotor reflex mechanism, have been shown to take place even in the dark, and in fact are mediated by feed-forward from the semi-circular canals.

The next stage—opening the loop—presents very great technical difficulties. One must first identify the principal elements of the loop to discover where to break the control sequence, and to ensure that no unwanted breaks in other control systems are caused. Sometimes, as mentioned earlier, the loop includes a relatively inert element such as a limb which is driven with one physical quantity (e.g. force) by an effector, while another quantity (e.g. position) is monitored by a receptor; this may form a convenient point at which to open the loop (Robson, 1962). An input signal can be applied in the form of a displacement of limb position by means of a very rigid transducer, and the force which is invoked can be measured as the output signal. This force cannot affect the position of the limb due to the rigid clamping; the loop is therefore opened. It has sometimes proved convenient (Robson, 1962) to take the output from one element farther back, by using information from the electromyogram as an indication of muscle activity. This can then be related in a separate experiment to the corresponding muscle tension. Some very elegant techniques of loop-opening have been used in the investigation of the operation of ocular feedback mechanisms (Stark & Sherman, 1957; Fender & Nye, 1961), but outside these two fields there has been remarkably little work on the open-loop response of biological systems.

Whether transient or sinusoidal signals are used for testing is largely a matter of experimental convenience. For slowly responding systems such as osmoregulators the technical problem of producing, say, sinusoidal changes of concentration with periods of hours is clearly formidable; it is hardly surprising that step-function signals have been used exclusively. At the other end of the scale, the production of very fast step-function changes is difficult, and sinusoidal signals may be the simpler to generate. If there are no experimental difficulties, sinusoidal analysis should be chosen, for it has a number of advantages. With phase-sensitive devices to measure output it is possible to discriminate against noise, harmonics and other interfering signals (Machin & Pringle, 1960; Robson, 1962). It is therefore possible to work with inputs which are sufficiently small that non-linear distortion is negligible. Furthermore, it is much easier in practice to fit a transfer function to the results of a sinusoidal analysis than to those of a transient analysis. If purely qualitative results about a simple system are required, they may be obvious from, say, a step-function response; but if detailed information is required about a system with more than three terms in its transfer function, the sorting out of the various components of a transient response can be very difficult.

Once the transfer function has been evaluated, what sort of elements are we likely to find in it? The most common element is the *exponential lag* $1/(1+\tau s)$. This always arises when there is interaction between the flux of some quantity and its concentration. Thus the diffusion of a substance across a thin barrier between two compartments leads to an exponential lag, as does the conduction of heat between two bodies separated by a thin, poor conductor. In mechanical systems the interaction between a perfect spring and a perfect dashpot can lead to an exponential lag.

The *exponential lead* $1/[1+(1/\tau s)]$ occurs when the output of some device momentarily follows its input, but when the input remains constant the output decays exponentially to zero. It is the characteristic shown by a completely adapting receptor. The decay of output of such receptors is commonly not perfectly exponential, but this may be the result of non-linearity and of an over-large signal. If the response of the receptor is partially tonic and partially phasic, its transfer function will be of the form $k[(1+\tau s)/(1+k\tau s)]$, *the phase advance*. Those familiar with the design of feedback amplifiers will know that the phase-advance network is useful in achieving stability in a feedback loop, and there are reasons to believe that it is used in this way in biological systems (Merton, 1951).

Elements of the type $1/[(1+\tau_1 s)+(1/\tau_2 s)]$ arise when a transfer of energy is possible between a potential store and a kinetic store. This occurs with mechanical elements such as limbs; apart from this it is unusual to find this transfer function represented in a biological system. It is, however, possible to simulate such a transfer function by the use of a subsidiary feedback loop within a main loop, using only exponential leads or lags. If this element were to turn up in the analysis of a non-mechanical system, it may well have arisen in this way.

All the elements considered so far are of the minimum phase type. When the gain plot has been fitted with minimum phase elements, it is quite usual to find that a linear increase of phase with frequency, characteristic of a finite delay, is left over. If nerve transmission is involved, such a delay can readily be explained. But there are a large number of occasions in which no such transmission is present, or where the delay time is very much larger than would be expected from the propagation velocity (Stark & Sherman, 1957; DeVoe, 1961; Stark & Hermann, 1961). It is possible here that the time delay has been simulated by a large number of minimum-phase elements in cascade, and that the frequency range of the experiments is insufficient to show this. If the range were extended it might be found that the amplitude no longer remained constant, but fell off very rapidly; the rate of the fall would be proportional to the phase accumulated up to the cut-off frequency. Methods exist to simulate the effect of a finite delay with lumped elements (Truxal, 1955), and it would seem worth while trying to fit some of the observed finite delays in this way. The difficulty of doing so increases with the amount of 'excess phase' which is found, but more than 270° of excess phase has never yet been observed.

In at least one analysis of a biological feedback system (Fender & Nye, 1961) it has been found that the phase shift was *less* than that due to minimum phase elements. This can occur if within the main feedback loop there are subsidiary loops which, when isolated, are unstable. As mentioned earlier, the overall system can be stable; as engineers have found relatively recently, a useful improvement in the performance of a feedback system can be achieved by this technique.

Two curious elements, whose transfer functions are not usually found in the armoury of the control engineer, must be mentioned. The first, the *distributed delay*, has the transfer function $e^{-\sqrt{(\tau s)}}$; it is characteristic of the conduction of heat through a thick conductor, and of diffusion through a thick barrier. The other, for which no name is accepted but which has been called 'fractional differentiation' by Chapman & Smith (1963), who first drew attention to it, has the transfer function s^k, where k lies between 0 and 1. These authors found that many mechanoreceptors obeyed this equation. It also describes the properties of many substances with rubber-like elasticity, and probably arises from the interaction of many exponential delays with a broad spectrum of time-constants.

By the time the physiologist has reached the stage of fitting the elements of the transfer function to the structure of his biological system, he is beyond the point where feedback theory can help him. A knowledge of current engineering techniques may guide analogistic thinking, and some idea of what is not possible may prevent wrong guesses. A. V. Hill has said (1956) '... every muscle, or group of muscles, will show, qualitatively and quantitatively, the sort of properties that a very intelligent engineer, knowing all the facts, would have designed for them in order to meet, within wide limits, the requirements of their owners'. There are only two snags: we only know a minute fraction of the facts, and our most intelligent engineers are at least a hundred million years behind the times.

REFERENCES

BODE, H. W. (1945). *Network Analysis and Feedback Amplifier Design.* New York: Van Nostrand.

BROWN, G. G. & CAMPBELL, D. P. (1948). *Principles of Servomechanisms.* New York: Wiley.

CHAPMAN, K. M. & SMITH, R. S. (1963). A linear transfer function underlying impulse frequency modulation in a cockroach mechanoreceptor. *Nature, Lond.* **197**, 699–700.

DEUTSCH, J. A. & CLARKSON, J. K. (1959). Nature of the vibrato and the control loop in singing. *Nature, Lond.* **183**, 167–8.

DEVOE, R. D. (1961). Electrical responses to flicker in the eye of the wolf spider. Ph.D. Thesis. Rockefeller Institute.

DONALDSON, P. E. K. (1960). Error decorrelation: a technique for matching a class of functions. *Proc. 3rd int. Conf. Med. Electronics*, pp. 173–8.

EVANS, W. R. (1954). *Control-system Dynamics.* London: McGraw-Hill.

FENDER, D. H. & NYE, P. W. (1961). An investigation of the mechanisms of eye movement control. *Kybernetik*, **1**, 81–8.

GABOR, D., WILBY, W. P. L. & WOODCOCK, R. (1961). Universal non-linear filter, predictor and simulator which optimizes itself by a learning process. *Proc. Instn elect. Engrs*, **108**B, 422–35.

HILL, A. V. (1956). The design of muscles. *Brit. med. Bull.* **12**, 165–6.

MACHIN, K. E. & PRINGLE, J. W. S. (1960). The physiology of insect fibrillar muscle. III. The effect of sinusoidal changes of length on a beetle flight muscle. *Proc. roy. Soc.* B, **152**, 311–30.

MERTON, P. A. (1951). The silent period in a muscle of the human hand. *J. Physiol.* **114**, 183–98.

PRINGLE, J. W. S. (1962). Prologue: the input element. *Symp. Soc. exp. Biol.* **16**, 1–11.

PRINGLE, J. W. S. & WILSON, V. J. (1952). The response of a sense organ to a harmonic stimulus. *J. exp. Biol.* **29**, 220–34.

ROBSON, J. G. (1959). The effect of loading upon the frequency of muscle tremor. *J. Physiol.* **149**, 29*P*–30*P*.

ROBSON, J. G. (1962). An analysis of a human stretch reflex. Ph.D. Thesis. Cambridge.

STARK, L. (1959). Stability, oscillations and noise in the human pupil servomechanism. *Proc. Inst. Radio Engrs, N.Y.* **47**, 1925–39.

STARK, L. & HERMANN, H. T. (1961). The transfer function of a photoreceptor organ. *Kybernetik*, **1**, 124–9.

STARK, L. & SHERMAN, P. M. (1957). A servoanalytic study of consensual pupil reflex to light. *J. Neurophysiol.* **20**, 17–26.

TIZARD, R. H. (1961). Discussion on paper by Gabor, Wilby and Woodcock. *Proc. Instn elect. Engrs*, **108**B, 436–7.

TRUXAL, J. G. (1955). *Automatic Feedback Control System Synthesis*, pp. 546–51. New York: McGraw-Hill.

WESTCOTT, J. H. (1960). Estimation of values of parameters of a model to conform with observations. *Symp. Soc. exp. Biol.* **14**, 102–21.

WILLIAMS, W. T. (1960). The problem of communication in biological teaching. *Symp. Soc. exp. Biol.* **14**, 243–9.

COMMENTARY

By B. R. WILKINS
Department of Anatomy, University College London

It is difficult to discuss a paper which is a statement of well-established, if unfamiliar, facts, rather than a description of experimental results, because of the shortage of controversial material.

I have however found one point of disagreement with Dr Machin: he defines s, the transfer function variable, as being the operator d/dt. While there is not universal agreement amongst control engineers as to the 'correct' definition of a transfer function, opinion is now overwhelmingly in favour of the use of Laplace transforms. This may cause some alarm to the non-specialist, but in practice the transforms are obtained from tables and are as harmless as logarithms.

There are several advantages in defining the transfer function in this way. s is now the Laplacian variable, which is an algebraic quantity so that the schizophrenia complained of by Dr Machin is no longer necessary. More important, perhaps is that the differential operator definition will not work except in simple cases (what are we to make, for example, of a transfer function $s^{0.5}$ or $\epsilon^{-\tau s}$ if $s = d/dt$?). Finally, knowing that s is an algebraic variable, the restriction on the form of the poles and zeros of the transfer function follows immediately, since all the coefficients $a_0, a_1, \ldots, a_n, b_0, b_1, \ldots, b_m$, being derived from the physical system must therefore necessarily be real. A well-known (?!) property of a polynomial with real coefficients is that its roots are either real or occur in complex conjugate pairs.

As an engineer dabbling on the fringes of biology, I naturally rejoice to see this attention being paid to mathematical descriptions of biological systems. One particular advantage is that it fosters mathematical theories of system organization: such theories can be formulated and tested much more rigidly than the vague generalizations which are so often voiced, particularly in such fields as learning theory. Thus methodological improvement may well be the principal benefit to accrue from this kind of study.

Author Citation Index

Subject Index